SURFACES, INTERFACES, AND COLLOIDS

Principles and Applications

"For Graciela"

SURFACES, INTERFACES, AND COLLOIDS

Principles and Applications

DREW MYERS

D. Myers
C. C. 90
5850 Rio Tercero (Cordoba)
Republic of Argentina

Library of Congress Cataloging-in-Publication Data

Myers, Drew, 1946-
  Surfaces, interfaces, and colloids: principles and applications /
  by Drew Myers.
     p. cm.
  Includes bibliographical references.
  Includes index.
  ISBN 1-56081-033-5
  1. Surface chemistry. 2. Colloids. I. Title.
QD506.M39   1990
541.3'3--dc20   90-13112
                                            CIP

British Library Cataloguing in Publication Data

Myers, Drew *1946-*
  Surfaces, interfaces and colloids.
  1. Colloids & surface phenomena
  I. Title
  541.345

  ISBN 1-56081-033-5

©1991 VCH Publishers, Inc.

Printed in the United States of America.

ISBN 1-56081-033-5 VCH Publishers
ISBN 3-527-28091-X VCH Verlagsgesellschaft

Distributed in North America by:

VCH Publishers, Inc.
220 East 23rd Street, Suite 909
New York, New York 10010

Distributed Worldwide by:

VCH Verlagsgesellschaft mbH
P.O. Box 1260/1280
D-6940 Weinheim
Federal Republic of Germany

TABLE OF CONTENTS

PREFACE

After much neglect over the years as a "stepsister" of physical chemistry and physics, the "twilight zone" of surface and interfacial science is now coming to be generally recognized as a vital, if not *the* vital, component of many of our most important and complex technological and biological processes. Surface effects, including colloidal systems, have been recognized for thousands of years as being of great potential use in many areas of the lives of humankind.

Historically, improvements in systems involving interfacial and colloidal phenomena, and the development of new applications, have been a result of repetative processes of trial and error and the application of empirical rules developed over long periods of time --- in other words, the area has been perhaps more art than science. Current economic and social conditions, however, tend to make the old "artistic" approaches to invention, development, and production much less desirable and tolerable. In today's technological climate, a small advantage or disadvantage can be the difference between survival and extinction in the economic jungle.

True to historical form, our educational institutions in general continue to gloss over or ignore completely the subject of surface and colloid science. Probably greater than 95% of graduates in chemistry, physics, biology, engineering, materials science, etc, enter their careers totally ignorant of even the most basic concepts of surfaces, interfaces, and colloids -- this, despite the fact that literally trillions of dollars in economic capacity are directly or indirectly involved with this scientific area. As a result, scientists and technologists (and their employers) risk loosing large amounts of time and material (ie, money!) in search of solutions that, in some cases, may be obvious or at least more easily found by the application of the fundamental principles of surface and colloid science.

While there are a number of excellent standard reference texts available in the area of surface and colloid science, they are often found to be overly daunting to temporary visitors who hesitate to wade into the jungle of theory in order to find the basic concepts they seek. Intended primarily as texts for the training of surface and colloid specialists, those standard works often pose a significant barrier for someone with a limited familiarity with the finer points of thermodynamics, quantum mechanics, solution theory, electrical phenomena, and the like.

This book is intended to serve as a narrow footbridge for scientists, technologists, and students who may use or need to use some aspect of surface and colloid science in their work, or who want to attain some familiarity with the area during their training process. It is designed to provide a general introduction to concepts, rather than a strong theoretical background. While some theory must be included for clarity, details are left for the interested (or needy) reader to pursue in the cited references. In most cases, discussions center primarily on conceptualization, with semiquantitative or qualitative illustrations serving to highlight the principles involved.

Although a minimum amout of space has been dedicated to theory, the quantitative nature of the subject requires that certain mathematical formulas be introduced. However, formulae are basically tools to be used when needed and stored away when not needed. For that reason, large portions of this work will be found to be devoid of mathematics.

Likewise, while extensive references to the original literature may occassionally be useful, most readers probably prefer not to chase down such details. For that reason, very few such references are provided here. Instead, various works that do contain many original references are cited in the Bibliography for each chapter.

I wish to thank Drs. G.H. Pearson, B.W. Rossiter, and D.A. Smith for their strong professional support over many years, and Drs. H.D.Bier and W.P. Reeves for that first important push. In addition I want to salute the faculty and staff of the surface and colloid group in the School of Chemistry, the University of Bristol, England, who, along with a few other groups throughout the world, strive to maintain a high standard of excellence in this field.

Drew Myers
Rio Tercero (Córdoba)
Argentina

CHAPTER 1

SURFACES AND COLLOIDS:
THE TWILIGHT ZONE

In 1915 Wolfgang Ostwald described the material generally covered in colloid and surface science as a "world of neglected dimensions." Science has taken a firm theoretical and experimental hold on the nature of matter at its two extremes --- at the molecular, atomic, and subatomic level, and in the area of bulk materials, their physical strengths and weaknesses, their chemical and electrical properties, etc. Vast numbers of chemists, physicists, materials scientists, engineers, etc, are continuously striving to improve on that knowledge in academic and industrial laboratories around the world. In between those two extremes, however, lies the world referred to by Ostwald, the "twilight zone" of surfaces and colloids, which has generally lagged behind other areas in the development of theoretical and practical understanding of the nature of the "beast" and phenomena involved in that dimension.

This is not to say that we are completely devoid of understanding when it comes to interfacial and related colloidal phenomena. Great strides were made in the theoretical understanding of the interactions at surfaces in the late 19th and early 20th centuries. With modern computational and analytical techniques and equipment, the last few years have brought giant strides toward a more complete theoretical understanding of the unique nature of interfaces. However, because of the unique and sometimes complex character of interfaces and associated phenomena, the development of satisfying theoretical models for the various phenomena has been slow. By "satisfying" is meant a theory that produces reasonable agreement between theory and experiment in situations that are somewhat less than "ideal" or "model" systems. That, of course, makes the definition arbitrarily personal in many cases, so that there exists a great deal of controversy in many areas. For the surface and colloid scientist, such controversy is not necessarily bad, since it represents the fuel for continued fundamental research. For the practitioner who needs to apply the fruits of fundamental research, however, such uncertainty can only add to his work in solving practical surface and colloidal problems.

As a rather wild guess, one may estimate that for every trained surface and colloid scientist in academia and industry, there are probably 100 or more scientists and technicians whose work directly or indirectly involves some surface and colloidal phenomena. And it is very likely that of those 100 workers, a relatively small percentage were formally introduced to surface and colloid science during their scientific training. It would be almost impossible to name all of the human activities (both physiological and technological) that involve surface and colloidal phenomena, but a few examples have been listed in Table 1.1.

For purposes of illustration, the examples have been divided into four main categories, each of which is further divided (somewhat arbitrarily, in some cases) according to whether the main principle involved is "colloidal" or "surface." More exact definitions of what those two terms imply will be given in the appropriate chapters; however, for present purposes one can think of a *colloid* as being a state of subdivision of matter in which the particle (or molecular) size of the basic unit varies from just larger than that of "true" molecular solutions to that of coarse suspensions (eg, 1--1000 nm). *Surface* phenomena may be defined, in this context, as phenomena related to the interaction of at least one bulk phase (solid or liquid) with another phase (solid, liquid, or gas) or a vacuum in the narrow region in which the transition from one phase to the other occurs. As will quickly become apparent, the two classes of phenomena are intimately related and often cannot be separated. In fact, many workers in the field prefer to lump all under one term, as either colloidal or surface phenomena. For present purposes (and according to the author's preference) the examples have been divided according to those definitions based on the principle phenomenon involved.

By examining each subdivision, one can quickly get the idea that such systems are ubiquitous: we and our world simply could not function or even exist in the absence of interfacial and colloidal phenomena. Surprisingly, however, this "neglected dimension" has historically been undermanned in terms of scientists and technicians formally trained in the theoretical and experimental aspects of the discipline. As a result, one can speculate that untold amounts of time, money, and other resources have been wasted over the years simply because chemists, physicists, biologists, engineers, and numerous classes of technical operators were ignorant of certain basic ideas about surfaces and colloids that could have solved or helped solve many practical and theoretical problems.

AN HISTORICAL PERSPECTIVE

While the sciences of surfaces and colloids are now known to be intimately related, such was not always the case. In the history of their development, the two branches evolved from somewhat different sources and slowly grew together as it became obvious that the basic laws controlling phenomena related to the two were, in fact, the same.

Table 1.1. Some common examples of the importance of surface and colloidal phenomena in industry and in nature.

Surface phenomena	Colloidal phenomena

A. Products manufactured as colloids or surface-active materials:

Surface phenomena	Colloidal phenomena
Soaps and detergents	Latex paints
Drugs	Aerosols
Emulsifiers and stabilizers	Foods, eg, ice cream, butter, mayonnaise
Herbicides and pesticides	Cosmetics
Fabric softeners	Pharmaceuticals
	Inks
	Laquers, oil-based paints
	Oil and gasoline additives
	Adhesives

B. Direct application of surface and colloidal phenomena:

Surface phenomena	Colloidal phenomena
Lubrication	Control of rheological properties
Adhesion	Emulsions
Foams	Emulsion and dispersion polymerization
Wetting and waterproofing	Drilling muds
	Electrophoretic deposition (cataphoresis)

C. Use for the purification and/or improvement of natural or synthetic materials:

Surface phenomena	Colloidal phenomena
Tertiary oil recovery	Mineral ore separation by flotation
Sugar refining	Grinding and communition
	Sintering
	Sewage and wastewater treatment

D. Physiological applications:

Surface phenomena	Colloidal phenomena
Respiration	Transport of lipophilic components in blood
Joint lubrication	Emulsification of nutrients
Capillary phenomena in plant liquid transport	Enzymes
Arteriosclerosis	Cell membranes

Although the formal studies of surface and colloid science began in the early 19th century, humans observed and made use of such phenomena thousands of years earlier. The Bible and other early religious writings refer to strange clouds and fogs, which are colloidal in nature. Ancient Egyptain hieroglyphic paintings show scenes of slaves lubricating great stones being moved to build pyramids and other monuments. Hebrew slaves made bricks of clay, a classic colloid, while many ancient seafaring cultures recognized the beneficial effect of spreading oil on storm-tossed waters in order to help protect their fragile crafts. The preparation of inks and pigments, baked bread, butter, cheeses, glues, etc, all represent surface and colloidal phenomena of great practical importance to ancient cultures.

In more modern times, such notables as Benjamin Franklin began to take formal notice of surface and colloidal phenomena in philosophical discussions of, for example, the amount of oil required to cover a small pond in London completely with the thinnest possible layer (*monolayer* coverage). The first important quantitative analyses of surface phenomena were probably the works of Young,[2] Laplace,[3] Gauss,[4] and Poisson.[5] From the middle of the 19th century on, understanding of the phenomenology of surfaces and interfaces became more and more clear at the molecular level, although the nature of the forces involved remained uncertain until the advent of quantum mechanical theory in the 1930's. The study of colloidal phenomena followed a similar track in that certain characteristics of colloidal systems were recognized and studied in the last certury (and before), but a good quantitative understanding of the principles and processes involved remained elusive.

The main reason for the delay in developing good quantative theories for surfaces and colloids was the lack of good, well-characterized systems which gave consistent and reproducible results in the hands of different investigators (and sometimes those of the same one). System purity is a key requirement for understanding the performance of most surface and colloidal phenomena. Contaminants at levels of small fractions of a percent can significantly affect the performance of a system. In the pioneering days of surface and colloid science, such low levels of impurities often went undetected, so that correlations between theory and experiment were sometimes less than optimal. Modern techniques of purification and analysis have made it possible for us to work with much more knowledge of the systems involved, with the result that experimental data can be used with more confidence in trying to correlate results with theory. Even now, however, the sensitivity and complexity of surfaces and colloids causes difficulties in developing complete understanding of many phenomena of academic and practical importance.

A FUTURE PROSPECTIVE

Because modern, quantitative surface and colloid science is a relatively new development, a great deal remains to be done in terms of

Table 1.2. Some important areas of surface and colloid science inviting future research.

Theoretical studies	Surface energies of solids, surface and interfacial tensions and the interfacial region, thermodynamics of colloidal systems, improved electrical double layer theory, adsorbed polymer layers and steric stabilization, relationships between surface energies and bulk properties
Surface chemistry	Equilibrium wetting and spreading processes, adhesion, dynamic wetting, physical adsorption, chemisorption and heterogeneous catalysis, spectroscopic and optical studies of surfaces, flow through porous media
Interparticle interactions	Measurement of forces between surfaces, effects of adsorbed layers, the role of solvation
Colloidal stability	Hydrodynamic and solvation factors, emulsion stability, microemulsions, multiple emulsions, coagulation and flocculation theory, foam stability, demulsification and defoaming, the effects of adsorbed polymers on stability and flocculation
Colloidal properties	Optical and spectroscopic properties of model colloids, rheological properties, electro-phoretic properties
Chemical reactions at surfaces	Heterogeneous catalysis by colloids, chemical and biological reactions in colloidal systems, interfacial reactions, including those between body fluids and foreign surfaces
Lyophilic colloids	Association colloids, gels, studies of polymers in solution and adsorbed onto surfaces, microgels, liquid crystals
Aerosols	Methods of formation, stabilization, and destruction
Biocolloids	Membranes, cell and particle adhesion, cell--antibody interactions, drug delivery, transport phenomena

extending the basic ideas and concepts stemming from the "classical" period to include new information and theories. Not only the desire for improved theories, but technological innovation demands that our understanding of surface and interfacial phenomena be improved. Just a few of the general areas that warrant close, specific attention are listed in Table 1.2. The list is obviously incomplete, but may serve to lead the student into new and important areas of research. It is hoped that the following material, as limited as it is in its coverage of a large subject, will assist the "needy" or just the interested in finding a door into the "twilight zone" of surfaces and colloids.

CHAPTER 2

SURFACES AND INTERFACES :
GENERAL CONCEPTS

The subject matter to be covered in the following chapters is concerned with the regions of our physical world that lie between two distinct and identifiable phases of matter. The bulk characteristics of various phases will not, for the most part, be considered, except insofar as they affect interphase interactions. Our primary area of interest lies in that region of space in which the system as a whole undergoes a transition from one phase to another. For purposes of terminology, it is common practice to refer to that nebulous region as a surface or an interface. As will become evident, the exact definition of what constitutes a surface or an interface is not always unequivocally clear. While the two terms are often used to indicate distinct situations, they are in practice interchangable, exact usage depending more on personal preference than on any physically definable differences. In general, however, one usually finds that the term "surface" is applied to the region between a condensed phase (liquid or solid) and a gas phase or vacuum, while "interface" is normally applied to systems involving two condensed phases. Where complete generality is implied,"interface" is probably the better term. That convention will generally be employed in the material to follow. However, no guarantees of complete consistency are offered.

There are several types of surfaces or interfaces that are of great practical importance and that will be discussed in turn. These general classifications include: liquid--vacuum, solid--vacuum, liquid--gas, liquid--liquid, liquid--solid, solid--gas, and solid--solid. From a practical standpoint, solid-- and liquid--vacuum interfaces are of little concern. They are most often encountered in the context of theoretical derivations, since the absence of a second phase simplifies matters greatly, or in studies of high-vacuum processes such as deposition, sputtering, etc. The true two-phase systems (assuming that a vacuum is not considered to be a true "phase") are the ones which are of most importance in practical applications and that are addressed in most detail here. A list of commonly encountered examples of these interfaces is given in Table 2.1.

Table 2.1. Common interfaces of vital natural and technological importance.

Interface type	Occurance or application
Solid--vapor	Adsorption, catalysis, contamination, gas chromatography.
Solid--liquid	Cleaning and detergency, adhesion, lubrication, colloids.
Liquid--vapor	Coating, wetting, foams.
Liquid--liquid	Emulsions, detergency, tertiary oil recovery.

THE NATURE OF INTERFACES

In order for two separate phases to exist in contact, there must exist an interface through which the intensive properties of the system change from those of one phase to those of the other, as for example in the boundary between a solid and a liquid. In order for such a boundary to be stable it must possess an *interfacial free energy* such that work must be done to extend or enlarge the boundary or interface. If such is not the case, and if no other external forces such as gravity act to separate the phases by density, etc, then no energy will be required to increase the interfacial area and random forces will distort, fold, and convolute the interface until the phases become mixed. In other words, if the interface does not have a positive free energy, it cannot exist as a stable boundary between two phases.

"Stable," however, can be a relative term in surface and colloid science (as will become apparent later). For that reason, one should always have clearly in mind just what is intended by the term in a given situation. Our "chemical" world is one of both thermodynamics and kinetics, so that even if a system is energetically unstable, it may require a rather long time for it to go from its unstable to its stable configuration. Such systems are commonly referred to as being *metastable* . While thermodynamics is an essentially irresistable drive to a final energetic state, over which we sometimes have little control, we can sometimes use kinetics as a tool to slow that drive for periods of time sufficient to achieve a particular goal

In order to define an interface and show in chemical and physical terms that it exists, it is necessary to think in terms of energy, keeping in mind that nature will always act so as to attain a situation of minimum total free energy. In the case of a two-phase system, if the presence of the interface results in a higher (positive) free energy, the interface will be reduced to a minimum --- the two phases will separate to the greatest extent possible within the constraints of the container, gravitational forces, etc. If the composition of the system is altered the energetic situation at the interface may also be altered, possibly resulting in a lower interfacial energy or some other effect that results in an increase in the time required for complete phase separation. That is, the change may alter the energetic drive to phase separation (ie, change the *thermodynamics* of the situation) or it may alter the *rate* at which the phase separation occurs (ie, its *kinetics*), or both. Overall the interfacial energy will probably still be positive, but the alterations caused by the additive have served the purpose of prolonging the "life" of the dispersed system.

There exist innumerable practical situations in which the energetics of interfacial regions must be controlled in order to make use of the unique characteristics of a system. The primary purpose of this work is to present in a "bare bones" way the fundamental natures of various interfaces and illustrate how those characteristics can (in principle) be manipulated to the best advantage of people and their environment.

In the discussions to follow, the concepts of interfacial regions will be presented from a molecular (or atomic) perspective and from the viewpoint of the thermodynamics (energetics) involved. In this way one can obtain a mental picture of the situation and events occurring at an interface and also have a set of basic mathematical tools for understanding the processes and to aid in manipulating the events to best advantage. The "pictures" presented are intended to be qualitative in nature and do not necessarily represent reality in every detail. Similarly, the mathematical tools will be, for the most part, the basic elements necessary for accomplishing the purpose, with little or no derivation presented. More elaborate and sophisticated treatments of the subjects will be referenced but left for the more adventurous reader to pursue as needed.

SURFACE FREE ENERGY

Before beginning any discussion of surfaces and interfaces, it is important to have a clear concept of just what is meant by surface free energy. As will be seen throughout, the unique characters of surfaces and surface-related phenomena stem from the fact that atoms and molecules at surfaces and interfaces, because of their special environment, possess energies and reactivities significantly different from those of the same species in a bulk or solution situation. If one visualizes a unit (an atom or molecule) of a substance in a bulk phase, it can be seen that, on average, the unit experiences a uniform force field due to its interaction with neighboring units (Figure 2.1a). If the bulk phase is cleaved along a plane that just touches the unit in question (Figure 2.1b), and the two new faces are separated by a distance H , it can be seen that the forces acting on the unit are no longer uniform. Instead, it will continue to "feel" the presence of the adjacent units in its bulk phase, while feeling less interaction with those units being removed in the other section.

Because the unit at the new surface is in a different environment relative to its nearest neighbors, its total free energy must change. In this

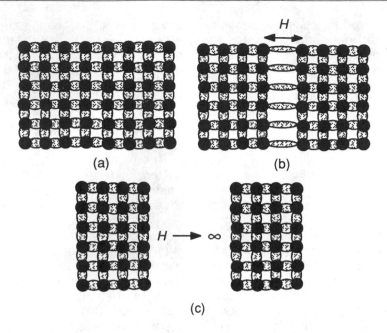

Figure 2.1. Schematic representation of changes in interatomic or intermolecular forces during formation of a new surface: (a) equilibrium position of bulk units; (b) at separation distance H, incipient surface units continue to interact, but to a reduced extent; (c) at separation distance of infinity (effectively) surface units interact only with adjacent bulk units, giving rise to the existence of an excess surface energy.

case, since the interactions in the bulk phase produce a net lowering of the free energy of the units, the removal of those interactions by the formation of a new surface results in an increase in the free energy of the units at or near the surface.

The increase in free energy of the system as a whole resulting from the new surface will be proportional to the area, A, of new surface formed and the surface density of the units. The actual change in system free energy will also depend on the distance of separation, since unit interactions will generally fall off by some inverse power law. When the two new surfaces are separated by what can be termed practical infinity, the additional free energy of the system becomes constant. The additional energy is then termed the "surface excess free energy," given by

$$\Delta G = \Delta W = 2\sigma A \qquad (2.1)$$

where the proportionality constant σ is termed the *surface* or *interfacial* tension or energy. W, in this case, is the amount of *reversible* work necessary to overcome the attractive forces between the units at the new surface or interface. At this point it is convenient to introduce two terms related to eq. 1.1 that will appear in various contexts in later chapters, namely, the work of cohesion and the work of adhesion.

The *work of cohesion* , W_c, is defined as the reversible work required to separate two surfaces of unit area of a material with surface tension σ (Figure 2.2a). Since the process involves the creation of two unit areas of fresh surface, and since the work required for that process is the surface tension, the work of cohesion is simply

$$W_c = 2\sigma \qquad\qquad (2.2)$$

It must be remembered that W_c is a reversible thermodynamic function and represents a minimum amount of work for carrying out the process. Additional work may be expended in associated <u>irreversible</u> processes such as heat generation. Related to W_c is the *work of adhesion* , $W_{a(12)}$, defined as the reversible work required to separate unit area of interface between two different materials or phases (1 and 2) to leave two "bare" surfaces of unit area (Figure2. 2b). The work is given by

$$W_{a(12)} = \sigma_1 + \sigma_2 - \sigma_{12} \qquad\qquad (2.3)$$

where the subscripts refer to the two phases being separated, and the σ's are the respective surface or interfacial tensions.

The nature of the environment of the newly formed surfaces will affect the actual surface free energy of the system, as illustrated in Figure 2.3. If a vacuum separates the surfaces, there are, naturally, no atoms or molecules present to interact with the surface units. Those units, therefore, can be considered to have "bare" areas exposed which

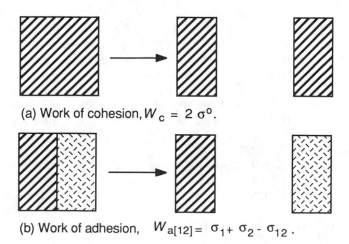

(a) Work of cohesion, $W_c = 2\sigma^0$.

(b) Work of adhesion, $W_{a[12]} = \sigma_1 + \sigma_2 - \sigma_{12}$.

Figure 2.2. Schematic illustration of the works of cohesion and adhesion.

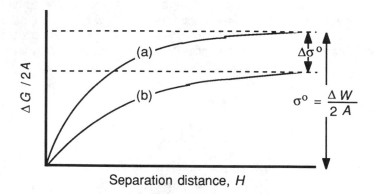

Figure 2.3. Energetic consequences of distance of surface separation in (a) a vacuum and (b) in the presence of an intervening fluid phase.

represent a very high energy situation (relative to the bulk). When in contact with an adjacent fluid phase (liquid or gas), surface units can interact to some extent with units of the fluid phase and thereby dissipate some of the excess energy they have gained by being at the surface (Figure 2.3). The greater the interaction between surface and adjacent phase, the lower the energy of the surface units due to asymmetric forces. The net result is that the energy of a surface will be greater in a vacuum than in the presence of a fluid. In some cases, eg, liquid surface tensions, the difference between vacuum and vapor may be negligible. For high- energy solid surfaces, however, the difference can be significant, as will be seen in later chapters. Conversely, for liquid--liquid and solid--liquid interfaces where significant interactions take place, the interfacial tension can be quite low.

Since the result of the formation of new surface is an increase in the free energy of the system, it should not be surprising to find that most systems will be thermodynamically driven to minimize surface area. A vivid illustration of the effect of surface thermodynamics is the picture of a blob of liquid forming itself into an almost perfect sphere when left to its own devices --- that is, when no mechanical agitation, gravitational effetcs, etc, are present. The technological consequences of surface thermodynamics are far reaching. As will be seen, it is the control of those surface effects that makes the study of surface science such an economically important undertaking.

In addition to the tendency for liquids to form spherical droplets in order to minimize their surface area, it can be easily demonstrated that liquid surfaces have other properties that can be traced back to the concept of the work necessary to form new surface area. For example, if one takes a clean needle and carefully places it on the surface of pure distilled water, the needle will float on the surface, even though the needle has a density many times that of the water. In order for the needle to sink, it must penetrate the surface of the water. Penetration of the liquid surface must increase the surface area of the water to allow the liquid to engulf the needle. The process involves increasing the water surface area with respect to the vapor phase and with respect to the interface between water and the needle. The force driving the needle to sink is the mass times the acceleration due to gravity. Opposing it is the reversible work necessary to create new interface as given by eq. 2.1.

One way to visualize the phenomenon is to think of the liquid surface as having a membrane under tension stretched across the surface and holding the needle up. The idea of the stretched membrane gave rise to the concept of a surface of tension running parallel to the surface along the bulk phase --- the *surface tension* . In fact the operative phenomenon is more accurately an energy term, so that the surface tension is more correctly a *surface energy* . The two terms are often interchanged and for liquids are numerically equal. The units employed are different, being milliNewtons per meter (mN m^{-1}) in SI units or dynes per cm for tensions and milliJoules per meter^2 (mJ m^{-2}) or ergs per cm^2 for energy.

If one thinks of the forces acting between molecules as being springs, it is easy to visualize the situations as follows. In the bulk phase, molecules are being pulled and pushed from all sides by vibrating springs of equal strength. The time-average result is some equilibrium position for a specific molecule (Figure 2.4). At an interface, the pull of springs into the bulk phase is stronger than that due to springs on the "outside." The result is that the surface molecules are pulled into the bulk (to the extent allowed by their finite size and repulsive interactions) and the net density of molecules in the surface region is decreased. There is more space between surface molecules and the springs acting between them are therefore stretched beyond their equilibrium length, creating a tension pulling along the surface working to keep the molecules together. The force of the springs pulling along the surface, then, is the surface tension or surface energy.

The application of the above concept to solid--vacuum or solid--vapor interfaces is not quite as straightforward. While it is certainly true that forces and stresses experienced by atoms and molecules near a solid surface differ greatly from those in the bulk, in general those stresses will not be isotropic, as is (or is assumed to be) the case for more mobile liquid systems. If one defines the surface tension of a solid in the same way as that of a liquid, the tension must be expected, at least for a crystalline material, to depend on the direction in the surface being considered as well as the exact crystal structure of the surface. It should be immediately obvious that for a solid the idea of a homogeneous surface tension (in the sense of the spring analogy) can become quite complicated, and a completely satisfactory definition in those terms difficult to achieve.

For that reason, it is more convenient when talking about solid--vacuum or solid--vapor interfaces to speak directly in terms of energy and to avoid completely the concept of tensions. In that way the various conceptual problems associated with the normally

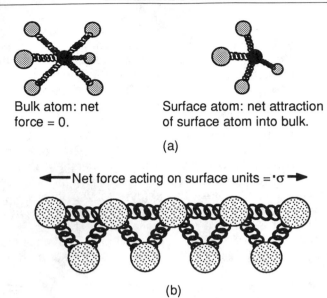

Bulk atom: net force = 0.

Surface atom: net attraction of surface atom into bulk.

(a)

←—Net force acting on surface units = ·σ—→

(b)

Figure 2.4. Schematic representation of the spring model of surface energy: (a) for the individual atom, location at the surface results in an unbalanced force pulling it into the bulk; (b) for the surface in general, the summation of the individual attraction for the units produces the net effect of surface tension or surface energy.

heterogeneous nature of solid surfaces can be avoided.

In summary, the concept of "tension" is normally applied to the interface between two fluid phases, while "energy" is most often employed with respect to systems involving at least one solid phase. The reason for the distinction is that in solid systems, the actual surface will not, in general, be a molecularly smooth. Rather, it will be irregular with different surface units being located in distinct environments relative to neighboring units (Figure 2.5a). The result is that the free energies of the surface units will vary from unit to unit and the total excess surface free energy of the system will be *history dependent* and not uniform over the entire surface. Also, it is often conceptually difficult to visualize a solid surface "tension" in a way analogous to that of a fluid surface. The surface "energy" and "tension" for solids, then, are not necessarily equivalent and the energy term is most often used.

The presence of an asymmetric force field at a phase boundary is, for liquids and fluids at least, the apparent presence of a tension at the interface acting tangent to that region at the point of interest. It is conventional to consider that this surface or interfacial tension resides in a narrow monomolecular region between the two phases. However, experimental evidence indicates that contributions can arise from second, third, and possibly even deeper molecular layers. For that reason it is most convenient at times to refer to the *interfacial region* with the implication that more than one molecular layer must be considered.

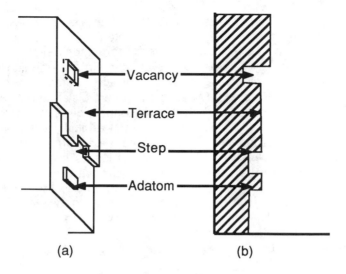

(a)   (b)

Figure 2.5. Schematic illustration of a "typical" solid interfacial region: (a) physical profile and (b) concentration profile.

Such an approach can sometimes cause "philosophical" problems in the discussion of an interface using certain models and mathematical approaches. In reality, however, since we still do not fully understand the exact nature of molecular interactions in interfacial regions, it is best not to concern ourselves too much with apparent philosophical contradictions. Nature is full of apparent contradictions resulting from our own ignorance of the true situation. For the time being we must use what tools we have that seem to work and hope for further enlightenment in the future.

Standard Reference States

The simplified description of surface energy given above is far from sufficient to explain all of the surface and interfacial phenomena such as wetting, adhesion, and colloidal stability, that are of theoretical and practical importance. In fact, depending on the specific situation, it is often necessary, or at least convenient, to approach the question of surface interactions from completely opposite points of view. For example, when one is considering a question of colloidal stability, in which the desired effect is to prevent two surfaces from interacting in an attractive way, it is convenient to think in terms of imposing a barrier, either energetic or physical, between the two interacting species which prevents the metastable dispersed state from converting into the energetically more favorable state of phase separation. For the case of adhesion, on the other hand, it is convenient to think in terms of increasing the net attractive interactions between the surfaces to be joined, so that it may be conceptually easier to consider the situation in terms of decreasing the interfacial energy between the surfaces.

In chemistry and physics it is customary to discuss energies with reference to some specified state. That is, instead of stating an absolute energy (which may be difficult or impossible to determine) for a system, the *change* in energy relative to a standard state is measured. For example, the preceeding discussion of surface energy was given in implied terms of an initial state of zero separation distance between two surfaces, going to a state of some separation distance H. It may, however, be of more utility to think in terms of an initial state of infinite separation, and measure energy changes as a function of the approach of two surfaces. Because each situation carries its own requirements there can be no set rules governing the zero point of reference for all interfacial interactions. In each of the specific areas to be discussed in the following chapters, the "zero-energy" or preferred initial state will be specified or will be obvious from the evolution of the discussion.

The Molecular Nature of the Interfacial Region

It was stated above that the free energy of a surface or interface arises due to asymmetric forces acting on atoms or molecules at or in the boundary region between phases. While the quantitative nature of those forces will be addressed in Chapters 4 and 5, it will be useful to develop the qualitative picture of the situation a bit more at this point. To begin with, let us assume that there are only three phases with which we need be concerned --- solid, liquid, and vapor. We will for the moment neglect the vacuum "phase" and ignore the existence of the various classes of solids, including crystalline, quasicrystalline, liquid crystalline, glass, and amorphous. In a practical context, the differences between the classes of solid surfaces cannot be ignored because that nature will greatly affect it surface properties. For now, however, we will keep life simple.

When two phases are in contact there is a transition region of molecular dimensions in which the composition of the system changes from that of one phase to that of the other. In the case of a nonvolatile solid surface in contact with a vacuum or an inert gas, the transition region will be essentially one molecule in thickness. That is, there will be a very sharp boundary in which the composition will change abruptly from molecules of the solid to molecules of the gas. For a molecularly smooth solid, that transition region will be molecularly smooth. For a more common irregular surface, the transition region will be characterized by the physical irregularities of the surface (Figure 2.5a). A concentration profile of the region will reflect the presence or absence of solid phase units resulting from vacancies, steps, adatoms, etc, present on the surface (Figure 2.5b).

For a pure liquid in contact with its pure vapor, the transition will be much less abrupt, going from a molecular density corresponding to the bulk material, through a zone where the unit concentration gradually decreases until the density reaches that of the pure vapor. In such a case, the transition region may be found to be several molecular diameters thick (Figure 2.6). At a mixed liquid--vapor interface, each component will have its own concentration profile depending on such factors as volatility and miscibility. A similar situation would hold for the interface between two liquid phases.

In the case of fluid systems, critical phenomena require that the interfacial region become thicker as the temperature of the system is increased, until the point where the critical temperature is reached and the two phases cease to exist as such. Solid--liquid systems will also exhibit the characteristics of the above concentration profile, although its detailed nature will depend on the solubility of the solid in the liquid (and vice versa).

A consequence of the nature of the interfacial region is that the total concentration of a given component in a system of a given volume and interfacial area will be determined by the shape of the concentration profile at the interface. For example, if a component in a fluid--fluid

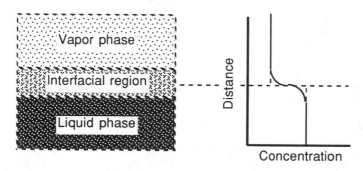

Figure 2.6. Schematic illustration of the change in concentration in going through a liquid--vapor interface.

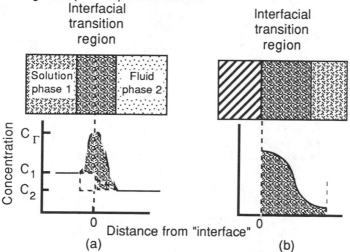

Figure 2.7. Schematic illustration of positive solute adsorption at (a) the solution--fluid and (b) solid--solution interfaces.

system is preferentially adsorbed at the interface, its concentration profile resembles that depicted in Figure 2.7a. In a solid--liquid system Figure 2.7b would be expected. The concept of the concentration profiles illustrated have been around in theory for many years. Unfortunately, procedures for the direct quantitative verification of the concept have, until very recently, eluded even the best experimentalists. A number of indirect methods for obtaining the interfacial concentration profile have been suggested, but the one most often used in that of Gibbs.[1] Although much criticized, the approach of Gibbs has so far stood the test of time based upon its ability to fit experimental data, its generality, and the relative ease of obtaining useful results.

The Gibbs Surface Excess

The Gibbs approach is quite simple and straightforward. Consider a system containing a substance i in one or both of two phases α and β. If the unit concentration of i in phase α ($C_i{}^\alpha$) is uniform throughout and that in β ($C_i{}^\beta$) is likewise uniform, for given volumes of α (V_α) and β (V_β), the total amount of i, n_i is given by

$$n_i = (C_i{}^\alpha V_\alpha + C_i{}^\beta V_\beta) \tag{2.4}$$

However, since the local value of C_i varies going through the interface (except in very unusual circumstances), there will generally be a different amount of i present in the interfacial region than that given simply by eq. 2.4. That difference, defined as the *surface excess amount* of i ($n_i{}^\sigma$), is given by

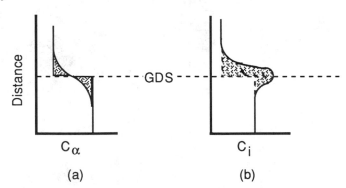

(a)   (b)

Figure 2.8. Illustration of the Gibbs approach to location of the dividing surface: (a) component α (solvent) and (b) component i (solute). The Gibbs dividing surface (GDS) is defined as the point where the shaded areas in (a) are equal --- ie, the adsorption of component α at the interface = 0.

$$n_i{}^\sigma = n_i - (C_i{}^\alpha V_\alpha + C_i{}^\beta V_\beta) \qquad (2.5)$$

The quantity $n_i{}^\sigma$ results from the presence of the interface and is dependent upon the shape of the concentration profile of i in the transition region between α and β. From a practical standpoint, the surface excess can also be considered to be the amount of i *adsorbed* at the interface. The major theoretical problem with this approach is the question of exactly how to define or locate the interface between α and β. Gibbs' approach was to define the interface (or the *Gibbs dividing surface*) as the region in which the concentration of one component, for example substance α, becomes zero. In a solution, it is standard practice to define α as the solvent. The concentration profiles shown in Figure 2.8 illustrate graphically the situation under consideration. Figure 2.8a gives the concentration profile for component α, the solvent. The shaded area represents the excess concentration of α in the region. The profile for i is given in Figure 2.8b and shows how, at $C_\alpha{}^\sigma = 0$, there is a relative excess amount of i with respect to α. When the system is limited to an interface of area A^σ, one obtains a concentration termed the *surface excess concentration* of i with respect to α, $\Gamma_i{}^{(\alpha)}$, where

$$\Gamma_i{}^{(\alpha)} = n_i{}^\sigma / A^\sigma \qquad (2.6)$$

It should be kept in mind that the Gibbs approach is a model that facilitates the handling of data mathematically, and does not imply that the surface excess of i is actually physically located in the region of the Gibbs dividing surface. The dividing surface is a mathematical plane with zero dimension in the third direction (into phases α and β), while the molecules of i are three dimensional and cannot occupy such a mathematical plane.

As mentioned, it is extremely difficult to measure a surface excess quantity directly. Such measurements have been made at solid--liquid interfaces, but liquid--liquid and liquid--vapor interfaces still pose extreme experimental difficulties. That is not to say, however, that the Gibbs approach is not applicable in such systems. To the contrary, fundamental thermodynamics allows one to use the Gibbs approach to determine surface excess concentrations in fluid systems from relatively simple and straightforward determinations of interfacial tensions via the *Gibbs adsorption equation* .

ADSORPTION

A topic of particular importance in surface and colloid science is that concerned with the adsorption of atoms and molecules at interfaces. The process of adsorption is one of the principle ways in which high-energy interfaces can be altered to lower the overall energy of a system. While the phenomenon can be quite complex from the point of view of molecular theory, the classical approaches to understanding the various processes have been founded mostly on empirical observations and conceptual insight rather than fundamental first principles. Modern computers allow one to attack the problems more rigorously, but with little gain from the standpoint of practical applications.

Adsorption can be most simply defined as the taking up of one substance at the surface of another. It can occur at any type of interface; however, the distinct characteristics of solid versus liquid interfaces makes the analysis of the phenomenon in each case somewhat different. For that reason, the discussion of each situation is presented in the general context of specific interfaces. In many practical systems, all four of the principle interfaces may be present, leading to a complex situation that may make analysis very difficult or impossible.

Interfaces containing only liquids and vapors exhibit somewhat simpler adsorption characteristics (in principle, at least) than those containing solid surfaces, because in liquid surfaces, the complications arising due to specific structures and surface heterogeneities can be ignored. The adsorption characteristics of solid surfaces, on the other hand, can be very much history dependent. Such differences aside, there are several underlying principles of adsorption which represent a bedrock for understanding all adsorption phenomena --- the Gibbs equation, the Kelvin equation, and the Young-Laplace equation.

Under the general topic of adsorption phenomena, there are a few terms that must be kept clearly in mind to avoid confusion, particularly when solid surfaces are involved. When *adsorption* occurs on a solid, the solid is referred to as the *adsorbent* and the adsorbed material the *adsorbate*. In some cases it may occur that another phenomenon, *absorption*, the incorporation or penetration of a substance into the body of another, can also occur. It can be difficult to distinguish between the effects of the two phenomena if they occur together, so that the general term for the two is *sorption* .

In the consideration of adsorption processes, there are two aspects that must be addressed: (1) the effect of the adsorbed species on the final equilibrium interfacial energy of the system, and (2) the kinetics of the adsorption process. For the most part, the discussions to follow will be concerned only with equilibrium conditions, and dynamic processes will not be addressed. For many applications, such a restriction will not result in significant limitations to the validity of the concepts involved. For others, however, the kinetics of adsorption can play a very important role. Some of those situations will be addressed in later chapters.

The topic of adsorption is of great importance to the overall understanding of interfacial phenomena, both theoretically and practically. For now, however, it will be discussed only in so far as it pertains to the conceptual and mathematical development of the Gibbs adsorption equation, a relationship that is fundamental to understanding many of the most important phenomena in the field of surface and colloid science.

The Gibbs Adsorption Equation

In order to define unambiguously the state of a bulk phase, α, it is necessary to specify the values of several variables associated with that state. Those variables include the temperature, T^α, its volume, V^α, and its composition, n_i^α. Alternatively, the system pressure, P^α, may be specified. In terms of the Helmholtz free energy F the system can be specified by

$$F^\alpha = -S^\alpha T^\alpha - P^\alpha V^\alpha + \Sigma \mu_i^\alpha n_i^\alpha \qquad (2.7)$$

or in differential terms (with constant P)

$$dF^\alpha = -S^\alpha dT^\alpha - P^\alpha dV^\alpha + \Sigma \mu_i^\alpha dn_i^\alpha \qquad (2.8)$$

For a system of two phases, a similar equation can be written for the second or β phase. At equlibrium, the two phases will have the same temperature T, the same pressure, P, and the same chemical potential, μ, for all components. The complicating factor in a system of two phases in contact is that the presence of the α--β interface that may be considered to be a third "phase," σ, and which will make a separate contribution to the overall energy of the system. The total energy, then, will be given by

$$F^T = F^\alpha + F^\beta + F^\sigma \qquad (2.9)$$

where F^σ is the interfacial free energy. In "normal" systems in which the interfacial area is small relative to the bulk volumes, the contribution of F^σ is usually ignored. However, in many contexts, such as adsorption phenomena, catalysis, membrane activity, and especially colloidal systems, the free energy of the surface may be the primary factor determining the overall molecular, microscopic, and macroscopic properties observed. Analogous to eq. 2.8, the derivative of the surface free energy may be given by

$$dF^\sigma = -S^\sigma dT + \sigma dA^\sigma + \Sigma \mu_i dn_i^\sigma \qquad (2.10)$$

In eq. 2.8, σ is the surface or interfacial tension between α and β, and μ_i has the same value as that in the bulk phases. The term σdA^σ replaces the $P dV$ term from eq. 2.8 because the surface (as the Gibbs dividing surface) is a mathematical plane having only two dimensions and is therefore an area rather than a volume term. Similarly, the sign of the σdA^σ term is positive rather than negative because σ is a tension (pulling) rather than a pressure (pushing).

In the area of bulk thermodynamics, one can derive the Gibbs-Duhem equation by integration of eq. 2.8 while holding the intensive properties T, P, and μ_i constant to give

$$F^\alpha = P V^\alpha + \Sigma \mu_i^\alpha n_i^\alpha \qquad (2.11)$$

Differentiation then yields, at equilibrium,

$$dF^\alpha = V^\alpha dP - S^\alpha dT + \Sigma n_i^\alpha d\mu_i^\alpha = 0 \qquad (2.12)$$

An identical process for the interfacial "phase" yields the Gibbs adsorption equation

$$- A^\sigma d\sigma - S^\sigma dT + \Sigma n_i^\sigma d\mu_i = 0 \qquad (2.13)$$

At constant temperature, eq. 2.13 reduces to

$$- d\sigma = \Sigma n_i^\sigma d\mu_i / A^\sigma \qquad (2.14)$$

or, where $\Gamma_i = n_i^\sigma / A^\sigma$,

$$- d\sigma = \Sigma \Gamma_i d\mu_i \qquad (2.15)$$

For a two-component liquid--vapor system where the Gibbs dividing surface is defined so that the surface excess concentration of the solvent is zero ($\Gamma_\alpha = 0$), the summation in eq. 2.15 is no longer necessary and a simple relationship between the surface tension of the liquid phase, σ, and the surface excess concentration of solute i, Γ_i, is obtained. It is therefore possible to employ experimentally accessible quantities such as surface tension and chemical potential to calculate the surface excess concentration of a solute species and to use that information to make other indirect observations about the system and its components.

A more general form of the Gibbs adsorption equation is

$$d\sigma = - \Gamma_2^{(1)} d\mu_2 \qquad (2.16)$$

in which 2 designates the solute dissolved in bulk phase 1. At equilibrium, the chemical potential of each component is equal in all phases, so that μ_i at the interface can be taken as that value in either of the adjacent bulk phases. The chemical potential of 2, then, can be related to its concentration in either of the bulk phases by

$$d\mu_2 = RT \; d\ln a_2[1] = RT \; d\ln x_2\gamma_2 \qquad (2.17)$$

where $a_2[1]$ is the activity of 2 in [1], the bulk phase, x_2 is its mole fraction, and γ_2 its activity coefficient. These manipulations lead to the relationship

$$d\sigma = -RT \; \Gamma_2^{(1)} \; d\ln x_2\gamma_2 \qquad (2.18)$$

In any equilibrium situation, the chemical potential of a species will increase with its concentration, although not necessarily in a linear fashion. As a result, an increase in Γ_2 with μ_2 (positive adsorption) will produce a decrease in the interfacial tension, σ.

A material which is strongly adsorbed at an interface, a *surface-active material* or a *surfactant* , will normally produce a dramatic reduction in interfacial tension with small changes in bulk phase concentration. In dilute solution, it is customarily assumed that the activity coefficient of a material, γ_2, can be approximated as unity so that the last term in eq. 2.18 can be substituted for by the molar concentration, c_2. The practical applicability of this relationship is that the relative adsorption of a material at an interface, its surface activity, can be determined from measurement of the interfacial tension as a function of solute concentration:

$$\Gamma_2^{(1)} = -1/RT \; [d\sigma/d\ln c_2] \qquad (2.19)$$

The preceeding discussion of the Gibbs adsorption equation was referenced to a fluid--fluid interface in which the surface excess, Γ, is calculated based on a measured quantity, σ, the interfacial tension. For a solid--fluid interface, the interfacial tension cannot be measured directly, but the surface excess concentration of the adsorbed species can be, so that the equation is equally useful. In the latter case, eq. 2.19 provides a method for determining the surface tension of the interface based on experimentally accessible data.

The principles given above allow one to derive an expression relating theoretical concepts of surface excess concentration and adsorption to experimentally obtainable quantities. But what is the practical importance of those ideas? In fact, the phenomenon of adsorption at interfaces, tied to the resultant effects of such adsorption, carries with it a multitude of important consequences (some good and some bad) for many technological and biological processes.

Some of those effects will be discussed in detail in following chapters dealing with specific interfacial situations and interactions. For now, suffice it to say that the ability to understand and control interfacial tensions (or energies) through adsorption processes, and the secondary electrical and thermodynamic effects of the adsorbed species constitute the bedrock of modern technology related to colloidal stability (and instability), wetting phenomena, emulsification and demulsification, foam formation and destruction, adhesion, lubrication, fluid displacement in capillary systems, and many more areas. Specifically, the areas of pharmaceuticals, cosmetics, food preparation, inks, paints, adhesives, lubricants, crude oil recovery techniques, mineral ore separations, wastewater treatment, heterogeneous catalysis, lithographic and xerographic printing techniques, microelectronics fabrication, and photographic and magnetic recording media, live and die by the effective control of interfacial interactions through the preferential adsorption of surfactants, polymers, and colloidal solids. With so much riding on our ability to understand and control interfaces, it is important that the people involved in research and development in those areas have at least a basic working knowledge of the phenomena involved.

Before diving into the thick of the twilight zone of colloids and interfaces, it will be important to know more about the "actors" that really control events in that narrow region of space, in terms of both the chemistry (eg, surface-active materials) and the fundamental forces involved.

CHAPTER 3

THE MOLECULAR BASIS OF SURFACE ACTIVITY

Throughout the wide range of topics related to surfaces and colloids one encounters reference to chemical species that have a special propensity to locate (ie, adsorb) at interfaces, or to form colloidal aggregates in solution at very low molar concentrations. Such materials are given the general name of *surface -active agents* or *surfactants* . The physical chemistry of surfactants, in the specific contexts of interfaces and colloids, will be covered in subsequent chapters. This chapter is devoted to a description of the structural aspects of surfactant molecules, that is, the atomic compositions and groupings which produce the observed physicochemical characteristics of such materials.

BASIC STRUCTURAL REQUIREMENTS
FOR SURFACE ACTIVITY

Surface-active materials possess a characteristic chemical structure that consists of: (1) molecular components that will have little attraction for the solvent, normally called the *lyophobic group*, and (2) chemical units that have a strong attraction for the solvent, called the *lyophilic group* (Figure 3.1). Although, in principle, surface activity and related concepts are applicable to any system composed of at least one condensed phase, the bulk of the scientific and technological literature is concerned with aqueous solvents and their interaction with a second phase. As a result, the term *hydrophobic* will quite often be employed in place of the more general "lyophobic"; analogously, *hydrophilic* will be employed instead of "lyophilic." It should always be kept in mind, however, that generality is implied in most discussions, even when the specific terms applicable to water-based systems are used.

Materials that possess chemical groups leading to surface activity are generally referred to as being *amphiphilic* ("liking both"), indicating that they have some affinity for two essentially immiscible phases. When

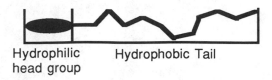

Hydrophilic   Hydrophobic Tail
head group

Figure 3.1. The basic molecular structure of surface-active materials.

a material exhibiting the characteristics of surface activity is dissolved in a solvent (whether water or an organic liquid), the presence of the lyophobic group causes an unfavorable distortion of the liquid structure, increasing the overall free energy of the system. In an aqueous surfactant solution, for example, such a distortion (in this case ordering) of the water structure by the hydrophobic group decreases the overall entropy of the system. That entropy is regained when surfactant molecules are transported to a surface or interface and the associated water molecules released. The surfactant will therefore preferentially adsorb at interfaces, or it may undergo some other process to lower the energy of the system (eg, micelle formation). Since less work is required to bring surfactant molecules to an interface relative to solvent molecules, the presence of the surfactant decreases the work required to increase the interfacial area.

The amphiphilic structure of surfactant molecules not only results in the adsorption of surfactant molecules at interfaces and the consequent alteration of the corresponding interfacial energies, but it also results in the *orientation* of the adsorbed molecules such that the lyophobic groups are directed away from the bulk solvent phase (Figure 3.2). The resulting molecular orientation produces some of the most important macroscopic effects observed for surface-active materials. Energetic considerations aside for the moment, it is important to understand the qualitative relationships between the nature of interfaces and the general chemical structures required for a molecule to exhibit significant surface activity.

The chemical structures having suitable solubility properties for surfactant activity vary with the nature of the solvent system to be employed and the conditions of use. In water, the hydrophobic group (the "tail") may be, for example, a hydrocarbon, fluorocarbon, or siloxane chain of sufficient length to produce the desired solubility characteristics when bound to a suitable hydrophilic group. The hydrophilic (or "head") group will be ionic or highly polar, so that it can act as a solubilizing functionality. In a nonpolar solvent such as hexane the same groups may function in the opposite sense. As the temperature, pressure, or solvent environment of a surfactant varies, significant alterations in the solution and interfacial properties of the system may occur. As a result, changes in conditions may require modifications in the chemical structure of the surfactant to maintain a desired degree of surface activity.

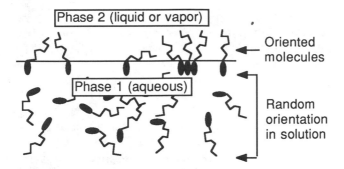

Figure 3.2. Schematic illustration of the preferrential orientation of surfactant molecules at interfaces.

SURFACTANT STRUCTURES AND SOURCES

To understand the relationship between the surface activity of a given material and its chemical structure, it is important to understand the chemistry of the individual chemical components which in concert produce the observed phenomena. The following discussion will introduce many of the structural aspects of surfactants, ranging from basic raw materials and sources to the chemical group combinations that result in the observed surface activity. Since the chemical compositions and synthetic pathways leading to surface active molecules are limited primarily by the creativity and ingenuity of the synthetic chemist and production engineer, it is not possible to discuss all chemical classes, their preparation, and subtle variations. However, the majority of surfactants of academic and technological interest can be grouped into a rather limited number of basic chemical types and synthetic processes. The chemical reactions that produce most surfactants are rather simple, understandable to anyone surviving the first year of organic chemistry. The challenge to the producer lies in the implementation of those reactions on a scale of thousands of kilograms, reproducibly, with high yield and high purity (or at least known levels and types of impurity), and at the lowest cost possible. With very few exceptions, there will always be a necessity to balance the best surfactant activity in a given application with the cost of the material that can be borne by the added value of the final product.

Building Surfactant Molecules

Synthetic surfactants and the natural fatty acid soaps are amphiphilic materials that tend to exhibit some solubility in water as well as some affinity for nonaqueous solvents. As a basis for understanding the relationship between surfactant structures and surface activity, it is useful to work through the simple example of how changes in the polarity (ie, the head group) for a specified hydrocarbon chain affects its solubility and surface activity. As an illustration, consider the simple, straight-chain hydrocarbon dodecane, $CH_3(CH_2)_{10}CH_3$, a material that is, for all practical purposes, insoluble in water.

　　If a terminal hydrogen in docecane is exchanged for a hydroxyl group (---OH), the new material, n -dodecanol, $CH_3(CH_2)_{10}CH_2OH$, still has very low solubility in water, but the tendency toward solubility has been increased substantially and the material begins to exhibit characteristics of surface activity. If the alcohol functionality is placed internally on the dodecane chain, as in 3-dodecanol, $CH_3(CH_2)_8CH(OH)CH_2CH_3$, the resulting material will be similar to the primary alcohol but will have sightly different solubility characteristics (slightly more soluble in water). Those differences will generally be carried over in other functional modifications. The effects of the position of substitution on surfactant properties can be quite large and will be discussed in more detail later.

　　If the original dodecyl alcohol is oxidized to dodecanoic acid (lauric acid) $CH_3(CH_2)_{10}COOH$, the compound is practically insoluble in water ; however, when the acid is neutralized with alkali it becomes water soluble --- a classic soap. The alkali carboxylate will be a reasonably good surfactant. If the hydrocarbon chain length were increased to 16 or 18 carbons, its solubility would decrease, but many of the surfactant properties (eg, foaming and detergency) would improve, illustrating the importance of obtaining the proper balance between the hydrophilic and hydrophobic portions of the molecule.

　　A major drawback to the use of the soaps has always been their great sensitivity to their aqueous environment. The main components of "hard water" are calcium, magnesium, and other di- and trivalent salts. In the presence of such materials, the carboxylic acid soaps form salts of low water solubility, which precipitate to produce scummy deposits, commonly encountered as "bathtub ring." The effectiveness of the polyvalent carboxylate salts as surfactants may not be greatly reduced from that of the monovalent salt on a molecular basis; however, their solubility in water is simply too low for the system to attain a high enough concentration to produce optimum results. In nonaqueous solvents, on the other hand, the polyvalent salts of carboxylate soaps show a significantly enhanced solubility and perform admirably at many surfactant functions. The inability of the carboxylate soaps to tolerate the presence of commonly encountered ions, as well as their sensitivity to pH changes, was one of the major driving forces for the development of synthetic surfactants that would not be so adversely affected by the common circumstances of hard water and cool washing temperatures, each detrimental to the effectiveness of the soaps.

　　Having the alcohol in hand, it can now be sulfated to produce the dodecane sulfuric acid ester, $CH_3(CH_2)_{10}CH_2OSO_3H$, a compound with very high water miscibility. When the sulfuric acid ester is neutralized with alkali, certain alkaline earths, or organic amines, the material becomes highly soluble in water and an excellent surfactant. It is, in fact, probably the most extensively studied and best understood surfactant known to science, sodium dodecylsulfate (SDS).

　　If the dodecyl alcohol is treated with ethylene oxide (OE) and base, the material obtained is an alkyl polyoxyethylene (POE) polyether,

$$C_{11}H_{23}CH_2\text{---}O\text{---}CH_2CH_2(OCH_2CH_2)_n OCH_2CH_2OH$$

which can have widely varying solubility characteristics, depending upon the value of n , the number of OE groups added to the molecule. If $n = 10$, the material will be completely soluble in water and will show good surfactant properties. If n is as little as 5, its water solubility will decrease significantly, as will its usefulness as a surfactant. If n is taken to 20 or higher, high water solubility will be maintained, but most of the good surfactant qualities will again be lost. For n less than 5, the material will have little significant water solubility.

By going down a slightly different road, the parent hydrocarbon can be sulfated to yield dodecane sulfonic acid, $CH_3(CH_2)_{10}CH_2SO_3H$, which closely resembles the sulfuric acid ester and has similar miscibility with water. When neutralized with the proper base, the resulting material is an excellent surfactant. It should be noted that while the sulfonic acid is related to the ester, their solution and surfactant properties are not identical, so that their potential applications may be different as well.

If the original dodecane molecule were terminally chlorinated and reacted with trimethylamine, the resulting compound would be dodecyltrimethylammonium chloride, $CH_3(CH_2)_{10}CH_2N^+(CH_3)_3Cl^-$, a water-soluble compound exhibiting some surfactant properties, but not generally as useful as the anionic analogs. The utility of such compounds is limited not so much by their surface activity as by their interaction with various oppositely charged components found in practical systems .

To this point we have covered the first three of the four general classes of surfactants defined previously. To produce an example of the fourth class, an amphoteric or zwitterionic surfactant, it is only necessary to react the dodecyl chloride prepared above with a difunctional material such as N ,N -dimethyl-3-aminopropane-1-sulfonic acid

$$(CH_3)_2NCH_2CH_2CH_2SO_3H$$

The result is just one of several possible chemical types that possess the amphoteric or zwitterionic character of this class of materials.

$$CH_3(CH_2)_{10}CH_2N^+(CH_3)_2CH_2CH_2CH_2SO_3^-$$

The number of modifications of the dodecane molecule that can lead to materials with good surfactant characteristics is limited primarily by the imagination and skill of the organic chemist --- and by the time and money available for indulgence in creative molecular architecture. In each example discussed, a solubilizing group has been added to the basic hydrophobe to produce materials with varying amounts of useful surfactant characteristics. When one considers the wide variety of hydrophobic groups that can be coupled, as was dodecane, with just the hydrophiles discussed, the number of combinations becomes impressive. When viewed in that light the current existence of 1500 or so distinct surfactant structures doesn't seem quite so outrageous.

THE CLASSIFICATION OF SURFACTANTS

With all of the possible chemical structures available for surfactant synthesis, it is necessary to have some system of classification to guide the user to the material best suited to immediate and future needs.

Surfactants may be classified in several ways, depending on the intentions and preferences of an author or user. One of the more common schemes relies on classification by the application under consideration, so that surfactants may be classified as emulsifiers, foaming agents, wetting agents, dispersants, etc. For the user whose work is confined to one type of application, such a classification scheme has certain obvious advantages. It does not, however, tell much about the specific chemical nature of the surfactant, nor does it give much guidance as to other possible uses of a material.

Surfactants may also be generally classified according to some physical characteristic such as water or oil solubility or stability in harsh environments. Alternatively, some specific aspect of the chemical structure of the materials in question may serve as the primary basis for classification, an example being the type of linking group between the hydrophile and the hydrophobe (eg, oxygen, nitrogen, amide, sulfonamido). Perhaps the most useful scheme from a general point of view, however, is that based upon the overall chemical structure of the materials in question. In such a classification system, it is easier to correlate chemical structures with interfacial activity, and thereby develop some general rules of surfactant structure--performance relationships.

The simplest structural classification procedure is that in which the primary type is determined by the nature of the solubilizing functionality (the lyophilic group or the hydrophile in aqueous systems). Within each primary classification by solubilizer there will exist subgroups according to the nature of the lyophobic moiety. It is possible to construct a classification system as complex as one might like, breaking down the lyophobic groups by their finest structural details such as branching and unsaturation. Such extremes, however, can introduce unnecessary complications in any discussion of structure-performance relationships, especially since industrially important surfactant systems often consist of several isomers or homologs, or other complex mixtures.

In aqueous systems, which constitute by far the largest number of surfactant applications, the hydrophobic group generally includes a long-chain hydrocarbon radical, although there are examples using halogenated or oxygenated hydrocarbon or siloxane chains. The hydrophilic group will be an ionic or highly polar group that can impart some water solubility to the molecule. The most useful chemical classification of surface-active agents is based on the nature of the hydrophile, with subgroups being defined by the nature of the hydrophobe. The four general groups of surfactants are defined as follows:

1. Anionic, with the hydrophilic group carrying a negative charge such as carboxyl (RCOO⁻), sulfonate (RSO₃⁻), or sulfate (ROSO₃⁻).

2. Cationic, with the hydrophile bearing a positive charge, as for example, the quaternary ammonium halides ($R_4N^+Cl^-$).

3. Nonionic, where the hydrophile has no charge but derives its water solubility from highly polar groups such as polyoxyethylene (---OCH_2CH_2O---) or polyol groups.

4. Amphoteric (and zwitterionic), in which the molecule contains, or can potentially contain, both a negative and a positive charge, such as the sulfobetaines, $RN^+(CH_3)_2CH_2CH_2SO_3^-$.

Surfactant Solubilizing Groups

The solubilizing groups of modern surfactants fall into two general categories: those that ionize in aqueous solution (or highly polar solvents) and those that do not. Obviously, the definition of what part of a molecule is the solubilizing group depends upon the solvent system being employed. For example, in water the solubility will be determined by the presence of a highly polar or ionic radical, while in organic systems the active group (in terms of solubility) will be the organic moiety. It is important, therefore, to define the complete system under consideration before discussing surfactant types.

The functionality of ionizing hydrophiles derives from a strongly acidic or basic character, which permits the formation of true, highly ionizing salts on neutralization. In this context, the carboxylic acid group, while not generally considered as such, is classed as a strong acid. A weak acid would be an alcohol or phenol. The nonionizing, or nonionic hydrophilic groups, on the other hand, have functionalities or element groups that are individually rather weak hydrophiles but have an additive effect so that increasing their number in a molecule increases the magnitude of their solubilizing effect.

The most common hydrophilic groups encountered in surfactants today are illustrated in Table 3.1 where R designates some suitable hydrophobic group that imparts surface activity, M is an inorganic or organic cation, and X is an anion (halide, acetate, etc). The list is in no way complete, but the great majority of surfactants available commercially fall into one of the classes.

It is possible, and sometimes even advantagous, to combine two or more of the above functionalities to produce materials with properties superior to a monofunctional analog. Prime examples of that would be the alcohol ether sulfates in which a POE nonionic material is terminally sulfated $R(OCH_2CH_2)_nOSO_3^-$ M^+ and, of course, the zwitterionic and amphoteric materials noted, which often exhibit the advantages of both ionic and nonionic surfactants while having fewer of their potential

Table 3.1. The most commonly encountered hydrophilic groups in commercially available surfactants.

Sulfonate	$R\text{-}SO_3^-\ M^+$
Sulfate	$R\text{-}OSO_3^-\ M^+$ *anionic*
Carboxylate	$R\text{-}COO^-\ M^+$
Phosphate	$R\text{-}OPO_3^-\ M^+$
Ammonium	$R_xH_yN^+X^-$ (x = 1-3, y = 4-x)
Quaternary ammonium	$R_4N^+X^-$ *cationic*
Betaines	$RN^+(CH_3)_2CH_2COO^-$
Sulfobetaines	$RN^+(CH_3)_2CH_2CH_2SO_3^-$
Polyoxyethylene (POE)	$R\text{-}OCH_2CH_2(OCH_2CH_2)_nOH$
Polyols	Sucrose, sorbitan, glycerol, ethylene glycol, etc
Polypeptide	$R\text{-}NH\text{-}CHR\text{-}CO\text{-}NH\text{-}CHR'\text{-}CO\text{-}...\text{-}CO_2H$
Polyglycidyl	$R\text{-}(OCH_2CH[CH_2OH]CH_2)_n\text{-}...\text{-}$ $OCH_2CH[CH_2OH]CH_2OH$

nonionic (handwritten brace for the last four rows)

drawbacks. The "hybrid" classes of surfactants, while not yet composing a large fraction of total surfactant use, can be particularly useful because of their flexibility and, especially in personal care items such as shampoos, because their low level of eye and skin irritation.

 Building on the basic hydrophilic functionalities, we will now turn our attention to some of the specifics of the structural subgroups based on the nature of the hydrophobic groups commonly encountered.

Common Surfactant Hydrophobic Groups

 By far the most common hydrophobic group used in surfactants is the hydrocarbon radical having a total of 8 to 20 carbon atoms. Commercially there are two main sources of supply for such materials that are both inexpensive enough and available in sufficient quantity to be economically feasible: biological sources such as agriculture and fishing, and the petroleum industry (which is, of course, ultimately biological). Listed below are the most important sources of hydrophobic groups, along with some relevant comments about each. There are, of course, alternative synthetic routes to the same basic molecular type, as well as other surfactant types that require more elaborate synthetic schemes. Those shown, however, constitute the bulk of the synthetic materials used today. Each source of raw materials may have its own local geographic or economic advantage, so that nominally identical surfactants may exhibit slight differences in activity due to the subtle influences of raw materials variations. Such considerations may not be important for most applications but should be kept in mind in critical situations.

 Natural fatty acids. Obtained primarily from triglyceride esters, the most useful members of the group have from 12 to 18 carbon atoms. The chains are usually saturated, although some unsaturated examples are employed (especially oleic acid). The chains usually have very little branching or heteroatom substitution.

Paraffins. These materials are obtained from petroleum distillates boiling higher than gasoline. They are normally saturated materials with 10 to 20 carbon atoms. The mixture will normally contain many branched isomers, some cyclic materials, and aromatics.

Olefins. The surfactant range olefins (C_{10}--C_{20}) are generally prepared by oligomerization of ethylene or propene, or by cracking of higher molecular weight petroleum fractions. They may be terminal (alpha) olefins or with internal unsaturation, depending of the particular process and conditions used.

Alkyl benzenes. These materials are prepared by the Friedel-Crafts reaction between olefins and benzene. The reaction may be carried out using surfactant-range olefins, or by using olefins such as propene or butene under conditions where a limited amount of oligomerization accompanies the alkylation. Generally the alkyl group will contain an average of 8 to 12 carbon atoms. The products will be highly branched, with the benzene being substituted randomly along the hydrocarbon chain.

Alcohols. Long-chain alcohols (C_8 --C_{18}) may be prepared by the catalytic reduction of fatty acid esters or by the oxidation of oligomers of ethylene. They may have an even or odd number of carbon atoms, with significant amounts of secondary alcohols usually being produced.

Alkylphenols. These materials are produced by the reaction of phenol with olefins. The products are a mixture of branched alkyl substituents with random substitution of the ring along the chain and mixed-ring substitution with respect to the hydroxyl (*ortho* , *meta* , and *para*).

Polyoxypropylenes. This is the most important class of hydrophobe containing noncarbon atoms in the chain. They are prepared by the base-catalyzed oligomerization of propylene oxide. They are particularly important in the preparation of block copolymer surfactants with ethylene oxide.

Fluorocarbons. Fluorocarbons are prepared primarily by the electrolytic substitution of fluorine for hydrogen on the carbon chain of carboxylic acid fluorides or sulfonyl fluorides. They may be completely fluorinated (perfluoro-) or have a terminal hydrogen atom. In that respect, it is important to know which type of chain is present in a material, since the properties of the two may differ significantly in critical applications. They may also be prepared by the oligomerization of tetrafluoroethylene. Linkage to many hydrophilic groups is accomplished through a short-chain hydrocarbon unit.

Silicones. These are generally oligomers of dimethylsiloxane that may then be functionalized with an appropriate solubilizing group.

Table 3.2. Some of the major modern applications of surfactants.

Industrial	Consumer Goods
Agricultural crop applications	Adhesives
Building materials	Dry cleaning fluids
Cement additives	Foods and beverages
Coal fluidization	Household cleaning and
Coating and leveling additives	laundering
Electroplating	Pharmaceuticals
Emulsion polymerization	Photographic products
Industrial cleaning	Soaps, shampoos, creams
Leather processing	
Lubrication	
Mold release agents	
Ore flotation	
Paper manufacture	
Petroleum recovery	
Surface preparations	
Textiles	
Waterproofing	

THE ECONOMIC IMPORTANCE OF SURFACTANTS

Economic considerations, in addition to the chemical questions that must be addressed when choosing a surfactant system, often play an important role in the selection process. Unless the cost of the surfactant is insignificant compared to the rest of the system, the least expensive material producing the desired effect usually will be chosen. Economics, however, cannot be the only factor in the choice, since the final performance of the system may well be of crucial importance. To make a rational selection, without resorting to an expensive and time-consuming trial-and-error approach, the formulator must have some knowledge of: (1) the characteristic chemical and physical properties of the available surfactant choices; (2) the surface and interfacial phenomena which must be controlled to achieve the desired results; (3) the relationships between the structural properties of the available surfactants and their effects on the pertinent interfacial phenomena; and (4) any restrictions to the use of certain materials, as in, for example, foods, cosmetics, or pharmaceuticals.

The applications of surfactants in science and industry are legion, ranging from primary processes such as the recovery and purification of raw materials in the mining and petroleum industries, to enhancing the quality of finished products such as paints, cosmetics, pharmaceuticals, and foods. Table 3.2 lists some of their major areas of application. As the technological demands placed upon product and process additives such as surfactants increase, it seems obvious that our need to understand the relationships between the chemical structures of those materials and their physical manifestations in particular circumstances becomes more important.

For many of the applications noted in Table 3.2, the desired properties will vary significantly. For that reason, such characteristics as solubility, surface tension reducing capability, critical micelle concentration (*cmc*), detergency power, wetting control, and foaming capacity may make a given surfactant perform well in some applications and less well in others. The "universal" surfactant that meets all of the varied needs of surfactant applications has yet to emerge from the industrial or academic laboratory.

There have been developed over the years a number of useful generalizations relating surfactant structures to their activity in a given application. Some of those generalizations are pointed out in the appropriate context in later chapters. For now, it is enough to remember each application has specific requirements that determine the utility of a particular structure in a given system. Some of the fundamental characteristics that must be evaluated for a surfactant proposed for some specific applications are given in Table 3.3.

When discussing the commercial aspects of surfactant technology, especially with regard to the raw materials sources for surfactants, it is common to refer to materials based on their original starting materials. While such classifications are useful from economic and technological points of view, the complex natures of such materials make it very difficult to illustrate the role of chemical structures in determining surfactant properties. Later discussions, therefore, will be couched more in terms of structure than of source. It should always be kept in mind, however, that nominally identical surfactants derived from different raw materials may exhibit significant differences in activity due to different isomer distributions.

Table 3.3. Typical (but not all) characteristics for surfactants which must be evaluated for various applications.

Application	Characteristics
Detergency	Low cmc, good salt and pH stability, biodegradability, good foaming properties
Emulsification	Proper HLB, environmental and biological (safety) aspects for application
Lubrication	Chemical stability, adsorption at surfaces
Mineral flotation	Proper adsorption characteristics on the ore(s) of interest, low cost
Petroleum recovery	Proper wetting of oil bearing formations, microemulsion formation and solubilization properties, ease of emulsion breaking after oil recovery
Pharmaceuticals	Biocompatibility, toxicity

SURFACTANTS IN THE ENVIRONMENT

The use of surfactants throughout the world is increasing at a rate in excess of the population growth because of generally improved living conditions and processed material availability in the less industrially developed Third World countries. Hand in hand with increased surfactant use go the problems of surfactant disposal. As the more developed nations have learned by painful and expensive experience, the ability of an ecosystem to absorb and degrade waste products such as surfactants can significantly affect the potential usefulness of a given material.

Of particular importance are the effects of surfactants on ground water and waste treatment operations. Although it may be technologically possible to remove all detectable residual surfactants physically or chemically from effluent streams, the economic costs would undoubtedly be totally unacceptable. The preferred way to address the problem is to allow Nature to take its course and solve the problem by natural biodegradation mechanisms.

Biodegradation of Surfactants

Biodegradation may be defined as the removal or destruction of chemical compounds through the biological action of living organisms. For surfactants, such degradation may be divided into two stages: (1) *primary degradation* , leading to modification of the chemical structure of the material sufficient to eliminate any surface-active properties; and (2) *ultimate degradation* , in which the material is completely removed from the environment as carbon dioxide, water, inorganic salts, or other materials that are the normal waste byproducts of biological activity. Years of research indicate that it is at the first stage of primary degradation that the chemical structure of a surfactant molecule most heavily impacts biodegradability.

Some of the earliest observations on the biodegradability of synthetic surfactants indicated that linear secondary alkyl sulfates (LAS) were biodegradable, while the branched alkylbenzene sulfonates (ABS) in use at the time were much more resistant to biological action. Continued investigation showed that the distinction between the LAS and ABS surfactants was not nearly as clear as first thought; that is, the observed differences in biodegradability did not stem from the presence of the benzene ring in ABS systems. It was found that the biodegradability of a particular ABS sample depended to a large degree upon the source, and therefore the chemical structure, of the sample. Early producers of ABS surfactants used either petroleum-derived kerosene (largely linear) or tetrapropylene (highly branched) as their basic raw material, without great consideration for the structural differences between the two. As a result, great variability was found in the assay of materials for determination of biodegradability. In fact, those materials derived from tetrapropylene showed little degradation while the nominally identical materials based on the kerosene feedstocks were much more acceptable in that respect. The difference lay in the degree of branching in the respective alkyl chains.

It was subsequently shown conclusively that the resistance of tetrapropylene ABS surfactants to biodegradation was a result of the highly branched structure of the alkyl group relative to that of the kerosene-derived materials and the linear alkyl sulfates. As a result of extensive research on the best available model surfactant compounds it was concluded that it was the nature of the hydrophobic group on the surfactant that determined its relative susceptibility to biological action, and that the nature and mode of attachment of the hydrophile was of minor significance. Subsequent research using an increasingly diverse range of molecular types has continued to support those early conclusions.

Over the years, the following generalizations have developed which seem to cover the biodegradation of most surfactant types:

1. The chemical structure of the hydrophobic group is the primary factor controlling biodegradability; high degrees of branching, especially at the alkyl terminus, inhibit biodegradation.

2. The nature of the hydrophilic group has a minor effect on biodegradability.

3. The greater the distance between the hydrophilic group and the terminus of the hydrophobe, the greater is the rate of primary degradation.

For an in-depth discussion of the complex relationship between the chemical structure of surface-active materials and biodegradability the interested reader is referred to work of Swisher cited in the Bibliography.

CHAPTER 4

LONG-RANGE ATTRACTIVE FORCES

When two atoms bind to form a typical nonionic molecule, the forces involved in bond formation are referred to as *covalent forces*, and the resulting bonds, *covalent bonds*. In such bond formation, the electrons involved become shared by two (or more) atoms and the individual characters of the atoms are, to some extent, lost. The formation of *metallic bonds* also involves the sharing of electrons among a number of atoms so that the nature of the "bonded" atoms differs from that of a "free" atom of the same metal.

We know from the general principles of chemistry that the exact nature and number of covalent bonds formed by a given atom depends on its location in the periodic table, relative to the other atoms involved. In addition, covalent bonds have certain characteristic bond lengths and bond angles which depend on the atoms involved; that is, they are *directional*. The number of bonds an atom may form, its *valency*, is a fundamental property and controls how that atom contributes to the overall nature of the resulting molecule. The number, length, and direction of bonds to an atom will control how the atoms and resulting molecules can arrange themselves in three-dimensional patterns or lattices. For example, the special bonding properties of the carbon atom results in its ability to form the perfectly regular three-dimensional lattice structure of diamond or the more two-dimensional structure of graphite, as well as determining the three-dimensional structure of all carbon-based organic compounds.

Obviously, covalent bonding is of primary importance to the nature of things as we know them. However, covalent bonds are very localized in the bonding regions between atoms, and are *short range* in the sense that they act over bond distances of 0.1--0.2 nm. The energies of normal covalent bonds range from 150 to 900 kJ mol^{-1} ($\approx$ 100--300 kT), and generally decrease in strength as the bond length increases. Some typical covalent bond energies are given in Table 4.1.

While covalent bonds are strong, they are restricted in the range of their actions. They are, in effect, limited to the interactions between atoms involved in molecular formation and formal chemical reactions. They are,

Table 4.1. Characteristic strengths of covalent bonds.

Bond type	Strength (kJ mol^{-1})	Bond type	Strength (kJ mol^{-1})
F---F (F$_2$)	150	C---H (CH$_4$)	430
N---O (NH$_2$OH)	200	O---H (H$_2$O)	460
C---O (CH$_3$OH)	340	C=C (C$_2$H$_4$)	600
C---C (C$_2$H$_6$)	360	C=O (HCHO)	690
Si---O	370	C≡N (HCN)	870

therefore, *chemical* forces. In most systems involving surface and colloidal phenomena, on the other hand, one is not so much concerned with molecular transformations as with the interactions between discrete, nonbonded atoms or molecules over distances significantly greater than molecular bond dimensions. Such interactions are conventionally called *physical interactions* , resulting from *physical forces,* and producing *physical bonds* . While such physical interactions do not, in general, involve electronic transformations analogous to covalent bond formation, they can, under some circumstances, be equally strong. They are not usually referred to as formal bonds, however, because such interactions normally lack the specific characteristics of covalent bonds --- strong directionality, fixed stoichiometry, and high specificity. While physical interactions may perturb the electronic configurations of the atoms or molecules involved, the electrons themselves remain associated with their original system. (An exception to this would be heterogeneous catalytic reactions, to be addressed later.)

While physical interactions lack the qualifications to be considered as classical "bonds," it is through such interactions that all but the smallest atoms and molecules are bound together to form the soilds and liquids we normally encounter at room temperature. They also exhibit themselves as the fundamental factors involved in all colloidal systems, all biological assemblies, and all natural phenomena not involving chemical interactions. That covers quite a lot of territory! Because of their importance, physical interactions will be covered in somewhat more theoretical detail below than most topics to be presented in the following chapters.

THE ROLE OF LONG-RANGE PHYSICAL FORCES

We have seen from the discussion in Chapter 2 that as two surfaces in a vacuum are separated to an effectively infinite distance, their free energies increase to some maximum value characteristic of the system involved. This means that the net force acting between the two surfaces must be attractive. That will always be the case for pure substances in vacuum, but in reality, many situations exist in which a maximum in the free energy--separation distance curve is encountered.

Many important systems and processes, especially biological assemblies such as cell walls and protein secondary and tertiary structures form as a result of physical, inter- and intramolecular interactions. Such assemblies and conformations exist because the physical forces binding them together operate over distances greater than those of covalent bonds yet hold the various molecules at the proper distance and with the proper strength so that they can successfully carry out their vital functions.

Under certain circumstances it is possible to utilize physical interactions to maintain surfaces at some minimum distances of separation as a result of an energy maximum in the interaction energy. The practical result of such "long-range" energy maxima is that, properly utilized, they can prevent or at least retard the natural tendency of surfaces to approach and join spontaneously. This effect is especially imporatnt in colloidal systems, foams, emulsions, etc. Before considering the sources of the energy maxima, however, it will be useful to have a little more detailed understanding of the sources of the original attractive interactions that must be overcome.

Although our aim is to understand the interactions of various surfaces, it must be remembered that a surface is nothing more than a collection of individual atoms or molecules, and that the macroscopic properties of a surface will be a reflection of the interactions of all of the individual atomic or molecular interactions involved. Therefore, we will begin the discussion by addressing the question of the source and nature of the various types of interactions experienced by individual units (atoms or molecules), followed by an integration of those interactions over all of the units in the surface.

CLASSIFICATION OF PHYSICAL FORCES

The fundamental physical forces controlling the *non-chemical* interactions among atoms and molecules are of two kinds --- formal coulombic or elecrostatic interactions, and those lumped together under the general term of *van der Waals forces* . The coulombic interactions can be somewhat nebulous as to their classification as chemical or physical interactions because they are, of course, involved in chemical bonding in ionic molecules. However, they may also be classified as physical interactions because they function over distances much greater than those of covalent bonds and produce dramatic effects on the interactions between ions, ions and polar molecules, and ions and nonpolar molecules (and ultimately particles and surfaces).

Coulombic interactions are by far the strongest (in absolute terms) of the physical interactions, equaling and exceeding the magnitude of covalent bonds. They are not, however, the most widely encountered type of interaction, since they are present only in systems containing charged species. Nor are they even always the most important, since many circumstances, as will be seen, can greatly diminish or completely nullify their net effect in a system.

The term "van der Waals forces" is often encountered in contexts where it implies that only one type of interaction is involved. In fact, however, van der Waals forces include three separate types of atomic and molecular interactions, each of which has its own characteristics, its own theoretical basis, and its own limitations. Two of the three forces are reasonably easy to understand because they are based on relatively straightforward electrostatic principles similar to those used for the much stronger coulombic interactions. The third is sometimes less clear because it is quantum mechanical in origin, and quantum mechanics, even is its simplest form, seems to effect adversely the sanity of many who touch thereon. In any case, the following discussion will attempt to present the ideas involved as simply as possible while still conveying the essence of the subject.

For reasons of convenience, the discussion will begin with the "simplest" force, that of direct coulombic interaction between two charged species. The presentation of the material has been kept brief because the intention is to provide an understanding of the basis of charge--charge interactions, leaving the fine points and complications (of which there can be many) to more advanced texts on the subject referenced in the Bibliography. Other interactions involving ions will be addressed as well.

The interactions generally grouped as van der Waals forces will then be covered more or less in order of increasing complexity --- that is, interactions of permanent dipoles, induced dipolar interactions, and finally the quantum mechanical forces.

Coulombic or Electrostatic Interactions

The interaction between two charged atoms or molecules is potentially the strongest form of physical interaction to be considered in surface and colloidal systems. The basic concepts and equations involved are fundamental to many areas of physics and chemistry and will not be developed in detail here. More will be said about them in Chapter 5 in the context of repulsive interactions between surfaces. A few basic points of review, however, will be useful in order to facilitate reference to them in later discussions of dipolar interactions.

For two point charges Q_1 and Q_2, the free energy of interaction $w(r)$ is given by

$$w(r) = Q_1 Q_2 / 4\pi\varepsilon_0 \varepsilon r = z_1 z_2 e^2 / 4\pi\varepsilon_0 \varepsilon r \qquad (4.1)$$

where ε_0 is the permittivity of a vacuum or free space, ε is the relative permittivity or dielectric constant of the medium, and r is the distance between the two charges. The right-hand form of the equation is commonly used, where the value of Q can be readily specified in terms of the sign and valency of each ion, z, and the elementary charge, e (= 1.602×10^{-19} Coulombs, C).

The *force* of the coulombic interaction, F_c, is the differential with respect to r of the free energy

$$F_c = dw(r) / dr = Q_1 Q_2 / 4\pi\varepsilon_0\varepsilon r^2 = z_1 z_2 e^2 / 4\pi\varepsilon_0\varepsilon r^2 \quad (4.2)$$

For two charges of the same sign, both $w(r)$ and F will be positive, which means the interaction will be repulsive; for unlike charges they will be attractive. In terms of magnitude, the force (whether attractive or repulsive) is at a maximum when the distance of separation r is a minimum, that is, when the two ions are in contact and r equals the sum of the two ionic radii. For example, for a sodium and a chloride ion in contact, r will be ≈ 0.276 nm and the binding energy will be

$$w(r) = (-1)(+1)(1.602 \times 10^{-19})^2 / 4\pi(8.854 \times 10^{-12})(0.276 \times 10^{-9})$$

$$= -8.4 \times 10^{-19} \text{ J}$$

Throughout the following discussions, reference will be made to a *standard unit of thermal energy*, kT, where k is Boltzmann's constant and T is temperature (K). Thus, the reference energy at room temperature (≈ 300 K) will be $kT = (1.38 \times 10^{-23})(300) = 4.1 \times 10^{-21}$ J. The energy of the sodium and chloride atom interaction, then, is approximately $200 kT$. From Table 4.1, an average covalent bond energy may be estimated to be in the range of 9×10^{-19} J or $220\ kT$. Obviously, then, the coulombic interactions must be considered to be at least equal in strength to covalent bonds.

According to eq. 4.2 the magnitude of the coulombic interaction falls off as the inverse square of the distance between the charges. A quick calculation shows that, in a vacuum, the interaction energy will fall to kT only as the separation distance approaches 60 nm, a large distance in the normal world of atoms, ions, and molecules.

Whenever we talk about the free energy of a system, we are usually, in fact, talking about a *change* in free energy rather than an absolute energy. It is therefore necessary to keep in mind that such changes in free energy are normally compared to some *reference state* that is specified or understood. When considering intermolecular interactions, the usual reference state to be considered is that in which the ions or molecules involved start with a separation of $r = \infty$ and come together to form a condensed solid or liquid state. If the condensation occurs in the gaseous phase so that the surrounding medium is a vacuum, the values for the demoninator in eq. 4.2 will be $\varepsilon_0 = 8.854 \times 10^{-12}$ C^2 $J^{-1}m^{-1}$ and $\varepsilon = 1$, the dielectric constant of a vacuum. If the condensation occurs in a condensed liquid, then ε becomes the dielectric constant of the liquid. Normally encountered liquid media range from hydrocarbons ($\varepsilon \approx 2$) to water ($\varepsilon = 80$). Obviously, in media other than vacuum, the coulombic interactions to be considered will vary significantly with the nature of the intervening medium.

It is important to point out that there exists an apparent discrepancy between eq. 4.2 and experimental measurements of such interactions. It is generally found in experiments that all intermolecular forces decay faster than the inverse 4th power of distance (r^{-4}), while eq. 4.2 indicates a much slower inverse square relationship. The apparent contradiction is removed, however, if one remembers that the relationship is derived for two isolated ions, while in reality, ions will interact to some extent with all ions of opposite charge in their vicinity. The presence of these "associated" ions, which are not considered in simple one-to-one calculations of the interaction potential, acts to "screen" interactions between the charges and reduces the strength of their interaction and, therefore, the range over which its effects are significant. Even with that, however, coulombic interactions remain strong and significant over distances much greater than covalent bonds and are therefore considered to be long-range forces.

Other Interactions Involving Ions

Many commonly encountered materials include ionic and molecular units that may interact by mechanisms that may be considered to be "hybrids" between strictly coulombic phenomena and the van der Waals interactions to be considered below. Although weaker and perhaps more subtle, interactions involving a charged species and a neutral unit can make a significant contribution to the total interaction energy of a system.

Dipoles and Polarization Phenomena

Many molecules do not carry formal electrical charges, so that their mutual interactions do not involve the coulombic interactions discussed above. However, if one examines the chemical structures of many, if not most useful chemical species, including marcomolecules, proteins, and drugs, it is apparent that they often include bonds that can impart an overall polar nature to the molecule as *permanent dipoles* , or they can be *polarized* by the effect of neighboring electric fields producing *induced dipoles* . The presence of permanent or induced dipoles means that the molecules can become involved in specific interactions with charged species, other dipoles, or nonpolar molecules, and those interactions can significantly affect the physical characteristics of the system.

Some molecules have dipoles that result from differences in the electronegativity of the covalently bonded atoms, an example being the commonly encountered carbonyl group ($C^{\delta+}=O^{\delta-}$) found in many organic molecules. Other important molecules, for example the amino acids (and therefore the proteins they make up), contain acid and base functionalities that usually exist as *zwitterions* in which the two functionalities form, in effect, an internal salt. For glycine, the molecule takes the form

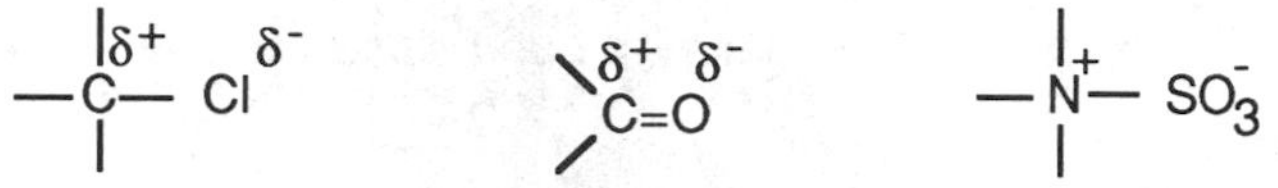

$$-\overset{\displaystyle |}{\underset{\displaystyle |}{C}}\!\!\overset{\delta^+}{}\!\!-Cl^{\,\delta^-} \qquad \overset{\displaystyle \diagdown}{\underset{\displaystyle \diagup}{C}}\!\!\overset{\delta^+}{=}\!O^{\,\delta^-} \qquad -\overset{\displaystyle |}{\underset{\displaystyle |}{N}}\!\!\overset{+}{}\!\!-SO_3^-$$

Figure 4.1. Typical dipolar molecular structures.

$$CH_2(NH_3^+)COO^-$$

In water at the isoelectric point, the molecule is electrically neutral, but the charge separation in the zwitterion produces a strong dipole that to a great extent governs the nature of the interactions of the molecule. At pH's other than the isoelectric point, the molecule becomes formally charged and coulombic factors prevail. Obviously, the interactions of molecules of these types can become quite complex and may exhibit a significant sensitivity to their solvent environment.

The Dipole Moment

A dipole arises due to the presence of an unsymmetrical distribution of electron density within a molecule, due either to a formal charge separation, such as in amino acids, or due to differences in the electronegativities of the atoms forming a covalent bond, as in carbonyl compounds or water. An isolated, neutral atom, of course, cannot have an unsymmetrical electron distribution; therefore atoms cannot be dipolar in nature. That is not to say, of course, that they cannot be *polarized* , or have their electron cloud distorted by an external electric field, but that subject is considered later. The *dipole moment* , μ, of a molecule is defined as

$$\mu = ql \qquad\qquad (4.3)$$

where l is the distance between the two charges, $+q$ and $-q$, of the dipole (or the positive and negative ends of the asymmetrical electron cloud in a covalent bond, Figure 4.1). For a zwitterionic species, for example, where $q = \pm e$ and the charge separation is $l = 0.1$ nm, the dipole moment will be

$$\mu = (1.602 \times 10^{-19})(1 \times 10^{10}) = 1.6 \times 10^{-29} \text{ C m} = 4.8 \text{ D}$$

where D is the Debye unit $= 3.336 \times 10^{-30}$ C m.

The magnitudes of the dipole moment for commonly encountered bonds and molecules range from approximately 1 D for slightly polar organic molecules, such as chloroform, to closer to 2 for water, carboxylic acids, and similar species. The dipole moments of complex molecules can, in principle, be calculated by taking a vectorial sum of all of the *bond moments* of the functionalities present. That can become quite complicated (and of marginal utility) in the case of complex organic molecules, especially since modern experimental techniques make their

Table 4.2. Characteristic molecular and bond dipole moments, μ (in Debye units).

Molecule	μ	Bond	μ
n -Alkanes	0	C---C	0
Bezene (C_6H_6)	0	C=C	0
Carbon tetrachloride (CCl_4)	0	C---N	0.22
Carbon dioxide (CO_2)	0	N---O	0.3
Chloroform ($CHCl_3$)	1.06	C---H	0.4
Hydrogen chloride (HCl)	1.08	C---O	0.74
Ammonia (NH_3)	1.47	N---H	1.31
Methanol (CH_3OH)	1.69	O---H	1.51
Acetic acid (CH_3COOH)	1.7	C---Cl	1.5-1.7
Water (H_2O)	1.85	F---H	1.94
Ethylene oxide (C_2H_4O)	1.9	N=O	2.0
Acetone (CH_3COCH_3)	2.85	C=O	2.3-2.7
N ,N -Dimethyl formamide	3.82		
Acetonitrile	3.92		

direct determination relatively easy. For reference purposes, the dipole moments and bond moments of several common molecules and functionalities are given in Table 4.2.

The Polarization of Nonpolar Atoms and Molecules

Molecules having no permanent dipole can also take part in electrostatic interactions as a result of deformations of their electron clouds by the presence of an external electric field. Polarization phenomena are referred to as *induced dipole* interactions and essentially involve the distortion of a normally symmetrical electron cloud as a result of the presence of some strongly polar unit in the vicinity. Theprocess is illustrated schematically in Figure 4.2.

All atoms and molecules are, in principle, polarizable. The ease with which the electron cloud of a given species can be distorted, its "polarizability," α, is defined by the strength of the induced dipole formed when placed under the influence of an electric field of strength E,

$$\mu_{ind} = \alpha E \qquad (4.4)$$

For a nonpolar molecule, the induced dipole results simply from the displacement of the center of mass of the negative electron cloud with respect to that of the positive nucleus. In polar molecules, the situation becomes more complex because the net induced dipole, μ'_{ind}, will

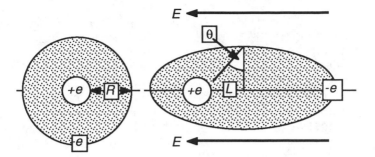

Figure 4.2. Schematic illustration of the polarization mechanism: (a) isolated atom with no external field, $\mu_{ind} = 0$; (b) with applied external field, E, the electrons are shifted by distance L, and $\mu_{ind} = Le = \alpha_o E$.

depend on the orientation of the induced dipole, μ_{ind}, with respect to the permanent dipole of the molecule.

To illustrate the point, imagine an atom with one associated electron of charge $-e$ in a spherically symmetrical orbital of "radius" R, the atomic radius. If the electron is subjected to an external field E, the orbit of the electron will be shifted by a distance l away from the nucleus. The induced dipole moment μ_{ind} is determined by the external force, F_{ext}, acting on the electron(s) as a result of the field E

$$F_{ext} = e\,E \tag{4.5}$$

balanced against the attractive force between the electron and the nucleus, the internal restoring force, F_{int}, where

$$F_{int} \approx (e\,/4\pi\varepsilon_o R^3)\,\mu_{ind} \tag{4.6}$$

At equilibrium $F_{int} = F_{ext}$ so that

$$\mu_{ind} = 4\pi\varepsilon_o R^3 E = \alpha_o E \tag{4.7}$$

where the polarizability, α_o is

$$\alpha_o = 4\pi\varepsilon_o R^3 \tag{4.8}$$

The phenomenon related to the displacement of electrons around an atom or molecule is termed *electronic polarizability* and has the dimensions $C^2\ m^2\ J^{-1}$. As a rule of thumb, it is usually found that the magnitude of the polarizability of a given atom or molecule will be of the order of, but slightly less than, its radius cubed multiplied by the term

$4\pi\varepsilon_0$. For water, for example, $\alpha_0/4\pi\varepsilon_0 = 1.48 \times 10^{-30}$ m^3 which is approximately equal to $(0.114$ nm$)^3$, some 15% less than the radius of water (0.135 nm).

The electronic polarizabilities of some characteristic atoms and molecules are given in Table 4.3. It has been found that the polarizability of a molecule can often be calculated with an accuracy of a few percent by summing the polarizabilities of isolated bonds within the molecule. While the procedure works well for isolated bonds, systems involving neighboring nonbonding electrons or delocalized structures, such as aromatic rings, are not so well behaved. In those circumstances, *group polarizabilities* have been assigned. Typical values of bond and group polarizabilities are also included in Table 4.3.

The Polarization of Polar Molecules

Equation 4.7 gives an expression for what is termed the *electronic polarizability* of a spherically symmetrical atom or molecule. If a molecule has a freely rotating permanent dipole, which has a time-averaged dipole moment of zero, in the presence of an external field E, there may develop an induced *orientational dipole*. This would then be related to the *orientational polarizability* of the molecule. If at some instant, the permanent dipole μ of the molecule is at an angle θ to the applied field, its energy in the field will be given by

Table 4.3. Electronic polarizabilities of typical atoms, molecules, bonds, and molecular groups, in units of $\alpha_0 / 4\pi\varepsilon_0$ (Å^3).

Atoms and molecules					
He	0.20	NH$_3$	2.3	CH$_2$=CH$_2$	4.3
H$_2$	0.81	CH$_4$	2.6	C$_2$H$_6$	4.5
H$_2$O	1.48	HCl	2.6	Cl$_2$	4.6
O$_2$	1.60	CO$_2$	2.6	CHCl$_3$	8.2
Ar	1.63	CH$_3$OH	3.2	C$_6$H$_6$	10.3
CO	1.95	Xe	4.0	CCl$_4$	10.5

Bond polarizabilities					
Aliphatic C---C	0.48	Aliphatic C---H	0.65	C=C	1.65
Aromatic C=C	1.07	N---H	0.74	C---Cl	2.60

Molecular groups			
C---O---H	1.28	CH$_2$	1.84
C---O---C	1.13	C=O	1.36

$$w(r, \theta) = -\mu E(r) \cos \theta \qquad (4.9)$$

so that the time-averaged induced dipole moment will be

$$\mu_{ind} = \mu^2 E /kT < \cos^2 \theta > = (\mu^2 / 3kT)E, \quad \mu E << kT \qquad (4.10)$$

From eq. 4.10, μ_{ind} is proportional to E, and $\mu^2 / kT = \alpha_{orient}$ represents an additional contribution to the overall polarizability of the molecule, the *orientational polarizability* . The total polarizability of the molecule, α, will be the sum of the electronic and orientational polarizabilities

$$\alpha = \alpha_o + \mu^2 / 3kT \qquad (4.11)$$

As an example, assume that a molecule has a permanent dipole moment of 1 D. At 300 K it will have an orientational polarizability of

$$\alpha_{orient} = (3.336 \times 10^{-30})^2 / 3(1.38 \times 10^{-23})300$$

$$= 9 \times 10^{-40} \, C^2 m^2 \, J^{-1} = (4\pi\varepsilon_o)8 \times 10^{-30} \, m^3$$

Such a value is comparable to the electronic polarizabilities of molecules as shown in Table 4.3. In the case of a very polar molecule in a very high field, as for example a water molecule next to a lithium ion, or at very low temperatures where molecular rotation is greatly limited, the polar molecule may become completely aligned with the field. At that point, the induced dipole moment will no longer be proportional to E, and the simple concept of the molecule's polarizability breaks down. Many of the unique, and sometimes perplexing, solvent characteristics of water can be traced to the ability of water molecules to be aligned and structured in solution by the presence of strong electric fields.

Ion--Dipole Interactions

Somewhere between the strong electrostatic and relatively weak dipole--dipole interactions to be discussed later lies an intermediate area involving ions and dipoles. If a charge Q lies at a distance r from the center of a polar molecule of moment μ and dipolar length l, with the orientation shown in Figure 4.3, the coulombic interaction between ion and dipole will be the sum of the interactions between Q and each end of the dipole with charges $\pm q$.

$$w(r) = - Qq /4\pi\varepsilon_o\varepsilon[1/AB - 1/AC] \qquad (4.12)$$

where

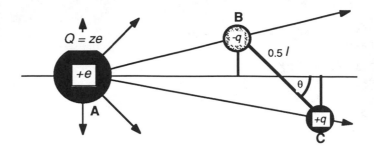

Figure 4.3. Schematic representation of ion--dipole interactions.

$$AB = \{[r - (1/2)\, l\, \cos\theta]^2 + [(1/2)\, l\, \sin\theta)]^2\}^{1/2} \quad (4.13)$$

and

$$AC = \{[r + (1/2)\, l\, \cos\theta]^2 + [(1/2)\, l\, \sin\theta)]^2\}^{1/2} \quad (4.14)$$

At distances greater than the dipole length the interaction energy can be estimated by

$$w(r,q) = -Q\mu \cos\theta / 4\pi\varepsilon_0 \varepsilon r^2 = -(ze)\mu \cos\theta / 4\pi\varepsilon_0 \varepsilon r^2 \quad (4.15)$$

Equation 4.15 gives the theoretical interaction between an ion of charge Q and a point dipole of moment μ, for which the length of the dipole is theoretically zero. In such a case, it is hard to visualize exactly what is meant by the statement that the dipole is pointing "toward" or pointing "away from" the interacting charged species. That point aside for the moment, when the dipole is pointing away from the charge ($\theta = 0°$, in Figure 4.3) the interaction will be at its most negative value (attractive). For $\theta = 180°$, the value will be its most positive (repulsive).

In fact, of course, point dipoles do not exist, so that some deviation from the simple theory provided here might be expected. It is only when the value of r approaches $2l$ that significant problems arise, however. Thus, for all reasonable ion--molecule separations and a dipole length of 0.1 nm or so, the use of eq. 4.15 produces no significant error. In zwitterions, however, in which the dipole arises from a formal separation of charges on different atoms in a molecule, the length of the dipole may be greater than 0.1 nm and larger deviations from eq. 4.15 are found. In that case, the total interaction must be calculated from each separate interaction between the ion and the ends of the dipole according to eq. 4.4. In those cases it is found that the total interaction is greater than that predicted by eq. 4.15.

Another consequence of the ion--dipole interaction is that it is often strong enough to bind polar molecules in the bulk liquid to the ion and establish an aligned structure within the overall matrix of the polar molecules. For the small monovalent cation Li^+ in the presence of water molecules (as opposed to *in* water), for example, the following data may be used: μ (water) = 1.85 D, r_w (the molecular radius of water) $\approx$ 0.14 nm, and r_{Li} (the ionic radius of Li^+) = 0.068 nm. Application of eq. 4.15 gives

$$w(r, q=0) = -\frac{(1.602 \times 10^{-19})(1.85 \times 3.336 \times 10^{-30})}{4\pi(8.854 \times 10^{-12})[(0.14 + 0.068) \times 10^{-9}]^2} = -2.05 \times 10^{-19} \text{ J}$$

which corresponds to a value of 123 kJ mol^{-1}, compared to the experimental value of 142 kJ mol^{-1}. Good correspondence is found between theory (eq. 4.15) and experiment for most of the small mono- and polyvalent cations. Because of their strong interaction with dipolar molecules such as water, cations are important in a number of phenomena steming from the structuring and aligning of water molecules, not the least of which is the nucleation of raindrops and snowflakes in cloud formations.

The above analysis assumes that the dielectric constant of the medium is unity. In *liquid* water, the interaction will be reduced by a factor of 80, the dielectric constant of water. Even there, however, the strength of the interaction can be significant with respect to kT for small di- and polyvalent ions and cannot be ignored for small monovalent ions.

Ion Solvation

In aqueous solution, when a bulk water molecule approaches an ion, it is not the same as when an isolated water molecule approaches an isolated ion. In the liquid phase, the molecule in question must replace a water molecule already associated with the ion, so that the net result, on a macro scale, is not an overall change in free energy, but simply an exchange of two water molecules. However, even here, eq. 4.15 suggests some finite interaction between the ion and individual water molecules in bulk water. The equation contains the orientational term cos θ which becomes important at small distances of separation.

When a water molecule is far away from the ion, it will be randomly oriented with respect to the ion, producing a spatial average for cos θ of zero. If the water molecule in question maintained that random orientation right up to the ion, the net interaction energy would, of course, be zero. However, as the approach takes place, the water molecule is no longer randomly oriented, but begins to assume orientations that produce a favorable (more negative) interaction. That is, the molecule becomes oriented relative to the random bulk molecules.

For small monovalent and polyvalent ions, the effect will produce a "shell" of oriented water molecules bound to the ion, with the orientation favoring θ = 0° for cations and 180° for anions. The shell of water molecules thus formed constitutes the *waters of solvation* or *hydration* of the ion. That is, the ions are *solvated* or *hydrated*. The number of water molecules associated with an ion (its *hydration number*) is characteristic of that ion but normally ranges between 4 and 6. Waters of hydration are not completely and irreversibly bound to a given ion. They are slowly (on a molecular scale) exchanged for other bulk water molecules.

Closely related to the hydration number is the *hydrated radius* of an ion in water, which is larger than its crystal lattice radius. Smaller ions, having the possibility of a closer approach to the water molecules and thus a stronger interaction, tend to have relatively larger hydrated radii than larger ions. On a molar basis they tend to structure the water more by orienting more molecules per ion. The effect of this hydrated radius is manifested in a number of solution physical properties including viscosity, conductivity, compressibility, diffusion, and a number of thermodynamic and spectroscopic properties of ionic solutions.

The structuring of polar solvent molecules, especially water, around an ion is not limited to the molecules directly "in contact" with the ion. The interaction is transmitted, to an increasingly reduced extent, to molecules in a second, third, etc, layer of surrounding solvent molecules. The structuring effect decays in an approximately exponential way but may extend several molecular diameters into the bulk liquid. This region of enhanced structuring in the solvent is referred to as the *solvation zone* around the ion. This *solvation* or *structuring* effect can have a number of important consequences for the interactions between ions, molecules, colloidal particles, and surfaces. If the orientation and mobility of solvent molecules near an ion differ from those in the bulk, one might expect, as is the case, that many physical properties of the solvation zone differ from those of the bulk. In particular, there are found to be differences in density, dielectric constant, conductivity, etc.

Of particular importance for present purposes is the fact that the dielectric constant of the solvation zone may differ significantly from the bulk value because the molecules there cannot respond to an imposed electric field in the same way as bulk liquid. Since the equations for calculating ionic, dipolar, and van der Waals interactions include the dielectric constant of the medium, differences such as those encountered in the solvation zone may have important consequences. In particular, when two ions approach close enough that their respective zones touch, their interaction energy may be much different than than predicted using the bulk value of ε. If the effective dielectric constant in the solvation zone is less than the bulk, as is most often the case, then the interaction will be stronger; if ε is increased (less common, but not unheard of) then the interaction will be decreased. Ion--solvent interactions are very specific and cannot be handled well by a general theory as yet. However, because of their potential importance in solvation-mediated processes (eg, nucleation, crystal growth), in colloidal stabilization, and in surface interactions, solvation effects or forces should be kept in mind to help explain seemingly anomolous experimental results.

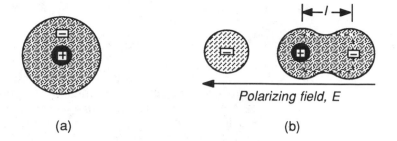

Polarizing field, E

(a)   (b)

Figure 4.4. Schematic illustration of the orientation of an induced dipole of length, l , relative to a polarizing field, E.

Interactions Between Ions and Nonpolar Molecules

When a nonpolar molecule is near an ion generating an electric field E, where

$$E(r) = (ze) / 4\pi\varepsilon_o \varepsilon r^2 \qquad (4.16)$$

the ion will induce a dipole moment in the molecule of

$$\mu_{ind} = \alpha E(r) = \alpha(ze) / 4\pi\varepsilon_o \varepsilon r^2 \qquad (4.17)$$

For a methane molecule with an electronic polarizability $\alpha_o = 2.6 \times 10^{-29}$ $C^2 m^2 J^{-1}$, located 0.4 nm from a monovalent ion ($E = 9 \times 10^9$ V m^{-1}), the induced dipole moment will be

$$\mu_{ind} = \alpha_o E = 4\pi\varepsilon_o(26 \times 10^{-30})(9 \times 10^9) = 2.6 \times 10^{-29} \text{ C m} = 0.78 \text{ D}$$

Considering the above induced dipole in a neutral molecule due to the presence of the ion, what will be the net interaction between the two species? If the ion involved is a cation, the induced dipole on the molecule will point away from the ion (Figure 4.4). An anion will induce a dipole pointing toward the ion. In either case, the overall interaction due to the induced dipole will be attractive. The induced dipole will interact with the ion with a "reactive" field given by

$$E_r = -2\mu_{ind} / 4\pi\varepsilon_o \varepsilon r^3 = -2\alpha E / 4\pi\varepsilon_o \varepsilon r^3$$

$$= -2\alpha(ze) / (4\pi\varepsilon_o \varepsilon)^2 r^5 \qquad (4.18)$$

The attractive force between ion and induced dipole will be

$$F = -2\alpha(ze)^2 / (4\pi\varepsilon_o \varepsilon)^2 r^5 \qquad (4.19)$$

for an overall interaction free energy of

$$w(r) = -\alpha(ze)^2 / 2(4\pi\varepsilon_0\varepsilon)^2 r^4 = -1/2\,\alpha E^2 \qquad (4.20)$$

The energy given by eq. 4.20 is one-half that expected for the interaction between an ion and an aligned, permanent dipole. The reason being that, in the process of inducing the dipole, some energy is used up in polarizing the nonpolar molecule. The energy used is that required to displace the center of the negative electron charge from the center of the positive nuclear charge.

The question of the full nature and consequences of coulombic interactions is, of course, much more complex than the material presented here. For excellent, more extensive discussions of the subject as it applies to surfaces and colloids the reader is referred to the works of Israelachvili, Adamson, and Kryuit in the Bibliography for this chapter. When the general topic of the stabilization of colloids is reached, the fundamental importance of coulombic interactions will become much more apparent. For now, however, our attention will turn to more subtle interactions involving ions and dipoles and those classed together as van der Waals forces.

VAN DER WAALS FORCES

It has been stated that there are four principle types of forces acting between atoms and molecules which control the fates of the units involved at the molecular and macroscopic levels. The first, coulombic interaction involving at least one formally charged species, was covered above. The remaining three forces make up what are commonly termed *van der Waals forces* and are comprised of three types of interaction. Of the three, two, those involving permanent and induced dipoles, are closely related to the coulombic forces, although they do not involve pure charge--charge interactions. The third is the most fundamental and universal force, and although generally the weakest of the three in absolute terms, is often the most important contributor to the total van der Waals interaction. This force is the so-called *London-van der Waals force* or *London dispersion force* first postulated by van der Waals in 1873[1] but actually explained by London in 1930.[2] Beginning with the simplest dipole--dipole interactions, each will be described in general terms.

Dipole-Dipole Interactions

If two polar molecules with dipole moments μ_1 and μ_2 approach in a vacuum there will develop a dipole--dipole interaction between the ends of the dipoles analogous to the interaction between the ends of two magnets. If the dipoles are oriented with respect to each other at a distance r as shown in Figure 4.5a, the interaction energy for the two dipoles will be given by

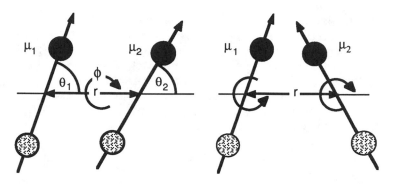

(a) Fixed dipoles.   (b) Freely rotating dipoles.

Figure 4.5. Schematic illustration of dipole--dipole interactions.

$$w(r,\theta_1,\theta_2,\phi) = -\{\mu_1\mu_2 / 4\pi\varepsilon_o\varepsilon r^3\}[\,2\cos\theta_1\cos\theta_2 -$$
$$\sin\theta_1\sin\theta_2\cos\phi] \qquad (4.21)$$

Maximum interaction will occur when the two dipoles are lying in line ($\theta_1=\theta_2=0°$), so that

$$w(r,0,0,\phi) = -2\mu_1\mu_2 / 4\pi\varepsilon_o\varepsilon r^3 \qquad (4.22)$$

Two like dipoles in a vacuum aligned for maximum interaction must, according to eq. 4.22, approach to within ≈ 0.36 nm for their inetarction energy to be equal to kT. If aligned parallel, the distance must be ≈ 0.29 nm. In a solvent medium, the interaction will be reduced so that the distance of separation for maximum interaction will be even less.

The expression in eq. 4.21 is an idealized expression that under many circumstances gives results which differ significantly from reality. That is found to be the case especially when the distance of separation is less than three times the length of the dipole. Since the distance of approach for maximum dipole--dipole interaction is of the same order of magnitude as normal molecular separations in condensed materials, it can be seen that dipolar interactions are of little practical significance (acting alone) in aligning or binding molecules to produce liquid or solid systems, or for imparting significant molecular structure in a system. Dipole--dipole interactions are usually significant only in systems involving very polar molecules. The exceptions involve small molecules with very large dipole moments such as water (O---H), hydrogen fluoride (F---H), and ammonia (N---H).

In those cases, the dipoles involve the small electron-defficient hydrogen atom bonded to a very electronegative atom. Here, the interactions become much more complex and are given the more specific descriptive name of *hydrogen bonding* interactions. In those cases the

electron-deff
icient hydrogen atom has such a small size that the
electronegative atoms can approach very closely and the hydrogen
experiences a much stronger electric field. This results in an enhanced
attractive interaction in the condensed state. Molecules undergoing this
special hydrogen bonding interaction form a special and very important
class of liquids called *associated liquids* . Their nature and the nature of
their interactions with other species is of great importance in many
practical areas of surface and colloid science, chemistry in general,
biology, etc. For now, we will be concerned with the more normal
molecules exhibiting weak dipolar interactions.

Angle-Averaged Dipolar Interactions

        To this point the discussion has been concerned with the
interaction between two dipoles effectively fixed in space by their mutual
interaction. However, at large separations, where the interaction energy
falls below the value of kT, or in a medium of high dielectric constant, ε,
the thermal energy of the system is such that the dipoles can no longer
be considered fixed; instead, they rotate or tumble with respect to one
another. Due to the random motion of the molecules, the overall
averages for the components of eq. 4.21 are zero. The angle-averaged
values of the interaction potential, however, will never be zero because
there will always exist a Boltzmann weighting factor that gives more
importance to some angles (or orientations) than to others. That is, those
orientations which produce more negative (lower energy) interaction
potentials will be favored over those with less negative or positive values.
For the dipole--dipole interaction, the Boltzmann angle-averaged
interaction will be given by the expression

$$w(r) = -\,(\mu_1\mu_2)\,/\,3(4\pi\varepsilon_0\varepsilon)^2kTr^{\,6} \quad \text{for} \quad kT > \mu_1\mu_2\,/\,4\pi\varepsilon_0\varepsilon r^{\,3} \quad (4.23)$$

The Boltzmann angle-averaged interaction potential is generally referred
to as the *orientation* or *Keesom interaction* and represents one of the
three 6th power of distance relationships involved in the total van der
Waals interaction.

Dipole-Induced Dipole Interactions

        The interaction between a polar molecule and a nonpolar
molecule is similar to that between an ion and a nonpolar molecule
except that the force field inducing the dipole arises as a result of a
dipole rather than a formal charge center. For a dipole of moment μ
oriented at an angle θ to the line joining the dipolar and nonpolar
molecules, the strength of the electric field acting on the polarizable
molecule is given by

$$E = \mu(1 + 3\cos^2\theta^2)^{1/2}\,/\,4\pi\varepsilon_0\varepsilon r^{\,3} \qquad (4.24)$$

The interaction energy, therefore, will be

$$w(r,\theta) = -1/2\,\alpha_o E^2 = -\mu\alpha_o(1 + 3\cos^2\theta)\,/\,2(4\pi\varepsilon_o\varepsilon)^2 r^6 \qquad (4.25)$$

For commonly encountered values of μ and α_o the strength of the interaction is not sufficient to orient the two molecules completely, as sometimes occurs in the case of ion--induced dipole and dipole--dipole interactions. The net effective interaction, $w_{eff}(r,\theta)$, will be the angle-- averaged energy. For the function $\cos^2\theta$, the angle average is 1/3, so that the interaction becomes

$$w(r) = -\mu^2\alpha_o\,/\,(4\pi\varepsilon_o\varepsilon)^2 r^6 \qquad (4.26)$$

For the situation in which two different molecules, each possessing a permanent dipole μ_1 and μ_2 and polarizabilities α_{o1} and α_{o2}, the net dipole-induced dipole interaction energy will be

$$w(r) = -[\mu_1^2\alpha_{o2} + \mu_2^2\alpha_{o2}]\,/\,(4\pi\varepsilon_o\varepsilon)^2 r^6 \qquad (4.27)$$

The interaction energy given by eq. 4.27, often referred to as the *Debye interaction* , represents the second of the three "inverse 6th power" contributions to the total van der Waals interaction between molecules.

The London-van der Waals (Dispersion) Force

As mentioned, the London-van der Waals or London dispersion force (hereafter generally referred to simply as *dispersion forces*) often makes the most important contribution to the total van der Waals interaction due to its universal nature, as contrasted to the dipolar and induced--dipolar forces, which depend on the exact chemical natures of the species involved, and may or may not be significant in a given case. Dispersion forces are important in a wide variety of phenomena, including the condensation of nonpolar molecules to the liquid and solid states; the boiling points, surface tensions, and other physical properties of condensed states; adsorption, adhesion, and lubrication processes; the bulk physical strength of primarily covalent materials; the aggregation and flocculation of molecular and particulate systems; and the structures and interactions of synthetic polymers, proteins, and other complex biological systems.

Although arising from complex quantum mechanical factors, dispersion forces have several easily understood characteristics; eg:

(1) They have a relatively long range of action compared to covalent bonds, their effect in some cases extending to a range of 10 nm or more.

(2) They may be attractive or repulsive, depending on the

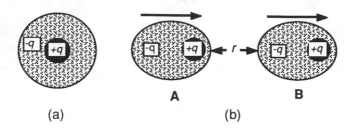

Figure 4.6. Schematic illustration of the source of dispersion forces: (a) isolated atom with a symmetrical electron cloud; (b) the "instantaneous" dipole in atom **A** induces a dipole in neighboring atom **B**.

situation, and generally do not adhere to simple power laws with respect to their dependence on separation distances.

(3) They are nonadditive, in that the interaction between any two atoms or molecules will be affected by the presence of other nearby atoms and molecules.

The London dispersion force is basically quantum mechanical in nature because it involves interactions between rapidly fluctuating dipoles resulting from the movement of the outer valence shell electrons of an atom or molecule. Rigorous derivations, therefore, can become quite complex and will serve little useful purpose in the present discussion. The interested reader is referred to the original work by London[2] and more recent treatments by McLachlan[3] and Israelachvili[4] for further enlightenment.

For a system of two isolated atoms (or molecules), one can visualize electrons around one atom as being particles (much like in the Bohr atom) which, although they travel close to the speed of light, can at any instant be located assymmetrically with respect to the nucleus with which they are associated (Figure 4.6). The asymmetric charge distribution produces an instantaneous dipole in the atom or molecule. The dipole generates a short-lived electric field that can then polarize a neighboring atom or molecule, inducing a dipole in the neighbor (Figure 4.6b). The result is a net coulombic attraction between the two species. Using the simple model of dispersion forces arising from fluctuating dipoles, their complex quantum mechanical origin can, for a first approximation, be forgotten and they can be treated as simple electrostatic interactions.

Theoretically, the strength of the attraction, F_{att}, between two such instaneous dipoles is found to be proportional to the inverse 7th power of the distance separating the two nuclei

$$F_{att} = -A'/r^7 \qquad (4.28)$$

where r is the distance between the nuclei or, for approximately spherical molecules, the centers of mass, and A' is a quantum mechanical constant related to the structure of the atom or molecule.

The amount of work required to separate reversibly a pair of atoms or molecules from a distance r to infinity is

$$\Delta W = -\int_r^\infty F_{att}\, dr = A'\int_r^\infty (1/r^7)\, dr = A'/6r^6 = A/r^6 \qquad (4.29)$$

If it is assumed that the interaction energy at infinite separation is zero, then the free energy of attraction will be

$$\Delta G_{att} = -A/r^6 \qquad (4.30)$$

The constant A ($= A'/6$) was given by London, for two identical units, as

$$A = (3/4)h\, v\, \alpha_o^2 \qquad (4.31)$$

where h is Planck's constant, α_o is the electronic polarizability of the atom or molecule, and v is a characteristic frequency identified with the first ionization potential of the atom or molecule, usually falling in the ultraviolet region. For two different interacting units, 1 and 2, the expression is

$$A_{12} = (3/2)[(v_1 v_2)/(v_1 + v_2)]\, \alpha_{o1}\alpha_{o2} \qquad (4.32)$$

The shape of the force--distance curve for eq. 4.30 is given schematically in Figure 4.7a. The free energy and attractive force between units becomes more negative as the separation distance decreases until the electron clouds of the respective units begin to interact. If no bonding interactions between the two units are possible, the interaction becomes repulsive (*Born repulsion*) and rises to infinity

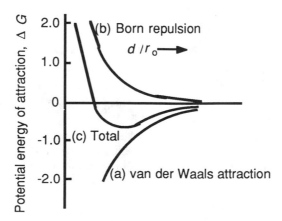

Figure 4.7. Force--distance relationships for total components of dispersion interactions between atoms and molecules.

as the electron clouds begin to overlap.

Theoretically, the Born repulsion function is takes the form

$$F_{rep} = B e^{-ar} \qquad (4.33)$$

where a and B are constants. This produces a contribution to the total interaction potential at distance r of

$$\Delta G_{rep} = (B/a) e^{-ar} \qquad (4.34)$$

For ease of manipulation, it is common practice to approximate eq. 4.34 by

$$\Delta G_{rep} = B'/r^{12} \qquad (4.35)$$

which produces a curve similar to (b) in Figure 4.7.

The total potential energy of interaction between two units will be the sum of the attractive and repulsive interactions and is illustrated by curve (c).

$$\Delta G = \Delta G_{rep} + \Delta G_{att} = (B'/r^{12}) - (A/r^6) \qquad (4.36)$$

Equation 4.36 is commonly referred to as the Lennard-Jones 6-12 potential. It may be noted that the attractive term in eq. 4.36 is an inverse 6th power relationship, the third such term in the overall van der Waals attraction.

One may estimate the strength of dispersion forces using the London equation for dispersion interactions between two small identical spherical atoms or molecules

$$w(r) = -(3/4) \alpha_o^2 h v / (4\pi\varepsilon_o)^2 r^6 \qquad (4.37)$$

The ionization potential of the atom or molecule (typically in the range of 2×10^{-18} J) can be substituted for hv. A good approximate value for the term $\alpha_o / 4\pi\varepsilon_o$ is 1.5×10^{-30} m^3. For two atoms in contact ($r \approx 0.3$ nm) at room temperature, the interaction potential will be $w(r) = -4.6 \times 10^{-21}$ J, which is approximately equal to 1 kT. It can be seen, then, that dispersion forces can be quite significant in magnitude considering the fact that they arise from little more than fleeting variations in the electronic balancing act between electrons and nuclei in the atom or molecule.

It is important to emphasize that eq. 4.37, while extremely useful for the type of systems specified, has several limitations when applied to larger and nonspherical molecules and to many-bodied systems such as condensed liquids and solids. These limitations include:

(1) The analysis neglects the effect of short-range repulsive forces

at small r which will work to lower the net attractive interaction between the atoms or molecules.

(2) Higher level quadrapole interactions, which tend to increase the attractive interaction, are not considered.

(3) Interactions with distant neighbors in condensed systems, which also tend to increase the attraction, are neglected.

By lucky chance, it appears that limitation 1 is pretty much cancelled out by 2 and 3 for simple systems, so that theoretical calculation based on eq. 4.37 are found to agree surprisingly well with experiment.

Total van der Waals Interactions Between Polar Molecules

We have seen, now, that there are three types of interactions that can be involved in the total van der Waals interaction between atoms or molecules: dipole--dipole (orientational or Keesome), dipole--induced dipole (induced or Debye), and dispersion interactions. The theories for all three interactions are found (to a first approximation) to involve an inverse 6th power of the distance separating the two interacting centers. The total van der Waals interaction potential, $w_{vdw}(r)$, can then be written as

Table 4.4. Typical values for the various partial contributions to the total van der Waals interaction in vacuum (10^{-79}J m^6) according to eqs. 4.40 and 4.41.

Interacting units	C_{ind}	C_{oriet}	C_{disp}	Theor.	Experimental (from gas law)	C_{disp} as % total
Ne / Ne	0	0	4	4	4	100
CH_4 / CH_4	0	0	102	102	101	100
HCl / HCl	6	11	106	123	157	86
HBr / HBr	4	3	182	189	207	96
HI / HI	2	0.2	370	372	350	99
CH_3Cl / CH_3Cl	32	101	282	415	509	68
NH_3 / NH_3	10	38	63	111	162	57
H_2O / H_2O	10	96	33	139	175	24
Ne /CH_4	0	0	19	19	---	100
HCl / HI	7	1	197	205	---	96
H_2O / CH_4	9	0	58	67	---	87

$$w_{vdw}(r) = -C_{vdw} / r^6 \qquad (4.38)$$

where C_{vdw} is the overall van der Waals constant for which

$$C_{vdw} = (C_{disp} + C_{ind} + C_{orient}) \qquad (4.39)$$

The individual constants are simply the coefficients of $1/r^6$ for each contributing factor (eqs. 4.23, 4.27, and 4.37).

For two identical molecules the full equation takes the form

$$w_{vdw}(r) = -[(3\alpha_o^2 h v_1/4) + 2\mu^2\alpha_o + (\mu^4/3kT)] / (4\pi\varepsilon_o)^2 r^6 \qquad (4.40)$$

For two unlike polar molecules, the full equation will take the form

$$w_{vdw}(r) = -\{[3\alpha_{o1}\alpha_{o2} h\ v_1 v_2 / 2(v_1+v_2)] + (\mu_1^2\alpha_{o2} + \mu_2^2\alpha_{o1})$$
$$+ (\mu_1^2\mu_2^2 / 3kT)\} / (4\pi\varepsilon_o)^2 r^6 \quad (4.41)$$

Some representative values for total van der Waals interactions are given in Table 4.4.

One can see from the table that the disprersion force contribution to the total van der Waals interaction can be quite significant. Except for very small and very polar molecules such as water, the dispersion force will exceed the Keesome and Debye contributions and dominate the character of the interaction, even in the case of two dissimilar molecules.

In many cases the interactions between dissimilar molecules cannot be calculated, so that some form of estimate must be employed. Experience has shown that in most cases, the experimentally determined value for the van der Waals interaction, and therefore the van der Waals constant C_{vdw}, falls somewhere between that for the two identical molecules. In practice, it is usually found that the geometric mean of the two values produces a theoretical result which agrees reasonably well with experiment. For example, the geometric mean of C_{vdw} for Ne (= 4) and methane (= 102) is

$$C_{vdw} = (4 \times 102)^{1/2} = 20$$

compared to the experimental value of 19. For HCl-HI (123 and 372) the geometric mean of 214 compares reasonably well with the measured value of 205.

Just as the van der Waals force theory to breaks down somewhat for the interaction between two water molecules, so the geometric mean approximation begins to lose its usefulness for interactions between water and dissimilar molecules. For the interaction between water and methane, for example, the geometric mean approach predicts a van der Waals constant of 120 while experimental values lie more in the range of 60--70. What that tells us is that methane molecules and water molecules greatly prefer to interact with molecules of their own kind, and explains the observed fact that methane (and other hydrocarbons) are immiscible with water. The same applies for other molecules such as fats and oils which may have some polar nature, but are predominantly hydrocarbon in makeup, as well as fluorocarbons, silicones, etc. Conventionally, this mutual "dislike" has come to be called *hydrophobicity* and the phenomenological effect the *hydrophobic* (water fearing) *effect* . We will see in the coming chapters that this "xenophobic" aspect of van der Waals forces has many important ramifications in real-world applications.

The simplified theory of van der Waals forces presented above is quite useful as a foundation for understanding the general principles involved in the physical interactions between atoms and molecules. As pointed out, it works well for many simple systems, in part due to the theory and in part due to good fortune, but begins to fail when the system of interest begins to deviate from one of small, spherically symmetrical units. Unfortunately, almost all practical systems deviate significantly from those restrictions. In addition, the theory assumes that atoms and molecules have only one ionization potential, and it cannot readily handle the presence of an intervening medium other than a vacuum between the interacting units. Theories that attempt to take such limitations into consideration have been developed but are presently of little practical use, except, of course, to the theoretician.

Effects of A Medium Other Than Vacuum

The equation for the total van der Waals interaction between two atoms or molecules (eq. 4.40) includes a factor for corrections due to changes in the dielectric characteristics of an intervening medium other than vacuum. That aspect of the theory can be of great importance both quantitatively and qualitatively and has significant ramifications in practical systems. A full discussion of the theoretical aspects of the effects of medium on van der Waals interactions would be beyond the scope of this book, but the reader is referred to the work by Israelachvili[4] for further enlightenment. From a practical standpoint, however, several important points arise from an analysis of the dispersion force equation for media of differing dielectric constants. A theoretical explanation of each will not be given here but can be found in the work by Israelachvili. The relevant points include:

(1) Since, for interactions in a vacuum, the ionization potential I (or $h\nu$) for most materials is much greater than kT for $\nu > 0$, the dispersion contribution to the total interaction is usually greater than the dipolar contributions for $\nu = 0$. This agrees with the results given in Table 4.4.

(2) The magnitude of the van der Waals interactions are greatly reduced in the presence of a medium other than vacuum. For example, in the case of two nonpolar molecules, the magnitude of their interaction in vacuum may be reduced by an order of magnitude in the presence of an intervening medium.

(3) Dispersion force contributions in a medium other than vacuum may be either attractive or repulsive, depending on the relative ionization potentials of the materials involved.

(4) In some cases where the interaction between two molecules in a solvent is very small, such as the lower molecular weight alkanes in water, the interaction becomes dominated, not by the dispersion force, but by an entropic term of the form

$$w(r)_{v=0} \approx -kTa_1^6/r^6 \qquad (4.42)$$

where a_1 is the radius of the small interacting molecules. In general terms, this means that there is an overall increase in entropy (of the water molecules) as the distance between two alkane molecules decreases. This agrees qualitatively with the observation that the interaction between "hydrophobic" species such as alkanes in water is primarily entropic in nature.

(5) For all but spherically symmetrical molecules, van der Waals forces are anisotropic. The polarizabilities of most molecules are different in different molecular directions because the response of electrons in a bond to an external field will usually be anisotropic. A consequence of this effect is that the dispersion force between two molecules will depend on their relative molecular orientation. In nonpolar liquids, the effect is of minor importance because the molecules are essentially free to tumble and attain whatever orientation is energetically favorable. However, in solids, liquid crystals, and polar media, the effect can be important in determining the relative fixed orientation between molecules, thereby affecting or controlling specific conformations of polymers or proteins in solution, critical transition temperatures in liquid crystals and membranes, etc. Repulsive forces in polar molecules are also orientation dependent, and are often of greater importance in controlling conformations and orientations.

(6) van der Waals forces are nonadditive and are affected by the presence of other interacting bodies in the vicinity. What this means is that the total interaction among a group of molecules or particles will not be a simple sum of the individual pairwise interactions. In fact, in most cases, a molecule interacting with a second molecule in a group not only will experience the force of interaction directly, but will also feel a "reflected" force due to the polarization of other neighboring molecules polarized by the first. The net result will be a total interaction somewhat greater than that which would be predicted by a simple summation. Although the effect is usually small (perhaps 30% of the "normal" interaction) it can be significant, especially in the case of relatively large particles interacting with a surface.

(7) Over relatively large distances, dispersion forces experience a *retardation effect* that results from the nature of the fluctuating dipoles which give rise to the interaction. When two molecules are an appreciable distance apart, the electric field generated by a fluctuating dipole takes a finite amount of time to reach a neighboring molecule. By

the time the second "induced" dipole can retransmit its effect back to the first, the first has had time to reoriente itself and may no longer have an orientation suitable for maximum interaction. Therefore the total interaction will be reduced in magnitude. Thus at large distances, the magnitude of the dispersion interaction is found to fall off faster than r^{-6}. In a vacuum, retardation effects begin to appear at distances of approximately 5 nm, which makes them of little practical importance. However, in a liquid medium retardation effects begin to be felt at shorter distances and become important, as in the interaction between particles and surfaces in liquid media. That is the case because in condensed media, the speed of light is reduced, thereby allowing more time between cause and effect, weakening the net interaction. Since it is only dispersion forces that suffer from retardation effects, as distances of separation increase, induced and orientational dipolar contributions become more important to the total van der Waals interaction.

INTERACTIONS BETWEEN SURFACES AND PARTICLES

The above discussion centered on the forces controlling the interactions between two isolated atoms or molecules. For multiunit systems it is assumed that the units interact mutually according to the Lennard-Jones potential, and that the total interaction is the sum of all indivudual interactions. For the repulsive term, it is common to neglect the repulsive component for units in one bulk phase and consider only repulsions between opposing surfaces.

Mathematically, the simplest situation to analyze is that involving two hard, flat, effectively infinite surfaces separated by a distance H in a vacuum. The free energy of attraction per unit area in such a case is approximated by

$$\Delta G^{att} = -A_H/(12\pi H^2) \qquad (4.43)$$

where A_H is termed the Hamaker constant. The value of A_H is related to A of eq. 4.31 by

$$A_H = (3/4)h\nu\,\alpha^2\,\pi^2 n^2 = A\pi^2 n^2 \qquad (4.44)$$

In eq. 4.44, n is the number of atoms or molecules in unit volume of the phase. For two identical spheres of radius a, where $H/a \ll 1$, a similar type of approximate equation is

$$\Delta G^{att} = -(A_H a/12H)[1 + (3/4)H/a + \text{higher terms}] \qquad (4.45)$$

In most practical instances, it is safe to neglect all of the higher terms.

A comparison of eqs. 4.30 and 4.43 will show that the free energy of attraction between two surfaces falls off much more slowly than that between individual atoms or molecules. This extended range of bulk interactions plays an important role in determining the properties of systems involving surfaces and interfaces. A combination of the attractive and repulsive forces between surfaces leads to a curve such as in Figure 4.7a.

Due to the fact that interactions between surfaces fall off much more slowly with distance than those for individual atoms or molecules, the retardation effect mentioned above becomes more significant. While the retardation effect is important in quantitative theoretical discussions of surface interactions, from a practical standpoint, it is still relatively insignificant compared to other factors.

Surface Interactions in Media Other Than Vacuum

The equations for surface interactions given above were derived for the situation in which the interacting units were separated by a vacuum. Obviously, for practical purposes, that usually represents a rather unrealistic situation. "Real life" dictates that in all but a few situations, interacting units be separated by some medium that itself contains atoms or molecules that will impose their own effects on the system as a whole. How will the relevant equations be modified by the presence of the intervening medium?

Surfaces interacting through an intervening fluid medium will experience a reduced mutual attraction due to the presence of the units of the third component. The calculation of interactions through a vacuum involves certain simplifying assumptions, therefore it is not surprising to find that models for three component systems are even more theoretically complex. Although a number of elegant approaches to the problem have been developed over the years, for most purposes a simple approximation of a composite Hamaker constant is found to be sufficient. When two surfaces of component 1 are separated by a medium of component 2, the effective Hamaker constant (A_H^{eff}) is approximated by

$$A_H^{eff} = [A_{10}^{1/2} - A_{20}^{1/2}]^2 \tag{4.46}$$

where A_{10} is the Hamaker constant for component 1 in a vacuum, and A_{20} is the same for component 2. A result of the relationship in eq. 4.46 is that as the vacuum Hamaker constants for 1 and 2 become more alike, the effective Hamaker constant tends toward zero, and the free energy of attraction between the two surfaces of component 1 is also reduced to zero. As we will see in Chapter 9, such a reduction in the attractive forces due to an intervening medium gives one a handle on ways to successfully prevent the spontaneous joining of surfaces thereby imparting a certain added stability to the separated system. Since eq. 4.46 involves the square of the difference between the Hamaker constants for components 1 and 2, the same will be true for surfaces of component 2 separated by a medium of 1. The form of the interaction curve for the above situation will be the same as that for the vacuum case, although the shape and values will differ because of the diferent value of the effective Hamaker constant.

Qualitatively, the above discussion of surface interactions tells us that free surfaces are inherently unstable and will usually experience a net attraction for similar surfaces in the vicinity. The practical repercussion is that if only the van der Walls forces were involved, systems involving the formation and maintanence of expanded interfaces would all be unstable and spontaneously revert to the condition of minimum interfacial area, thereby making impossible the preparation of paints, inks, cosmetics, many pharmaceuticals, many food products, emulsions of all kinds, foams, bilayer membranes, etc. It would be a decidedly different world we lived in. In fact, life as we know it would not exist! Obviously, something is or can be involved at interfaces that alters the simple situation described above and makes things work. In the following chapters we will introduce other "actors" that allow Nature (and humankind, when we're lucky) to manipulate surfaces and interfaces to suit our purposes.

Dipole, Induced Dipole, and Acid-Base
Interactions at Interfaces

In the preceeding sections the discussion centered on the source of attractive forces acting between atoms, molecules, and larger material units. The London-van der Waals forces were characterized as being universal and almost always attractive over relatively long distances. Many, if not most, practical systems, however, involve situations in which the substances and surfaces in question are composed of materials that can interact through forces other than London-van der Waals attractions. Such non-quantum mechanical interactions can be classed generally as electrostatic in nature. That is, they involve to one degree or another the interaction of partial or complete electronic charges. Because electrostatic interactions can occur between like charges (repulsive) or unlike charges (attractive), they can have different and significant effects on the characteristics of atomic, molecular, and interfacial interactions. The following chapter will introduce the basic concepts underlying the electrostatic nature of interfaces as related to some important colloidal phenomena.

CHAPTER 5

ELECTROSTATIC FORCES AND THE
ELECTRICAL DOUBLE LAYER

Most solid surfaces in contact with water or an aqueous solution will be found to develop some type of electrical charge. The magnitude of that charge may be quite small or very large, but it will almost always exist. In macroscopic systems, the presence or absence of a surface charge may often be overlooked. Such is not the case, however, in such systems as electrodes for electrolytic processes, systems subject to electrostatic deposition, and other sensitive situations. In what we might call the microscopic world of colloids, the presence or absence of a surface charge can have significant ramifications in terms of stability, sensitivity to environment, electrokinetic properties, etc. The previous chapter introduced the basic concepts governing coulombic interactions between atoms, molecules, and bulk materials. The following material will build on that material and lay the ground work for the application of those principles to the problem of the stability of colloidal systems discussed in Chapter 9.

SOME CONSEQUENCES OF SURFACE CHARGES

As a result of the presence of electrical charges, surfaces exhibit various properties that are not present in systems having no surface charge (eg, nonpolar solids dispersed in a nonpolar medium). Of particular interest are the so-called *electrokinetic properties* they impart to colloids, discussed later, and, perhaps more importantly, the fact that they provide the primary mechanims for the stabilization (and therefore existence) of many colloids, both natural and manmade. As a result of their unique characteristics, they provide a handle for manipulating a wide variety of multiphase systems that are of vital importance to life and technology. By the proper manipulation of surface charges, we may be able to transform an interesting but useless system into one of great practical importance. Likewise, we may be able to take a system that is a nuisance colloid and, by changing its electrical environment, remove it completely.

SOURCES OF INTERFACIAL CHARGE

A surface may acquire an electrical charge by one or more of several mechanisms, the most common of which include: (1) preferential (or differential) solution of surface ions, (2) direct ionization of surface groups, (3) substitution of surface ions, (4) specific ion adsorption, and (5) charges deriving from specific crystal structures. While other mechanims can be invoked, those five represent the most common and most important in terms of colloidal systems. Each type of surface charge carries with it certain characteristics that define, partially at least, the electronic nature of the resulting colloid. The five main classes are illustrated schematically in Figure 5.1.

Differential Ion Solubility

The preferential solubilization of ions from the surface of a sparingly soluble crystalline material represents one of the most common, and earliest recognized, mechanisms for the development of electrical charges on solid surfaces, especially colloids. A widely encountered example of such colloids is the silver halide colloids (incorrectly, but commonly referred to as "emulsions") used in photographic products.

The silver salts of chlorine, bromine, and iodine are of very limited solubility in water. When crystals of, for example, silver iodide are placed in water, ions dissolve from the surface until the product of the concentration of the two ions is equal to the solubility product of the material. For AgI, that would be $K_{SP} = [Ag^+][I^-] = 10^{-16}$ mol L^{-1}. If the two ions were dissolved equally readily, their concentrations in solution would be equal (10^{-8} mol L^{-1}), as would their occurance on the surface of the crystal, leaving a net surface charge of zero. One finds, however, that silver ions are more readily dissolved, so that their concentration in solution is greater than that of iodide, while the concentration of I^- on the surface of the crystal is enhanced, leading to the formation of a net negative charge on the surface. The situation is shown in Figure 5.1a.

If a soluble silver salt (eg, $AgNO_3$) is added to the solution, the solution of Ag^+ from the crystal surface will be suppressed (by the *common ion effect*) and the negative surface charge correspondingly reduced. At some characteristic silver ion concentration, the dissolution of silver ion from the crystal will be zero, and there will be no surface charge. That point is termed the *point of zero charge* (ZPC). If addition of Ag^+ is continued, a net positive surface charge will develop. If soluble iodide salts had been added to the solution, the solution of I^- would have been suppressed further, leading to an increase in the net negative surface charge on the crystal. The control of the magnitude and sign of the surface charge on materials such as the silver halides by controlling

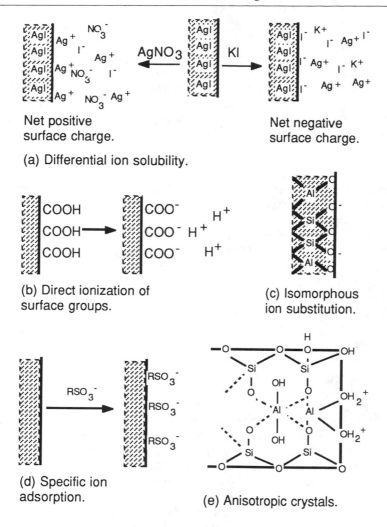

Figure 5.1. The principle sources of surface charge in solids.

the concentration of one of the two ions through the common ion effect gives one a useful handle for the manipulation of colloidal systems containing such materials.

Direct Ionization of Surface Groups

Materials containing surface groups that can be directly ionized, but in which one of the ions is permanently bound to the surface, represent a second important mechanisms for the development of surface charges. This group includes many metal oxides as well as many polymer latices (Figure 5.1b). Many metal oxides are amphoteric in that they can develop either negative or positive surfaces, depending on the pH of the solution. Such surfaces will obviously exhibit a characteristic point of zero charge such as that for the previous class, except that it will be controlled by pH rather than the concentration of a common ion.

Typical examples of ionizable polymer surfaces include those containing carboxylic and sulfonic acids and their salts (---COOH, ---COO⁻ M⁺, ---SO₃H, and ---SO₃⁻ M⁺), sulfuric acid esters and their salts (---OSO₃H and ---OSO₃⁺ M⁺), basic amino groups (---NR₃, R = H or an organic group), and quaternary ammonium groups (---NR₄⁺X⁻). In some of those cases, the degree of ionization will be determined by the pH and the acid (or base) strength of the ionizable group. For weak acids and bases, such as carboxyl and amino groups, the surface charge will be specific, either negative or positive, respectively, or zero. In the case of the sulfonates and sulfuric acid esters, since they are strong acids, the complete suppression of their charge will require a very low pH (pK_a's ≈ 1). Similarly, quaternary ammonium salts will be essentially independent of pH, giving a positively charged surface under all conditions, although the degree of ionization can be suppressed by high ionic strength in the solution. In the special case of surfaces containing amino acid groups (eg, ---C(NH₂)COOH) the surface may acquire a positive or negative charge, according to the pH.

Substitution of Surface Ions

Many minerals, clays, oxides, etc, can undergo what is termed *isomorphous substitution*, which means that structural ions are substituted by ions of valency one less than the original. For example, a silicon atom (valency = 4+) in clay may be replaced by aluminum (3+), producing a surface with a net negative charge (Figure 5.1c). Such a surface can be brought to its ZPC by lowering the pH.

Specific Ion Adsorption

Some surfaces that may not possess a direct mechanism for the formation of surface charges may do so by the adsorption of specific ions that impart a charge to the surface. Gold sols produced by the reduction of HAuCl₄ adsorb "free" chloride ions to produce a surface that apparently has a structure related to AuCl₄⁻. Of more practical importance is the adsorption of surfactant ions. In that case, the adsorption of an anionic surfactant produces a negatively charged surface; the adsorption of a cationic surfactant produces a positively charged surface (Figure 5.1d).

A variation on the theme of specific ion adsorption is seen when a surface charge arises from the dissociation of a salt, say a sodium carboxylate (-COO⁻ + Na⁺), to produce a negatively charged surface. If di- or trivalent ions are present in the solution, they may adsorb onto the surface in such a manner that the net result is a *charge inversion* from a negative to a positive surface charge. Such an *ion-exchange* mechanism has been seen in many systems, including biologically important bilayer membrane structures.

Anisotropic Crystals

Some important materials, for example kaolinite clay, are composed of aluminosilicates and have crystal structures that, when cleaved, can result in the production of both positively and negatively charged surfaces. Depending on the crystal face exposed, there may be positive $AlOH_2^+$ groups or negatively charged basal groups, or (more likely) both. As a result, such materials may exhibit very special properties, including the formation of characteristic open structures as illustrated in Figure 5.1e. They may also show the apparent existence of more than one ZPC. Typically, an aluminosilicate clay may be found to increase its volume tenfold on addition of water due to the special properties of its anisotropic crystal structure.

The above five mechanisms for the formation of surface charges cover the great majority of examples encountered in colloidal system. In order to appreciate the significance of their formation and activity in the context of colloidal systems, it is necessary to begin with some basic principles of electrostatics.

ELECTROSTATIC THEORY: COULOMB'S LAW

The fundamental law governing interactions between charged species was introduced in the preceding chapter. As a reminder, however, for the interaction of two charges q_1 and q_2 in a vacuum (F_{el}) separated by a distance r, the law takes the form

$$F_{el} = q_1 q_2 / (4\pi\varepsilon_0\varepsilon r^2) \qquad (5.1)$$

where the symbols are as defined previously. The work necessary to bring the two charges together from infinity to the distance r is

$$w(r) = -\int_\infty^r F_{el}dr = q_1 q_2 / (4\pi\varepsilon_0\varepsilon r) \qquad (5.2)$$

For charges of the same sign, $w(r)$ will be positive and the interaction will be repulsive. If of opposite charge it will be attractive.

Consider for a moment that a charge q_1 is isolated in space. It will produce an electric field at a point r, such that the work necessary to bring a unit electrical charge from infinity to distance r from q_1 will be equal to $q_1/(4\pi\varepsilon_0\varepsilon r)$. That quantity of work is defined as the *electrical potential* at r due to the charge q_1, and is given the symbol ψ. According to eq. 5.1, then, the force involved in bringing a charge q_2 to within a distance r of q_1 will be simply $q_2\psi$.

Boltzmann's Distribution and the Electrical Double Layer

While Coulomb's law is simple enough in isolation, in order to make use of it in more realistic situations the presence of all ions in the system must be taken into consideration. In practical applications, one is concerned with solutions containing many charges (ie, dissolved ions), particles, and surfaces that also contain charges of the same or different sign. In order to apply Coulomb's law to solutions of electrolytes and colloids it is necessary to employ *Boltzmann's distribution law* , which relates the probability of a unit (atom, molecule, ion, particle, etc) being at a certain point with a specified free energy (or potential energy), ΔG , relative to a specified reference state. The probability is generally expressed in terms of an average unit concentration, c, at the point r relative to a concentration, c_0, at some reference distance at which the energy is taken as zero. At a temperature T, the Boltzmann distribution is given as

$$c = c_0 \exp\ (-\Delta G\ /\ kT) \tag{5.3}$$

When applied to the situation involving charged particles and Coulomb's law, eq. 5.3 predicts that if there exists a negative electrical potential ψ at some point in an electrolyte solution, then in the region of that point the concentration of positive charges will be given by

$$c_+ = c_0 \exp\ (-z_+ e\ \psi/\ kT\ ) \tag{5.4}$$

where z_+ is the valency of the positive ion and c_0 is the concentration of the positive ion in a region where $\psi = 0$. A similar expression can be written for the negative ions in the solution

$$c_- = c_0 \exp\ (+\ z_- e\ \psi/\ kT\ ) \tag{5.5}$$

Although the solution taken as a whole will be electrically neutral, in the vicinity of the electrical potential there will exist an imbalance of electrical charges. That is, for a negative ψ, there will be more positive ions in the region than negative ions. For the case where $z_- = z_+ = 1$, the excess is given by

$$c_+ - c_- = c_0 [\exp(-e\, \psi / kT) - \exp(+e\, \psi / kT)] \qquad (5.6)$$

The region of excess charge of the opposite sign around a potential is commonly referred to as the *ionic atmosphere* or *charge cloud* associated with that potential.

Double Layer Thickness: The Debye Length

In the broad field of physical chemistry, the Boltzmann distribution law is fundamental to the derivation of the Debye-Hückel theory of electrolyte solutions. In the more narrow arena of surface and colloid science, it is applied to the determination of the ionic atmosphere around charged surfaces and particles. In that context, the charge cloud is more commonly referred to as the *electrical double layer* (EDL). The concept is illustrated schematically (Figure 5.2) for the situation in which a particle posseses an evenly distributed charge that is just balanced by the total opposite charge, the *counterions* in the electrical double layer.

The idea of the electrical double layer was first formally proposed by Helmholtz[1], who developed the concept of a system having charges arranged in two parallel planes as illustrated in Figure 5.2a. Such a situation describes, in essence, a molecular capacitor and is relatively easy to handle mathematically on that basis. In reality, of course, the thermal motion of ions in solution introduces a certain degree of chaos causing the ions to be spread out in the region of the charged surface, forming a *diffuse* double layer in which the local ion concentration is

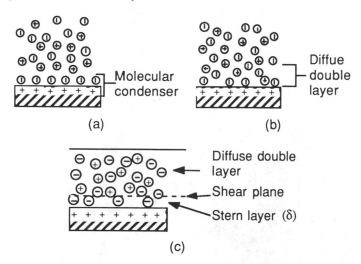

Figure 5.2. Models of the interface for charged surfaces: (a) the early Helmholtz model of a molecular capacitor; (b) the Gouy-Chapman model of the diffuse double-layer; (c) the Stern model.

determined by eq. 5.6. In that case, the analysis of the electronic environment near the surface is more complex and requires more detailed analysis. Such analysis gave rise to the more accurate Gouy-Chapman[2] model of the electrical double layer illustrated in Figure 5.2b. An additional consequence of the "reality" that charges occupy a finite amount of space and therefore have certain steric requirements is the existence of the so-called *Stern layer* . The Stern layer is a small space separating the ionic atmosphere around a surface, the actual diffuse double layer, from the steric "wall" of the charged plane just adjacent to the surface (Figure 5.2c). The thickness of the Stern layer, δ in Figure 5.2c, is usually on the order of a few angstroms and, as implied, reflects the finite size of charged groups and ions specifically associated with the surface.

While the Boltzmann distribution is relatively easy to evaluate for a single point charge, the situation can become quite complex in the case of a surface having many charges. For a detailed discussion of the finer points of double-layer theory, the reader is referred to the works of Verwey and Overbeek,[3] Kruyt,[4] and Hiemenez.[5] From a practical (and very simplistic) point of view, it is normally assumed that the electrical potential in the solution surrounding the surface in question falls off exponentially with distance from the surface (Figure 5.3) according to the *Debye-Hückel approximation*

$$\psi = \psi_0 \exp(-\kappa z) \qquad (5.7)$$

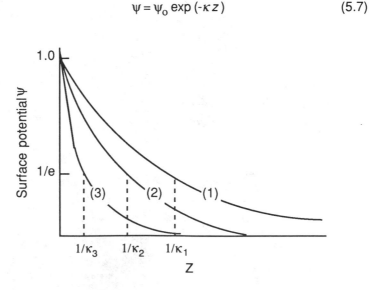

Figure 5.3. Illustration of the decay of surface potential with distance from the surface in various electrolyte concentration ranges: (1) low; (2) medium; and (3) high.

In eq. 5.7, κ is identified as the reciprocal of the thickness of the electrical double layer, also commonly referred to as the *Debye length*. Thus at a distance of $1/\kappa$ from a charged surface, the potential has fallen off by a factor of $1/e$. For low surface potentials (< 25 mV), the theoretical equation for the double-layer thickness, $1/\kappa$, is

$$1/\kappa = [\varepsilon_0 \varepsilon \, kT \, /(e^2 \sum_i c_i z_i^2)]^{1/2} \qquad (5.8)$$

One can see that the thickness of the electrical double layer is inversely proportional to the concentration of electrolyte in the system and to the square of the valency of the ions involved. In terms of colloidal stability, this means that the distance of separation between two particles that can be maintained under a given set of circumstances will depend on, among other things, those two factors. And their important effect gives one a strong handle for manipulating the characteristics and stability of many colloidal systems.

In Table 5.1 are given the values of $1/\kappa$ for various concentrations of a selection of electrolyte types in water at 25°C calculated according to eq. 8. One can see that the double-layer thickness drops off very rapidly as the concentration of electrolyte increases. This effect of electrolyte concentration has important ramifications in the world of practical colloids, as does the effect of the ionic charge (z) involved.

For ease of calculation of $1/k$, eq. 5.8 can be simplified to the following relationships:

For 1:1 electrolytes (eg, NaCl)
$$1/\kappa = 0.304 \, [MX]^{-1/2} \qquad (5.9)$$

For 2:1 and 1:2 electrolytes (eg, $CaCl_2$, Na_2SO_4)
$$= 0.178 \, [MX_2]^{-1/2} \text{ (or } [M_2X]^{-1/2}) \qquad (5.10)$$

For 2:2, 3:1, or 1:3 electrolytes (eg, $MgSO_4$, $LaCl_3$, Na_3PO_4)
$$= 0.152 \, [MX]^{-1/2} \text{ (or } [MX_3]^{-1/2} \text{ or } [M_3X]^{-1/2}) \qquad (5.11)$$

For 2:3 or 3:2 electrolytes [eg, $Ca_3(PO_4)_2$, $Al_2(SO_4)_3$]
$$= 0.136 \, [M_3X_2]^{-1/2} \text{ (or } [M_2X_3]^{-1/2}) \qquad (5.12)$$

The picture of the electrical double layer around a colloidal particle that arises as a consequence of eq. 5.8 is that of a cloud of ions dominated by charges opposite to that of the surface surrounding the particle. The distance over which this ionic "sheath" extends is a function

Table 5.1. Double-layer thickness ($1/\kappa$) for various electrolytes in water.

Electrolyte (mol L^{-1})	$1/\kappa$ (nm)				
	1:1(MX)	1:2(MX$_2$)	2:2(MX)	1:3(MX$_3$)	2:3(M$_2$X$_3$)
10^{-4}	30.4	17.6	15.2	15.2	13.6
10^{-3}	9.6	5.57	4.81	4.81	4.30
10^{-2}	3.0	1.76	1.52	1.52	1.36
10^{-1}	0.96	0.56	0.48	0.48	0.43
1	0.30	0.18	0.15	0.15	0.14

of the concentration and valency of the ions in the solution and the charge on the surface (Figure 5.3). According to eq. 5.8, the local concentration of ions near the surface varies as shown in Figure 5.4a, while the local charge density ($c_+ - c_-$) varies as shown in Figure 5.4b.

For the situation in which the surface charge density is constant, that is, progressive adsorption of ions with increased c_o does not reduce the surface charge (see "Charge Regulation," below), the area under curve b will be equal to the charge on the surface.

In practical situations, the stipulation of constant surface charge is often found to be invalid, especially in concentrated colloidal systems where the distance between interacting surfaces is relatively small. In those cases, a number of events can occur that will effectively result in changes in the net surface charge, and therefore the overall electrical characteristics of the system. The most important of these processes is the specific absorption of ions at the surface.

Specific Ion Adsorption and the Stern Layer

The derivation of equations related to the charges on surfaces have, to this point, been made with the aid of several important assumptions, most of which are valid only up to a certain point. The most significant of those included the following assumptions:

(1) The ions involved, both on the surface and in solution, are point charges --- that is, they have no finite volume. In fact, of course, ions possess characteristic radii which vary significantly with the ion involved. In addition, the *effective* radius of an ion will depend on the number of solvent ions associated closely with it (its *hydration number*).

(2) Charges on a surface are "smeared out" over the surface so that a uniform charge exists, when in fact charges arise from discrete

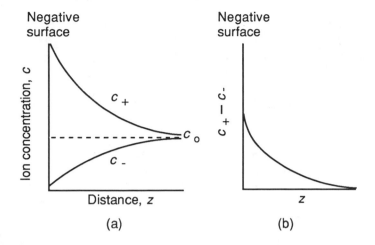

Figure 5.4. Schematic illustration of the distribution of charges in the diffuse double layer around a negatively charged surface: (a) the total concentration of ions near the surface and (b) the net local charge density.

points on the surface.

(3) Charges of opposite sign can approach infinitely closely, in obvious contradiction to (1).

(4) The solvent influences the double layer only through its dielectric constant and that the dielectric constant remains constant throughout the double layer. The latter is known not to be the case for water, in which the dielectric constant may be reduced by an order of magnitude near a surface due to orientation of the water molecules by charges in the area. In addition, as an ion is specifically adsorbed onto the surface, it is probable that it becomes dehydrated (desolvated), to some extent.

Luckily, most of those erroneous assumptions introduce only minor difficulties in the interpretation of electrostatic phenomena at surfaces, especially for practical purposes. Of perhaps more significance is the assumption that the charge density on a surface remains constant under all conditions. In fact, as the conditions of a surface are altered (eg, electrolyte concentration increases or two surfaces are brought into very close distances --- say 1--2 nm) the number of "free" charges on the surface may be reduced as a result of the specific adsorption of ions of opposite charge. Thus, as two charged surfaces are brought together, the surface charge density becomes a function of the distance of separation and tends toward zero at contact. That phenomenon is known as *charge regulation* . The end effect of charge regulation is to reduce the electrical potential of the surface, reduce the thickness of the electrical double layer, and reduce the effective repulsion between surfaces relative to that expected based upon the "theoretical" potential, ψ_o.

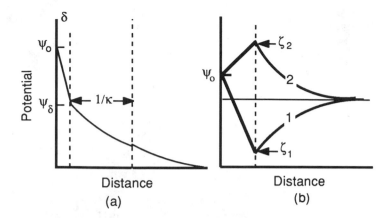

Figure 5.5. Schematic illustration of the electrostatic potential energy curve: (a) according to the Stern model and (b) the effect of adsorption of polyvalent or surface-active counterions (curve 1) and surface-active co-ions (curve 2).

Normally, a certain number of counterions will be strongly adsorbed in an area close to the actual surface in the Stern layer. Such ions will be adsorbed strongly enough that they will not be displaced (for a relatively long time, at least) by thermal Brownian motion. Because ions in the Stern layer are "fixed" relative to the ions further from the surface, in the so-called *diffuse double layer* , they effectively screen or neutralize a portion of the inherent surface charge. In that case, the surface potential ψ_o is replaced in the Gouy-Chapman treatment by ψ_s which is the Stern potential (Figure 5.5a). The curve is typical of a system with no specific adsorption; the electrical potential decreases rapidly from ψ_o to ψ_s within the Stern layer followed by the "normal" decay of ψ_s to zero in the diffuse double layer. The existence of the Stern layer should be considered as a general phenomenon in charged systems, as distinct from *specific adsorption* which can have special consequences not directly related to the Stern layer.

Figure 5.5b illustrates schematically the effects of two examples of specific adsorption on the electrical nature of the double layer. Curve 1 shows the effect of adsorption of a polyvalent counter ion or a surface active ion of opposite charge, leading to charge reversal (ie, ψ_s has the opposite sign to ψ_o). Curve 2 represents the situation in which the adsorption of ions or surface-active species of like charge causes the Stern potential to increase relative to that of the surface ($\psi_s > \psi_o$).

The complete mathematical expression for the double layer incorporating the Stern layer is quite complex and will not be given here. However, its existence and related effects are quite significant for

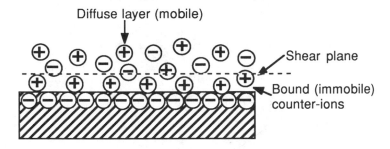

Figure 5.6. Schematic illustration of the physical meaning of the shear plane in relation to the surface potential and ion mobility in electrophoretic phenomena.

practical studies of electrokinetic phenomena discussed below because it is ψ_S that is actually being estimated in such procedures. When a charged particle (or surface) moves relative to an electrolyte solution, viscosity effects dictate that only that portion of the electrical double layer up to (approximately) the Stern layer will move. The ions in the Stern layer will remain with the surface. The dividing line between movement with the solution and that with the surface is referred to as the *shear plane* (Figure 5.6). The exact location of the shear plane, which is actually a very thin region in which viscosity effects change rapidly, is difficult to determine, but it is usually assumed to be just outside the Stern layer, which implies that the potential at that point will be slightly less than ψ_S.

The potential at the shear plane is termed the *electrokinetic* or ζ *(zeta) potential* and represents the actual value determined in the procedures discussed in the next section. It is generally assumed in tests of double-layer theory that the ζ potential and ψ_S are the same, since any error introduced will be small under ordinary circumstances. More significant errors may be introduced at high potentials, high electrolyte concentrations, or in the presence of adsorbed nonionic species that force the shear plane further away from the surface, reducing the ζ potential relative to ψ_S.

Considering all of the assumptions and approximations involved in the derivation of the Gouy-Chapman model of the double layer, it should be obvious that the real situation is likely to be much more complex. Nevertheless, results obtained based on that model have served (and continue to serve) well is furthering our understanding of electrical phenomena in colloidal systems. Further refinements of double layer theory have succeeded in explaining a number of bothersome observations in specific situations, especially very high surface potentials. However, the complications involved in their application, and the benefits derived, do not generally warrant such effort in most practical situations.

ELECTROKINETIC PHENOMENA

An important consequence of the existence of electrical charges on surfaces, whether they are formal colloids, porous materials, or some other system, is that they will exhibit certain phenomena under the influence of an applied electric field related to movement of some part of their electric double layer. Those phenomena are collectively defined as *electrokinetic* phenomena and include four main classes:

1. *Electroosmosis* : the movement of a liquid relative to a stationary charged surface under the influence of an electric field. The fixed surface will typically be a capillary tube or porous plug.
2. *Electrophoresis* : the movement of a charged surface (usually colloidal particles or macromolecules) plus its attached double layer relative to a stationary liquid, under the influence of an applied field. Electrophoresis is, of course, the complement of electro-osmosis.
3. *Streaming potential* : the electric field generated when a liquid is forced to flow past a stationary charged surface.
4. *Sedimentation potential* : the electric field which is produced when charged particles move relative to a stationary liquid.

The four phenomena are illustrated schematically in Figure 5.7.

Of the four, the phenomenon of greatest practical interest is electrophoresis. Over the years, several relatively easy techniques for the study and application of electrophoresis have been developed and today these are important tools in many areas of science and technology, including colloid science, polymer science, biology, and medicine. Of lesser practical importance, and less thoroughly studied, are eletroosmosis and streaming potential. Sedimentation potential has received relatively little attention because of experimental difficulties.

While a thorough discussion of the details of those techniques is not possible here, a brief conceptual description of the more important phenomena and their practical applications will be useful in guiding the interested reader to the method of choice for a specific application. Theoretical and experimental details can be found in the comprehensive colloid and surface chemistry works cited in the Bibliography.

Particle Electrophoresis

Particle electrophoresis, also sometimes referred to as *microscope electrophoresis* or *microelectrophoresis* , is one of the easiest and most useful techniques for investigating the electrical properties of colloidal particles. If the system of interest is in the form of a reasonably stable dispersion of particle size observable by light microscopy (say larger than 200 nm for practical application), the electrokinetic behavior of the

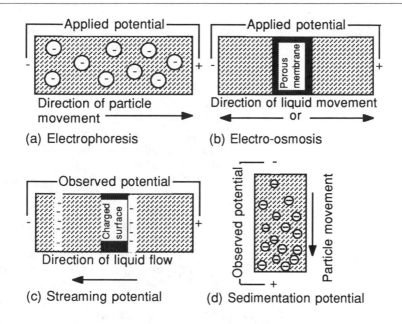

Figure 5.7. Four principle types of electrokinetic phenomena.

system can be observed and measured directly. Several commercial instruments are available for the purpose.

The measurement of the electrophoretic characteristics of a system can be very useful in evaluating the electrical nature of a surface --- its charge sign and, indirectly, its charge density --- and the effects of changes of pH, electrolyte content, electrolyte valency, etc, on the charge. It can also be used to evaluate the electrical nature of soluble species (eg, surfactants or polymers) that may be adsorbed onto the surface of a particle. For example, if a colloid is prepared that has little or no inherent charge and then placed in a solution of a polymer such as a protein, adsorbed protein will impart an electrical charge to the surface and measurably change the mobility of the particles in an electric field. From that information the electrical nature of the protein can be deduced, information which would be much more difficult to obtain otherwise.

Particle electrophoresis has proved to be very useful in many areas of theoretical and practical surface science, including "model" polymer latex and silver halide systems, and more practical problems related to water purification, detergency, emulsion science, the characterization of bacterial surfaces, blood cells, viruses, etc. With the advent of more sophisticated computer data analysis and laser light sources, the limits of resolution for particle sizes that can be analyzed has been, and is being, steadily reduced, so that with proper (and more expensive) instrumentation, the electrophoretic nature of particles in the size range of a few nanometers is now a possibility.

Moving Boundary Electrophoresis

An alternative to particle electrophoresis is *moving boundary electrophoresis* . The technique is used to study the movement of a boundary formed between a colloidal sol or solution and the pure dispersion medium under the influence of the electric field. The technique has found some application for determining not only electrophoretic mobility, but also for small-scale separation of species from a mixture for further identification. It found early application in the study of proteins and other dissolved marcomolecules.

If a protein solution contains a number of different species of different charge characteristics (and therefore different mobilities in a given charge field), the technique may be able to separate the fractions, or at least the peaks, sufficiently to indicate the number of distinct species present. The technique has largely been displaced by more sensitive, and experimentally easy, techniques; however, the inexpensive nature of the process still carries some weight in choosing an approach for some applications.

Gel (or Zone) Electrophoresis

Another alternative technique commonly used to investigate the electrophoretic properties of a material (especially soluble macromolecules) or mixtures thereof is *gel* or *zone electrophoresis* . The technique involves the use of a relatively inert solid or gel support for the solution of interest, which minimizes many of the experimental difficulties encountered in the moving boundary technique, especially convection and vibrational disturbances. It is also much simpler, since there now exist a number commercially available setups that require little manipulation and can be handled very easily by technicians with little advanced training or experience.

Gel electrophoresis requires very small sample sizes and can, in theory at least, give complete separation of a mixture of substances. While it cannot be used to determine electrophoretic mobilities, it allows for the separation and identification of components that would be extremely difficult or impossible to separate using other techniques. It is especially applicable to biological systems where sample availability may be a problem. It may even be used as a small-scale preparative procedure,

Zone or gel electrophoresis is limited in quantitative terms because it separates components according to two criteria --- charge and molecular weight or size. It may happen that two components coincidentally have the right combination of charge and size so that they move together under a given field. To overcome that problem, techniques have been developed that may be called "two-dimensional" electrophoresis in which an electric field is applied in one direction or a given period of time, followed by another field of different strength in a perpendicular direction. The net result is, hopefully, that components that fortuitously move together in the first field will be separated by the second, since only molecules of the same size and same charge would be expected to move together under two different electric fields.

Since the other electrokinetic phenomena are of significantly lesser practical importance, they will not be discussed here. For further theoretical and experimental details, the reader is referred to the works cited in the Bibliography. The true practical significance of the material presented in this chapter will become more apparent in the following chapter. Obviously, the more complex a colloidal system becomes, the more difficult it will be to pin down its behavior in terms of specific phenomena discussed so far. However, small pieces of information (such as electrophoretic mobility) can be important in helping to determine the performance of a given system.

CHAPTER 6

CAPILLARITY

Although defined in various ways depending on the context, the term *capillarity* for current purposes can be defined as the macroscopic motion of a fluid system under the influence of its own surface and interfacial forces. Such flow is similar to other types of hydraulic flow in that it results from the presence of a pressure differential between two hydraulically connected regions of the liquid mass. The direction of flow is such as to decrease the pressure difference. When the difference vanishes, or when there is no longer a mechanism to reduce the difference, flow ceases.

Capillary effects are encountered in many aspects of surface and colloid science, with its importance relative to other processes (eg, fluid dynamics) depending on the exact situation. For example, when two spherical drops of a liquid in an emulsion make contact and coalesce to form a larger drop (Figure 6.1a), the extent and duration of flow due to the capillary phenomenon is limited and fluid dynamics is of little practical importance. When there is an extensive amount of flow, on the other hand, such as in capillary imbibition, wicking processes, or capillary displacement (Figure 6.1b) fluid dynamics becomes important. Classical discussions of capillary action tend to concern themselves with the surface and interfacial driving forces behind the phenomena, with little attention given to the fluid dynamics aspect. In many important practical applications, however, fluid characteristics play an important role and must be considered. For the most part, the following comments will not take such factors into consideration but will address the capillary phenomena in isolation. For present purposes, the discussion will be limited to a qualitative and descriptive presentation rather than a rigorous derivation of the principles involved. A number of secondary references are provided that go into detail on the more theoretical questions.

It was stated above that capillary phenomena arise as a result of differences in pressure across a system containing at least one liquid phase and another liquid, vapor and/or solid phase. As will be illustrated below, that pressure difference arises as a result of differences in

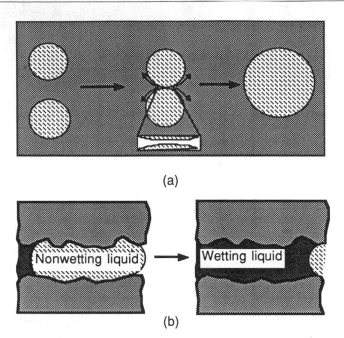

(a)

(b)

Figure 6.1. Important functions of capillary forces in practical situations: (a) as twoemulsion drops approach, the pressure at the nearest surfaces increases, deforming the drops and enlarging the radius of curvature in the immediate area. That deformation causes the capillary pressure in the regions outside that area to decrease in a relative sense, sunctioning continuous phase from between the drops and increasing the liklihood of contact and film rupture or coalescence. (b) in capillary displacement, the liquid that preferentially wets the solid will displace the less wetting liquid.

curvature of different regions of liquid--fluid phases in the system, and due to the presence of an effective mechanical tension in the interface, the surface or interfacial tension. The differences in curvature giving rise to the pressure differentials may result from various sources including the application of external forces, the contacting and coalescence of two masses of the liquid phase, or from the contact of the liquid phase with an second fluid phase and a solid surface.

A CAPILLARY MODEL

Capillary flow systems of most practical interest are those that involve a solid, a liquid, and a second fluid phase. In the absence of other external forces, the net driving force for capillary flow in such a system will be controlled by three basic quantities including the various interfacial tensions, the geometry of the solid--liquid--fluid interface, and the geometry of the solid surface at the three-phase boundary line. When a liquid phase contacts a second fluid phase and a solid surface

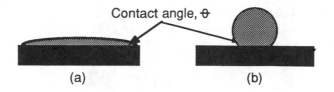

Contact angle, θ

(a)   (b)

Figure 6.2. Contact angles in solid--liquid systems: (a) a small angle measured through the liquid for a system with a low interfacial energy; (b) a large angle for a system with little favorable solid--liquid interaction, ie, a high interfacial energy.

there occurs a net effect of orienting the liquid--fluid interface by causing it to assume a characteristic equilibrium orientation with respect to the solid surface, reflected in the so-called *contact angle* (Figure 6.2). Under controlled, equilibrium conditions, that contact angle may be considered to be an intensive material constant depending only on the natures of the three component phases, but independent of the quantities present (within limits to the molecular level). Being an experimentally accessible quantity, the contact angle can be, as will be seen later, a very useful tool for studying interfacial effects.

Except under special circumstances (such as zero-gravity environments), most practical capillary systems will experience hydraulic pressure gradients in addition to those deriving from curvature, the most important of which is that due to gravity. Most treatments of capillary flow, then, should theoretically take into consideration the effect of the gravity-induced hydrostatic head. The gravitational pressure gradient must be included algebraically in many calculations of capillary flow, especially those containing a significant vertical component. For primarily horizontal systems, however, the pure capillary pressure contribution is so much greater than any gravitational effect that the latter may be neglected. In the discussion to follow, the effects of gravity have been neglected except where specifically indicated.

CAPILLARY DRIVING FORCES IN
LIQUID--FLUID SYSTEMS

Of the driving forces for capillary action mentioned above, the most fundamental is that of surface or interfacial tension and related effects (eg, contact angle). As pointed out in Chapter 2, a liquid--fluid interface behaves as if it is an elastic film stretched over (or between) the two phases and resisting any more stretching to produce greater interfacial area. The tension results fundamentally from the imbalance in the forces acting on the molecules in the surface, which tend to pull the molecules back into the bulk phase. At equilibrium, the surface tension is a material constant, the value of which can be determined by any convenient method and then be applied as needed in other situations.

Mathematically, an interface is a two-dimensional region. In reality, it will be three dimensional, although the third dimension may be of only one or two molecular thicknesses. Because it is three dimensional, the interfacial region may be treated in the context of hydrostatics, or in terms of molecular forces and distribution functions. Alternatively, a thermodynamic approach may be taken to arrive at the same conclusions.

The fact that a tension exists at a liquid--fluid interface implies that, if it is curved, there will be a difference in hydrostatic pressure across the interface. In 1806, Laplace[1] derived an expression for the pressure difference across a curved interface in terms of surface tension and curvature. The equation, referred to as the *Laplace Equation* , is

$$P_1 - P_2 = \Delta P = \sigma \left(1/r_1 + 1/r_2 \right) = P_{cap} \qquad (6.1)$$

In which P_1 and P_2 are the pressures in the two phases forming the interface, and r_1 and r_2 are the principal radii of curvature of the interface at the point in question. For a spherical surface $r_1 = r_2 = r$ and eq. 6.1 simplifies to

$$\Delta P = 2\sigma/r \qquad (6.2)$$

It is useful to work through the derivation of eq. 6.2 for a spherical surface to be certain of the relationship between surface tension and pressure, since pressure is the driving force for capillary action. If one takes a spherical drop of liquid of radius r and adds more liquid so that the radius increases by a factor dr, the surface area of the drop will increase by a factor $8\pi r\, dr$ (Figure 6.3). As seen in Chapter 2, the amount of work that must be done to expand a surface or interface is given by

$$W = \sigma\, dA \qquad (6.3)$$

so that for the drop in question, the work will be $\sigma \times 8\pi r\, dr$. Under conditions of mechanical equlibrium, that work will come from the pressure difference across the interface $(P_1 - P_2) \times 4\pi r^2$ acting through the distance dr, where P_1 is the pressure on the inside of the drop and P_2 is the pressure outside. If the two terms are equated, the result is

$$(P_1 - P_2) = 2\sigma/r \qquad (6.2a)$$

which is, of course, eq. 6.2. A similar analysis of nonspherical systems produces the same result, but in the more general form of eq. 6.1.

In a system involving purely capillary phenomena, the sole driving force is the pressure differential in various areas of the system. The sign of the pressure term in eq. 1 will depend on the assignment

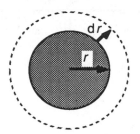

Figure 6.3. Illustration of the growth of a fluid drop as related to capillary pressure.

of P_1 and P_2; however, conventional procedure is to assign P_1 to the liquid phase of interest and P_2 to the adjoining fluid phase. In a flat surface where $r_1 = r_2 = \infty$, $\Delta P = 0$. Where the surface is concave inward (toward the liquid), $\Delta P > 0$; where it is convex inward, $\Delta P < 0$. Where a pressure differential exists, the liquid will flow from high to low pressure until the differential is decreased and ultimately eliminated. In situations where the interfacial tension is uniform from point to point, the capillary pressure will depend only on the curvature of the interface.

It has been noted that in most circumstances it is common to ignore gravitational effects for most calculations of capillary forces. The validity of such a procedure can be seen with a simple example. For example, suppose that a spherical drop of liquid of unit density and surface tension of 50 mN m^{-1} has a diameter of 0.1 cm. The hydrodynamic pressure difference due to gravity between the top and the bottom of the drop will be 98 mJ m^{-2}. The capillary pressure difference, on the other hand, will be, according to eq. 2, 2000 mJ m^{-2}. As the drop diameter and the radius of curvature decrease, the capillary pressure increases to approximately 1.01×10^5 J m^{-2} at $r = 10^{-4}$ cm and 1×10^7 J m^{-2} at $r = 10$nm. As the drop diameter approaches molecular dimensions, eqs. 6.1 and 6.2 should no longer be considered valid. Therefore the "calculation" of extremely high pressure in capillary systems of very small radius must be approached with great caution.

Solid--Liquid--Fluid Systems: The Effect of Contact Angle

The above discussion was concerned with capillary forces in a system containing only the liquid of interest and a second fluid phase. It has been stated, however, that the systems of most practical interest normally involve a third solid phase which will usually result in a three-phase boundary line. The situation can be represented as a drop of liquid resting on a flat solid surface and contacting the third liquid or vapor phase (Figure 6.4). If the drop is allowed to spread over the surface a small distance from point a to point b, the new liquid--vapor interface will remain essentially parallel to the old, but increase its area

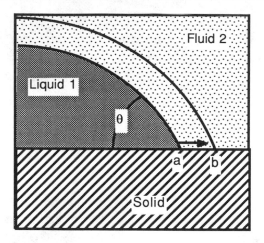

Figure 6.4. The wetting of a solid by a liquid in the presence of a second fluid phase.

by the amount bc = ab cos θ. In the process, the solid--liquid interface is increased by ab and the solid--vapor interface decreased by that amount. At equilibrium, the change in free energy will be zero, so that

$$\Delta G = \sigma_{SV} \Delta A_{SV} + \sigma_{SL} \Delta A_{SL} + \sigma_{LV} \Delta A_{LV} = 0 \qquad (6.4)$$

leading to the result

$$\sigma_{SV} = \sigma_{SL} + \sigma_{LV} \cos \theta \qquad (6.5)$$

which is generally known as *Young's equation* . While the exact physical interpretation of Young's equation is not clear for all situations, it is one of the fundamental equations of surface chemistry and will be encountered repeatedly. More will be said about its significance and interpretation below and in Chapter 17.

Capillary Flow and Spreading Processes

Before continuing with the specific topic of capillary flow, it may be useful to divert our attention to the broader topic of the spreading of a liquid on a surface. The topic of spreading is of great practical interest because the process is of importance in many applications. For that reason, another chapter has been devoted specifically to such phenomena. However, it will be useful to introduce some basic concepts at this point in order to clarify the some points in the current topic.

If a quantity of liquid is placed on a surface, which may be a solid or another liquid, there are two things that may occur. The liquid may spread across the surface to form a uniform duplex film, or the liquid may form a drop (on a solid) or lens (on a liquid) with a finite, nonzero contact angle. Thermodynamically, at constant temperature and pressure the change in the free energy of the system is given by

$$G = (\delta G /\delta A_A)\, dA_A + (\delta G /\delta A_{AB})\, dA_{AB} + (\delta G /\delta A_B)\, dA_B \quad (6.6)$$

where subscript A designates the substrate and B the liquid. Obviously, $dA_A = -dA_B = dA_{AB}$, so that

$$(\delta G /\delta A_A) = \sigma_A, \quad (\delta G /\delta A_B) = \sigma_B, \text{ and } (\delta G /\delta A_{AB}) = \sigma_{AB}$$

The term $(\delta G /\delta A_B)$ gives the free energy change for the spreading of liquid B over the surface A and is called the *spreading coefficient* of B on A, $S_{B/A}$, given by

$$S_{B/A} = \sigma_A - \sigma_B - \sigma_{AB} \quad (6.7)$$

From the definitions of the works of cohesion and adhesion given in Chapter 2, it can be seen that $S_{B/A}$ is the difference between the work of adhesion of B to A and the work of cohesion of B

$$S_{B/A} = W_{AB} - W_{BB} \quad (6.8)$$

From this analysis, the spreading coefficient will be positive if there is a decrease in free energy on spreading; that is, the spreading process will be spontaneous. If $S_{B/A}$ is negative, then the cohesive forces will dominate and a drop or lens will result.

What eq. 6.7 is saying, in general, is that when a liquid of low surface tension such as a hydrocarbon is placed on a liquid or solid of high surface tension such as clean glass or mercury, spontaneous spreading occurs. Conversely, if a liquid of high surface tension such as water is placed on a surface of low surface tension such as teflon or paraffin wax, drop or lens formation results.

Unfortunately, complications arise in spreading phenomena due to the unfortunate fact that liquids, solids, and gases tend to interact in bulk processes as well as at interfaces, and those bulk-phase interactions may have significant effects on interactions at interfaces. In particular, gases tend to adsorb at solid interfaces and change the free energy of those surfaces, σ_{SV}; they may also become dissolved in liquid phases and thereby alter the liquid surface tension. More importantly, liquids in contact with other liquids tend to become mutually saturated, meaning that the composition of the two phases may not remain "pure" and therefore no longer have the surface characteristics of the pure materials. Finally, liquids and solutes, like gases, can adsorb at solid interfaces to alter the surface characteristics of the solid and thereby change the thermodynamics of the spreading process. Most of these situations will be addressed in the context of specific areas of interest in later chapters. However, the classic example of benzene--water systems will serve as a useful illustration.

For a drop of pure benzene (σ_B = 28.9 mN m^{-1}) placed on a surface of pure water (σ_A = 72.8 mN m^{-1}) with an interfacial tension, σ_{AB} of 35.0 mN m^{-1}, eq. 7 predicts a spreading coefficient of

$$S_{B/A} = 72.8 - 28.9 - 35.0 = 8.9 \text{ mN m}^{-1}$$

The positive spreading coefficient indicates that benzene should spread spontaneously on water. When the experiment is carried out, it is found that after an initial rapid spreading, the benzene layer will retract and form a lens on the water. How can this seemingly anomolous result be explained?

In this and many similar cases, it must be remembered that benzene and many other such water-"immiscible" liquids have, in fact, a small but finite solubility and the water will rapidly become saturated with benzene. Benzene, having a lower surface tension than water, will adsorb at the water--air interface so that the surface will no longer be that of pure water but that of water with a surface excess of benzene. The surface tension of benzene-saturated water can be measured and is found to be 62.2 mN m^{-1}, which is now the value which must be used in eq. 6.7 instead of that for pure water, so that

$$S_{B/A(B)} = 62.2 - 28.9 - 35.0 = -1.7$$

where the subscript A(B) indicates phase A saturated with phase B. The negative spreading coefficient indicates that lens formation should occur, as is observed. The saturation process occurs, of course, in both phases. However, since water is a material of relatively high surface tension, it will have little tendency to adsorb at the benzene--air interface and will therefore cause little change in the surface tension of the benzene. In this case $\sigma_{B(A)}$ = 28.8 mN m^{-1} so that

$$S_{B(A)/A} = 72.8 - 28.8 - 35.0 = 9.0$$

If only the benzene layer were affected by the saturation process, spreading would still occur.Combining the two effects one obtains

$$S_{B(A)/A(B)} = 62.2 - 28.8 - 35.0 = -1.6$$

which illustrates the fact that it is the effect of benzene in water which controls the spreading (or nonspreading) in this system. The interfacial tension of water--benzene is unchanged throughout because it inherently includes the mutual saturation process.

Situations like that for benzene are very general for low surface tension liquids on water. There may be initial spreading followed by retraction and lens formation. A similar effect can in principle be achieved if a third component (eg, a surfactant) which strongly absorbs at the water-iair interface, but not the oil--water interface, is added to the system. Conversely, if the material is strongly adsorbed at the oil-iwater interface, lowering the interfacial tension, spreading may be achieved where it did not occur otherwise. This is, of course, a technologically very important process and will be discussed in more detail in later chapters.

Geometrical Considerations in Capillary Flow

When considering capillary flow problems there are several external factors that must be kept in mind, in addition to the question of pressure differentials due to surface curvature and surface tension effects. For a liquid--fluid system, if one assumes that σ_{LV} is constant and that there are no external factors inducing pressure differences in the system, then the capillary pressure, P_{cap}, is a function entirely of the curvature of the LV interface. Put another way, when the system is at mechanical equilibrium, P_{cap} will be constant and at its minimum value, and curvature will be constant in all parts of the system. If some external force induces a change in curvature at some point, the resulting increase in P_{cap} returns the system to its original state, provided that the disturbance has not been to great. If the perturbing force is great enough to remove the system significantly from equilibrium, the resulting P_{cap} may cause the liquid mass to be divided into separate masses rather than returning to the original configuration. This effect is often referred to as a *yielding* of the surface. Such an effect can be either advantageous or disadvantageous, depending on the situation.

A commonly encountered example of the yielding effect can be seen in the breaking of a foam formed in a pure liquid. It is generally observed that pure liquids do not foam; or that any transitory foam formed on agitation will dissipate very rapidly once agitation ceases. A schematic representation of an unstable foam system is given in Figure 6.5, where the liquid phase lies in the thin lamellae (L) between the vapor cells (V). Because of the large curvature differences in the *plateau regions* (P) relative to the lamellae there will exist a large pressure differential in the system. The P regions have a small convex radius of curvature, which leads to a large negative P_{cap}, while the L regions have a much larger radius of curvature and a correspondingly smaller (and less negative) P_{cap}. As a result of the pressure differential, liquid will flow from region L to region P , thinning the lamellae until the cell ruptures and the foam breaks. In unstable foams, this process occurs very rapidly. In systems containing various additives such as surfactants or polymers, the process can be slowed sufficiently to produce "stable" foams. That topic will be discussed in Chapter 12.

A potentially important disadvantage of surface yielding can be found in the process of secondary oil recovery. In many such operations,

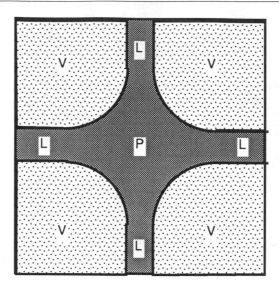

Figure 6.5. Schematic illustration of the situation for an unstable foam system.

the oil to be recovered is forced out of the porous rock formation by an aqueous "plug" which displaces the oil in the rock capillaries. If the system is not properly formulated and implemented, the force of the aqueous plug may cause the crude oil phase to undergo such yiedling processes, leading to its breakup into small particles and isolation in pores from which recovery will be difficult or impossible.

Measurement of Capillary Driving Forces

Classically, the approach used to calculate capillary flow has been to determine the curvature of liquid interfaces in the system and calculate P_{cap} from eq. 6.1. Those values could then be used to calculate the direction and magnitude of the driving forces. In systems of simple geometry such as liquids which form spherical interfaces and smooth cylindrical solid surfaces, the technique works out very well. The best known example of such a system is the capillary rise method for determining the surface tension of a liquid, illustrated in Figure 6.6. In this system, capillary forces cause the liquid to rise in the tube due to differences in curvature of the liquid--air interface within the tube (a small radius of curvature) and that in the reservoir ($R \approx \infty$). Here there is also a constant opposing force due to gravity that must be included.

If the contact angle of the liquid on a capillary surface is θ, the radius of the tube is r, and it is assumed that r is sufficiently small that the liquid surface in the tube is spherical, then the radius of curvature of the liquid--air interface, R, $= r$ /cos θ. According to eq. 6.1, the capillary pressure at point B will be $-2\sigma \cos \theta / r$ and the net driving force for

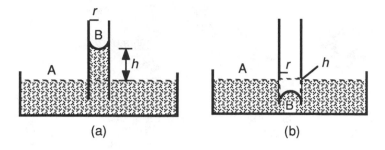

Figure 6.6. Capillary rise phenomena for (a) wetting and (b) nonwetting liquids.

capillary rise will be $P_{cap}(A) - P_{cap}(B)$, since at A $P_{cap} = 0$, the total driving force comes from the curvature of the interface in the tube. The movement will continue until the hydrostatic head of liquid in the tube, $\Delta\rho gh$, is equal to $P_{cap}(B)$, where $\Delta\rho$ is the difference in density between the liquid and the vapor, g is the gravity constant, and h is the height of the miniscus in the tube above the liquid level at A.

$$\Delta\rho gh = 2\sigma \cos\theta /r \qquad (6.9)$$

Equation 6.9 is, of course, the classic equation for determining the surface tension of a liquid by the capillary rise method. When $\theta = 0°$, the equation simplifies to

$$\sigma = \Delta\rho ghr /2 \qquad (6.10)$$

In the absence of gravity, a similar effect can be seen for two capillary tubes of unequal radius connected end-to-end as shown in Figure 6.7. In this case the net driving force for liquid flow arises due to the differences in curvature of the ends of the liquid mass in the two tubes at points A and B. The net capillary pressure is given by

$$P_{cap} = 2\sigma \cos\theta \, (1/r - 1/r') \qquad (6.11)$$

where r is the radius of the small and r' that of the large capillary. Beginning with the situation as illustrated in Figure 6.7a, the liquid will flow into the smaller capillary until it attains configuration (b), at which time all pressures will be equal and flow will cease.

The situation illustrated by Figure 6.7 is potentially important in many practical areas because it represents an idealized system for many wicking, blotting, and absorption processes. Unfortunately most real systems, such as textiles or paper products, have such complex geometries that it is not possible to determine simple values for r and r'. In such cases it is convenient to employ a thermodynamic approach

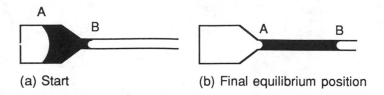

(a) Start   (b) Final equilibrium position

Figure 6.7. Capillary flow in a horizontal system of two joined capillary tubes of unequal diameters.

rather than use pressure differentials.

In such a thermodynamic analysis, it is necessary to use eqs. 6.4 and 6.5, and the fact that in the system the sum of the areas of solid--vapor (A_{SV}) and solid--liquid (A_{SL}) interfaces remains constant. Using the equations, it can be shown that the change in free energy dG caused by a change in position of the three-phase boundary by a distance ds can be represented

$$dG/ds = \sigma_{LV}\, dA_{LV}/ds - \sigma_{LV} \cos\theta\, dA_{SL}/ds \qquad (6.12)$$

The quantities σ_{LV} and θ are experimentally accessible and the area changes can be determined from the geometry of the system. According to eq. 6.12, the liquid will move in a capillary system if dG/ds < 0, dG/ds being numerically equal to the net driving force P_{cap}. The thermodynamic approach is in principle very general and has been successfully applied with a number of models including the pull of a liquid in a partially immersed rod (Figure 6.8a), the movement of a liquid in a notch, and the movement of a liquid on two closely spaced rods. Such models have been found useful for studying various types of woven or interlacing systems in the textile and paper industries (Figure 6.8b).

Complications to Capillary Flow Analysis

In the discussions of capillary flow so far, it has been assumed that the values of surface tension and contact angle are constant and that other noncapillary factors such as external forces (except gravity in some cases), liquid viscosity, electrical effects, etc, can be ignored. In practice, such is not always the case. The application of the principles of capillarity often involves the handling of variations in the surface tensions involved with time and location, changes in contact angles due to those surface tension gradients, and variations in contact angle due to compositional and geometrical changes in the solid surface, hysteresis, etc, not to mention the fluid dynamic and other external factors. In practice, it is important to keep such variables in mind in order to better understand the capillary characteristics of a particular system.

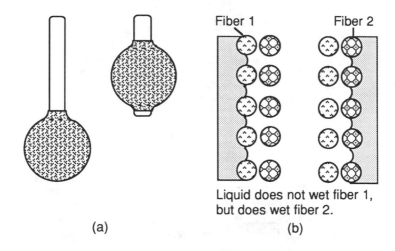

Liquid does not wet fiber 1,
but does wet fiber 2.

(a)   (b)

Figure 6.8. Schematic illustration of examples of capillary flow: (a) a liquid climbing a partially immersed rod; (b) wicking---the spontaneous movement of a liquid from a nonwetting to a wetting situation.

Surface Tension Gradients and Related Effects

In many practical systems, one of the most commonly encountered complications to the analysis of capillary flow is that arising from variations in the solid--liquid and/or liquid--vapor interfacial tensions in the system. Particularly, the value of σ_{LV} may vary significantly from point to point, leading to liquid flow unrelated to capillary phenomena. In a continuous liquid system in which surface tension gradients arise, liquid will spontaneously flow from regions of low to those of high surface tension. This flow is strictly a surface effect and is independent of curvature, which controls capillary flow. The rate of such flow will depend on the magnitude of the difference in σ_{LV} and the hydrodynamic characteristics of the liquid. Liquid flow at surfaces arising from surface tension gradients is commonly referred to as *Marangoni* flow and can be important in many systems other than capillary phenomena. Some of those systems will be encountered in later chapters.

Marangoni effects can be encountered in both single- and multicomponent liquid systems. In a pure liquid, surface tension gradients arise due to differences in temperature (or evaporation rate) from one point to another in the system. In almost all liquids, an increase in temperature lowers σ_{LV} so that where "hot spots" occur, liquid flows away to cooler regions of the liquid (Figure 6.9).

In multicomponent systems, surface tension gradients usually arise due to adsorption-related phenomena or, where possible, due to different rates of evaporation from the system (although temperature

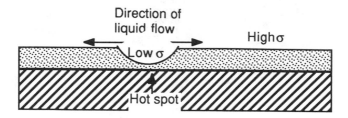

Figure 6.9. Schematic illustration of the Marangoni effect resulting from "hot spots" and surface tension gradients.

variations can also be important). If the system contains two liquid components of differing volatility, the more volatile liquid may evaporate from the LV interface. Normally, when two or more components are present, one will be preferentially adsorbed at the LV interface and lower σ_{LV}. If the "surface-active" component evaporates, the local surface tension of the liquid will rise and Marangoni flow toward the evaporation site will occur. If the evaporation process leads to an increase in σ_{LV} without significantly altering the contact angle, capillary flow can be much greater than that predicted based upon constant values.

Multicomponent systems may also involve the selective adsorption of one component at the SL interface. Since the component that lowers the surface tension will be preferentially adsorbed, the rate of the adsorption process can affect the local surface tension and the local contact angle. In many systems, the rate of adsorption at the solid surface is found to be quite slow compared to the rate of movement of the SLV contact line. As a result, the system does not have time for the various interfacial tensions to achieve their equilibrium values. Most surfactants, for example, require several seconds to attain adsorption equilibrium at a LV interface, and longer times at the SL interface. Therefore, if the liquid is flowing across fresh solid surface, or over any surface at a rate faster than the SL adsorption rate, the effective values of of σ_{LV} and σ_{SL} (and therefore θ) will not be the equilibrium values one might obtain from other measurements.

An effect closely related to that of varying adsorption rates is that resulting from changes in σ_{SL} due to direct chemical or physical interaction between the liquid and solid surface. Particularly important in the textile and paper industries would be the swelling of fibers on contact with water or other solvent liquids. As swelling occurs, the value of σ_{SL} will be continuously changing at an undetermined rate. Analysis of the flow then becomes difficult or impossible. In addition, some polymeric surfaces that do not swell on contact with aqueous systems have specific monomeric units that do interact with water (eg, acid groups). Such surfaces may experience changes in σ_{SL} or θ as a result of specific interactions at the molecular level (eg, ionization), again

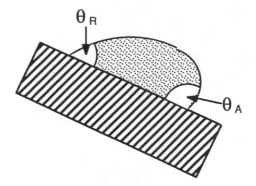

Figure 6.10. Schematic illustration of contact angle hysteresis of a liquid drop on an inclined surface.

affecting the observed capillary flow.

Contact Angle Effects

In practical capillary systems such as textiles, paper products, oil-bearing rock formations, etc, anomolous effects due to contact angle variations can almost be considered a certainty. The effects may be due to heterogeneities in the solid surface (compositional effects), geometry (surface roughness), or other dynamic or molecular factors. It was previously stated that the equilibrium contact angle of a given solid--liquid--fluid system could be considered as a material constant for the system. That is, in principle, true. However, it is commonly found experimentally that the equilibrium angle measured as a drop advances across a fresh solid surface may differ significantly from an angle measured as a drop moves across a previously contacted area. Conventionally, the former situation is referred to as the *advancing contact angle*, θ_A, and the latter as the *receding contact angle*, θ_R. The difference between the two angles is the *contact angle hysteresis* (see also, Chapter 17).

A commonly encountered illustration of contact angle hysteresis is that of a rain drop moving down a slanted car windshield (Figure 6.10). In that case, gravity is causing the front of the drop to advance across the fresh glass surface with a relatively large θ_A while the back moves with a much smaller angle. There is, of course, some distortion of the drop due to gravity, but that only exaggerates the effect somewhat.

Contact angle hysteresis measured experimentally may arise due to heterogeneities in the composition of the solid surface, surface irregularities, or dynamic effects due to adsorption or desorption phenomena, molecular reorientations, etc.

The cause of contact angle hysteresis on a chemically uniform but rough surface is illustrated schematically in Figure 6.11a. From the illustration it can be inferred that the extent of hysteresis observed on a

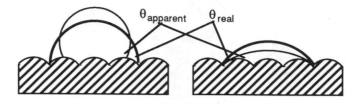

Figure 6.11. Contact angle hysteresis on a rough surface: (a) if $\theta_{real} \geq$ 90°, $\theta_{apparent} > \theta_{real}$; if $\theta_{real} < 90°$, $\theta_{apparent} < \theta_{real}$.

macroscopic scale may depend upon the scale of observation of the angle and the size of the liquid sample employed. Obviously, hysteresis can occur from point to point along the SLV contact line. The extent to which the curvature of the drop exhibits localized hysteresis will depend on the length of the moving boundary line and the area of the moving front. It can be shown that if the "true" contact angle of a liquid on a smooth solid surface is < 90°, its apparent contact angle on a rough surface of the same material will be smaller than the actual value. Conversely, if the true angle is > 90°, the apparent angle will be greater than the true value. A similar analysis of hysteresis on a nonuniform or *composite* surface has been employed to develop empirical relationships to quantitatively adjust observed contact angles for those complications. Those relationships and others concerning contact angles are discussed in more detail in Chapter 17.

Although hysteresis in many systems may be sufficiently small that it can be neglected, many important capillary systems are found to exhibit a hysteresis of 50--60°. Prime examples of such systems would be many textile and paper products that have θ_A quite large but θ_R at or near zero. The same can be expected in some rock and soil formations in which contact with water alters the hydration state or electrical characteristics of the surface and thereby θ. In such capillary systems simple empirical adjustments such as those available for planar systems do not suffice. In capillaries where the driving force for liquid flow must be calculated, it is most convenient to employ a modified form of eq. 6.12, dividing the final term on the right into terms for θ_A and θ_R so that

$$dG/ds = \sigma_{LV}\, dA_{LV}/ds - \sigma_{LV} \cos\theta_A\, dA_{SL(A)}/ds -$$
$$\sigma_{LV} \cos\theta_R\, dA_{SL(R)}/ds \qquad (6.13)$$

where the subscripts SL(A) and SL(R) refer to the area of the advancing and receding liquid fronts, respectively. If one applies eq. 6.13 to a fused capillary system such as that illustrated in Figure 6.7, where θ_A is sufficiently greater than θ_R, it will be seen that the liquid will remain stationary rather than move into the smaller capillary because the curvatures of the two ends of the liquid mass will be equal and P_{cap} will be zero.

Dynamic Contact Angle Effects

The contact angle hysteresis effects discussed above were taken in a context in which is was assumed that liquid movement was sufficiently slow that "equilibrium" or static values of θ_A and θ_R were involved. In capillary systems in which liquid flow is relatively fast, the effects of a *dynamic advancing* contatct angle, θ_{AD}, may become apparent. In such a situation the advancing contact angle measured will be greater than θ_A. The difference between θ_A and θ_{AD} will generally be found to increase with the speed of liquid flow. In systems of large-bore, short-path capillaries or those with high static θ_A, the effects of the dynamic contact angle on liquid movement may appear at relatively slow flow rates, even becoming self-limiting. Where relatively long capillaries are involved, so that the area SLV_A is small relative to the area of the SL and SLV_R interfaces, fluid dynamic factors such as viscosity become more important than θ_{AD}.

Phenomena related to dynamic contact angles are important in other, noncapillary applications, such as coating operations. Some examples will be addressed in Chapter 17.

Rates and Patterns of Capillary Flow

While it is important to analyze a capillary system with respect to the various factors affecting P_{cap} in order to understand liquid flow patterns, the effects of various fluid dynamic effects must also be kept in mind. While a thorough discussion of that topic is beyond the scope of this work, it may be useful to introduce a few simple ideas that can be applied in many situations.

When one considers the smooth, uniform (laminar) flow of a fluid in a narrow cylindrical tube, which is the classical model of a capillary, one can employ Poiseuille's equation to relate the volume rate of flow to various characteristics of the fluid and capillary system. The volume rate of flow, dv/dt (ml sec^{-1}) is given by

$$dv/dt = \pi r^4 P / 8\eta l \qquad (6.14)$$

where r is the radius of the tube, η the viscosity of the fluid, l is the distance of fluid movement in the tube in time, t, and P is the pressure drop across the distance l. For linear rates of flow the eq. 6.14 becomes

$$dl/dt = r^2 P / 8\eta l \qquad (6.15)$$

In capillary systems, P is replaced by P_{cap}.

If eq. 6.12 is employed to calculate the driving force dG/ds, it must be converted into an equivalent P_{cap} for use in eq. 6.14 or 6.15. For a relatively simple system such as that illustrated in Figure 6.7, the conversion is straightforward. For example, in the figure, if the normal projected cross-sectional area of the LV interface at point A (in cm^2) in the large end of the capillary system is significantly greater than that at point B at the small end, it can be assumed that P_{cap} at A will be negligable and only the contribution from P_{cap} at B need be considered. If ds in eq. 6.12 is measured at B, the net driving force $-dG/ds$ (in dynes) divided by the cross-sectional area at B will give the pressure drop from A to B (dynes cm^{-2}). In that case, eq. 6.15 becomes

$$dl/dt = 2\sigma_{LV} \, r \cos \theta / 8\eta l \qquad (6.16)$$

In practice, unfortunately, most capillary systems do not involve nicely uniform cylindrical tubes so that the above analysis will not apply in a pure quantitative way. In not overly complicated systems, reasonable and useful results can be obtained if r is replaced by the so-called *hydraulic radius* given by the volume of the liquid in the capillary section being considered, V, divided by the solid--liquid interfacial area, A, in the same section. In systems of still higher complexity calculation of a hypothetical value for r, or of a so-called *resistance factor*, r/η, becomes quite difficult and only qualitative or semiquantitative relationships can be expected.

Because many practical systems involving capillary flow involve irregular and ill-defined geometries, and therefore variations in curvature and P_{cap} from point to point the system, there has developed an area of investigation into hydraulics that envelops and combines aspects of purely capillary flow and the associated field of pure fluid dynamics (along with quite a bit of intuition and modeling). Fortunately, many complex systems such as filtration and wetting of woven textile and paper products, can be approximated using models with the liquid mass in a given starting location and configuration and using eq. 6.12 to calculate the driving force for movement and the final configuration. If reasonable values of P_{cap} are made, eqs. 6.14, 6.15, or 6.16 can be used to estimate initial rates of flow. From a comparison of predicted and measured initial flow rates, the validity of P_{cap} from the original assumptions can be evaluated and the model changed if necessary to improve the fit. By a series of such iterations one may arrive at a workable practical model for the system in question.

SOME PRACTICAL CAPILLARY SYSTEMS

While many practical capillary systems do not lend themselves to direct analysis according to the simple concepts presented above, there are areas in which those principles, along with some intuition and a

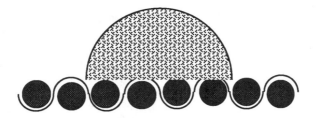

Figure 6.12. Schematic illustration of a drop of nonwetting liquid on an open woven capillary system.

dollop of luck, can be constructively employed to analyze and predict capillary phenomena. Several areas in which various degrees of success have achieved are wetting and repellency of woven fibers, paper products, and porous solids, wicking, and cleaning action in detergent baths. Approaches to some of those problems will be briefly discussed below.

Wetting in Woven Fibers and Papers

If one examines a woven fiber or paper system under a microscope, one will see an open capillary system that can, qualitatively at least, be approximated as shown in Figure 6.12. A typical fabric yarn will be composed of 100--200 approximately parallel fibers each being 15--20 μm in diameter and 3--8 cm in length. The yarn is then woven to produce the fabric. The spaces between the woven yarn will be several orders of magnitude larger than those between the fibers so that, in terms of capillary phenomena, the fibers act as the solid surface whose interaction at the SLV interface will determine the ultimate capillary driving force in the system.

In the treatment of woven fabrics for dyeing, waterproofing, sizing, etc, important criteria for evaluation are the speed and completeness of the wetting process. In the textile industry, "wetting" is taken to mean the submersion of the fabric in an aqueous treating solution, replacing all of the air in the cloth structure with the solution. Obviously, if the treatment is to produce a product with uniform characteristics, all portions of the cloth must contact the solution for an optimum period of time to allow completion of the desired process. If air bubbles remain in the cloth structure during the treatment, areas of fiber that were not wetted will not receive the desired treatment and will have characteristics distinct from those of treated areas. Of course, given time, all of the air may be removed from a system to produce the desired result. However, in industry, time is money, and it is usually desirable to have the wetting process occur evenly and rapidly. Since the interfiber spaces are the ones that dominate the capillary forces in the system, they will be the rate-determining factor in the overall process.

Neglecting the effect of gravity as an external force helping to remove air from the fiber, the remaining effect will be almost entirely due to capillary forces. Analysis of the pertinent equation for capillary flow rates (eq. 6.16) indicates that the desired conditions for rapid wetting include a small contact angle and high surface tension. Unfortunately, due to the nature of the relationship between those two quantities, the most desirable conditions are normally contradictory. High surface tension aqueous solutions tend to have large contact angles on most fibers. Contact angles, however, can be reduced significantly by the addition of *wetting agents* or surfactants (Chapter 17). Unfortunately, surfactants by nature also lower surface tensions. All is not lost, however.

In aqueous systems, the range of surface tensions that will be encountered using common surfactants and practical concentrations is somewhat limited. In extreme practical cases, one might expect σ_{LV} to be reduced to the range of 40 mN m^{-1}, a change of a factor of about two from that of pure water (72.8). Lower values are normally attained only with very high surfactant concentrations (which can introduce foaming problems), with purer, more expensive surfactants, or with special materials such as fluorocarbon or silicone surfactants. Significant contact angle reductions, on the other hand, can be achieved using much lower concentrations of less surface-active wetting agents. Table 6.1 presents hypothetical data illustrating the relative effects of changes in σ_{LV} and θ on potential wetting rates. One can see from the data that accessible changes in contact angle have a much greater relative (and positive) effect on the wetting process than the potentially detremental effects of large reductions in surface tension, which are in most cases difficult to achieve.

The presence of surfactants or wetting agents in textile treatment solutions can also introduce other complications in the understanding of the dynamics of the wetting process. Because surfactants adsorb at the SL interface as well as the LV interface, as the liquid front continuously moves across fresh solid surface, adsorption processes will tend to deplete the concentration of available surfactant and may cause localized changes in both σ_{LV} and θ. In many cases, however, adsorption rates at the SL interface is much slower than that at LV interfaces, so that such effects can be taken into consideration without too much difficulty.

Of more direct practical importance to textile processing is the fact that it becomes increasingly more difficult to remove the last traces of air from the fiber system. Fiber bundles in yarn can produce quite complex capillary systems that provide ample opportunity for air entrapment in very inaccessible nooks and cranies. It is found, for example, that yarns made of smooth essentially cylindrical fibers are much easier to wet completely than those composed of rough, irregular fibers. In any case, the last vestiges of air are probably not displaced by capillary process at all, but by direct solution of the small air bubbles in the aqueous solution. In that case, the dissolution process is probably "driven" by the high pressure in the bubble due to it small radius of curvature.

Table 6.1. The effects of changes in surface tension σ_{LV} (mN m^{-1}) and contact angle on the linear rate of flow in a hypothetical capillary system using eq. 6.16, where $r = 0.05$ cm, $\eta = 2.0$ cp, and $l = 5$ cm.

Situation	θ (°)	dl/dt (cm sec^{-1})	$\Delta(dl/dt)$ (x)
$\sigma_{LV} = 72$:			
1	89	0.0016	---
2	75	0.023	15
3	50	0.058	36
4	25	0.082	51
5	0	0.091	56
$\sigma_{LV} = 55$:			
6	89	0.0012	---
7	75	0.018	15
8	50	0.044	37
9	25	0.062	52
10	0	0.069	57
$\sigma_{LV} = 40$:			
11	89	0.001	---
12	75	0.013	13
8	50	0.032	32
9	25	0.045	45
10	0	0.05	50

Waterproofing or Repellency Control

In dyeing and other processing of textiles, complete wetting is of vital importance. However, once the final product has been obtained it is usually desirable to have a system that is no longer wetted by water or other liquid systems. Waterproofing or repellency, then, is the opposite of wetting and must be addressed with an essentially opposite approach. In this case, since the manufacturer will have no control over the surface tension of any contacting liquid, it is necessary to control wetting by attacking the problem from the aspect of the contact angle. Ideally, to have a completely nonwetting system for all possible liquids (water or oil) it is necessary to produce a fiber surface that exhibits a large θ for both classes of liquids. Achieving that end for water solutions is not all that difficult because water will almost always have a higher contact angle (in air) than oils on organically treated fiber surfaces, and even untreated ones in many cases. For the case of oily liquids, it is usually necessary to treat the fibers with a class of substances that produce high oil--solid θ's. Those materials, unfortunately, are quite limited and usually involve relatively expensive fluorocarbons or silicones.

Complete analysis of the effect of contact angle on a given system is complicated (naturally) by the possible existence of hysteresis, roughness effects, and surface compositional variations within the fibers. If θ_A for a fiber system is > 90° for a given liquid, that liquid will not spontaneously enter into the fiber network under the drive of capillary forces. Likewise, if θ_R is also > 90°, any liquid forced into the network by external physical forces will be spontaneously expelled from the system. Such a situation represents the "ideal" for most effective repellency control. Even if those conditions cannot be achieved, however, large values of θ_A less tha 90° will at least retard entrance of liquid into the fiber matrix or maintain a slow rate of movement.

If a liquid having θ < 90° is forced into a fabric it can , in principle, be removed relatively easily by contacting the fabric with another matrix of essentially the same texture and weave, but having a lower contact angle with the liquid. In that case, the liquid will flow or "wick" into the more easily wetted fabric, leaving the first essentially dry (see Figure 6.8b). For example, if one were interested in producing a material which would allow moisture to pass from the area contacting skin out to the air, but not allow movement in the opposite direction, a two layer system composed of an inner layer of polyester or nylon (high θ) contacting a layer of cotton (low θ) would, in principle, provide just such a wicking/repellent fabric. For simplier repellency control, surface treatment of the fibers to produce large θ's is normally used.

Many other processes related to textile and paper treatments, impregnation of capillary systems, and wicking are of great industrial importance. The analysis of such problems begins with the same basic ideas as those employed for understanding and controlling wetting and repellency. However, they can be further complicated by such factors as irregularities in the capillary system (producing unsymmetrical flow patterns), Marangoni effects due to variations in solvent evaporation rates and temperatures or swelling of fibers in contact with liquid. Such problems represent fruitful, but theoretically difficult areas for research and development.

Capillary Action in Detergency Processes

As a final topic for discussion we will briefly address the problem of capillary action in the context of cleaning and detergency. In its simplest form (Figure 6.13), *detergency* can be viewed as the process of separating a liquid (L_1) from contact with a solid by the action of a second liquid (L_2). The same principles can be applied to the case of separating

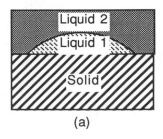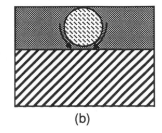

(a)   (b)

Figure 6.13. Schematic illustration of capillary action in detergency: (a) an oily soil (1) spread on the solid surface in contact with a better wetting liquid (2); (b) the wetting liquid penetrates between the soil and solid by capillary action, "rolling it up" and allowing it to be lifted off of the surface.

two solid surfaces by the action of a liquid, but for present purposes the discussion will center on SL_1L_2 systems. Probably the most familiar detergency system is that involving the removal of an oily soil from a fabric through the action of an aqueous wash. However, the removal of aqueous liquid from a solid surface (usually metal or ceramic) by the action of an organic liquid is also of great industrial importance.

Although mechanical agitation is important in detergency processes, the fundamental physical chemical process for removal of the soil is the capillary displacement of one liquid on the solid surface by a second liquid. In such a process, the contact angles of the two liquids on the solid surface are the primary factors controlling the rate of capillary displacement, although viscosity (lower is better) and σ_{LL} (higher is better) is also important. For an aqueous detergency system, the optimum condition will be where $\theta_{A/water}$ is as small as possible relative to $\theta_{R/oil}$. If $\theta_{A/w}$ is greater than 90°, water will not penetrate the fabric and no cleaning action will result. Conventionally, such situations are avoided by the addition of surfactants that lower $\theta_{A/water}$ and aid the capillary processes. Since most natural fabrics are slightly swollen by water but not by oils, soaking can allow time for the aqueous solution to penetrate not only into the capillary system, but also into the basic fiber network, swelling the fiber and further improving the contact angle situation in favor of oil removal. In nonaqueous detergency systems, of course, the thinking in terms of θ must be reversed.

This concludes the direct discussion of capillary phenomena as such. As has been noted, in many important applications, capillary action usually cannot be isolated from other important surface, hydrodynamic, or mechanical phenomena. Reference to the ideas presented above, therefore, will be encountered in many sections of chapters to follow.

CHAPTER 7

SOLID SURFACES

A solid is by definition a phase of matter that is rigid and resists stress. Like the liquid surfaces discussed previously, a solid surface must be characterized by some surface free energy and total free energy terms. It should be evident, however, that such free energies for solids cannot be characterized using the capillary and related methods so useful in the study of fluid surfaces. While liquid surfaces can, in general, be assumed to be in equilibrium and equipotential, a solid surface is generally of such a nature that those two assumptions will not be valid.

On a normal time scale, a liquid surface under stress will undergo *plastic flow* --- that is, as surface area is increased, molecules of the liquid phase will flow into the surface region from the bulk to maintain an equilibrium surface density (Figure 7.1a). A solid surface, on the other hand, will normally undergo *elastic flow* , in that as a stress is applied and the surface area is increased, there will not be a significant flow of molecules from bulk to surface. Instead, the distance between surface molecules will increase to produce a lower surface density (Figure 7.1b). As a result, since the distance between molecules increases, their lateral interactions decrease, resulting in a change in the energy of the stressed surface. Ultimately, events occur that bring the surface back to equilibrium; however, for many solids, the time scale of those events may be years. Therefore the nature of a solid surface, with some exceptions depending on the exact nature of the surface and environmental conditions, will be determined as much by its history as by equilibrium thermodynamics or surface tension forces. This "historical" effect on solid surfaces will be discussed further below.

SURFACE MOBILITY IN SOLIDS

For the purpose of placing events in the proper frame of reference, it is useful to do a simple calculation to determine the approximate mobility of atoms and molecules in a solid surface. To obtain such an approximation, one can view the surface as being in a dynamic state where there is a constant interchange of molecules between the surface,

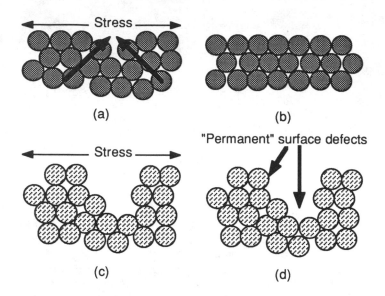

Figure 7.1. Schematic illustration of the responses of liquid and solid surfaces to stress: (a) a stressed liquid surface with "vacancies;" (b) molecules below the surface rapidly flow into the stressed area to "heal" the imbalance of forces; (c) a stressed solid surface with various defects; (d) due to lack of mobility, molecules cannot readily move into the stressed areas and the defects remain to produce a higher surface energy.

the bulk, and the vapor phase.

The number of moles of vapor hitting one cm^2 of surface per second Z is given by

$$Z = 0.23P \ (3/MRT)^{1/2} \tag{7.1}$$

where P is the vapor pressure of the material, M its molecular weight, and R and T have their usual meanings. If the equation is applied to a solid such as tungsten with a room temperature vapor pressure, P, of $\approx$ 10^{-37} mm Hg, then Z becomes about 10^{-20} atoms cm^{-2} s^{-1}, and the average lifetime of a surface atom becomes 10^{37} s! More volitile but still refractory solids such as copper will still have very long room temperature surface lifetimes.

For solids, there is defined a temperature, the so-called *Tamman temperature* , at which the atoms or molecules of the solid have acquired sufficient energy for their bulk mobility and reactivity (including sintering) to become appreciable. In general the Tamman temperature of a material will be approximately one-half its melting temperature (K). As the temperature at which noticible sintering occurs is approached, one begins to see dramatic changes in the average surface lifetime of the solid units. For example, at 725°C copper has a vapor pressure of about

10^{-8} mm Hg. Equation 7.1 gives $Z \approx 10^{15}$ atoms cm^{-2} s^{-1} for a surface lifetime of about 1 sec for copper atoms. From the point of view of bulk diffusion processes, under similar conditions, an average copper atom would travel approximately 100 Å in 0.1 s. At room temperature, that self-diffusion rate falls to the range of 10^{27} s for a distance of 100 Å.

Such calculations serve to illustrate the relatively low mobility of atoms or molecules in solids and help dramatize the differences that must be considered when discussing and comparing solid and liquid surfaces. They do not, however, exactly describe events when an atom or molecule moves within the surface region. For example, in bulk self-diffusion, one considers the ease with which a unit moves from one position in the bulk phase to another, but with a net result (to a first approximation, at least) of zero change in the total free energy of the system. Likewise, the calculation of Z from eq. 7.1 relates to the movement of atoms or molecules from the surface to the vapor phase, again with no net change in free energy of the system. For units in the surface, however, there is a third option for movement that is of more importance in the temperature region below the Tamman temperature. That movement is *surface diffusion* .

An atom or molecule undergoing bulk diffusion will experience no net change in the extent of its interaction with its neighboring units (assuming no change in the bulk structure). However, in order to move from one site in the structure to another, it must move past other units, meaning that there must be a significant activation energy which must be overcome for diffusion to occur. The process is illustrated (roughly!) in Figure 7.2a. For a unit in the surface (Figure 7.2b), however, one would expect a much lower activation energy since the unit, by its location in the surface, will usually have one or more "empty" sites into which it can move without requiring the movement of neighboring units. The surface unit, therefore, will face a much lower barrier to diffusion and will have much greater mobility at a given temperature. Surface diffusion, then, because of its lower activation energy, might be expected to represent an important phenomenon in temperature regions below that where sintering becomes important, especially with respect to surface chemical effects.

In summary, then, when one considers the nature of solid surfaces, one must always take into consideration the history (especially thermal) of the material. Since atoms and molecules in solid surfaces at room temperature have a very low mobility, they can be expected to retain the surface positions they acquired at the time of formation, even though they may be occupying positions of high relative energy. As a solid approaches its melting point, the surface units begin to acquire the properties of the bulk liquid phase, with overall greater mobility in terms of interchange with the vapor and bulk phases, and especially in lateral movement in the surface. In the interum temperature range, however, where bulk diffusion and evaporation--condensation are negligible, individual surface units can still move relatively rapidly toward more favorable energetic positions and thereby alter the surface chemical

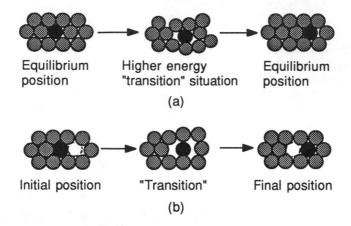

Equilibrium
position

Higher energy
"transition" situation

Equilibrium
position

(a)

Initial position

"Transition"

Final position

(b)

Figure 7.2. Schematic illustration of the comparative energetics of diffusion in bulk and in a surface: (a) bulk diffusion, illustrated in cross-section, involves the displacement of several nearest neighbor units, representing a relatively large energy barrier; (b) surface diffusion, shown in a top view, involves the displacement of fewer neighboring units, and therefore a relatively lower activation energy for the process. As a result, systems that undergo bulk diffusion with difficulty may exhibit orders of magnitude more surface diffusion under the same conditions.

characteristics of the system. Such relatively subtle surface changes may be reflected in more dramatic alterations in characteristics such as adsorption, wetting, adhesion, friction, or lubrication.

Sintering

Under some conditions, especially close to the melting point, many materials usually considered to be solids will exhibit sufficient plastic flow in the surface that capillary forces will slowly, but within a reasonable timeframe, come into play to move the surface toward equilibrium. A prime practical example of such action is the *sintering* of solids. If a solid powder --- metalic, crystalline, or amorphous --- is heated to some temperature below its melting point, usually, but not always, with some applied pressure, a fusion of adjacent particles will occur (Figure 7.3).

It is recognized that the main driving force for sintering is surface tension rather than external pressure. However, because of the above mentioned "historical" nature of solid surfaces, various parts of the powder surface will experience different net driving forces, leading to a rather complex situation during the sintering process. For example, atoms or molecules located at sharp asperities on the surface will have higher local excess surface energies than others located in situations more similar to the bulk phase. As a result, they will experience a greater surface tension force and have a greater surface mobility. In addition, the normally rough surfaces of solids mean that, due to the presence of

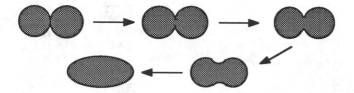

Figure 7.3. Schematic illustration of the stepwise process of particle sintering.

asperities, the actual area of contact between particles undergoing sintering will be small so that even at relatively low total pressures, the pressure applied at the points of contact will be multiplied so as to exceed the yield value of the material, producing plastic deformation and flow at the asperities (Figure 7.4).

In addition to local pressure effects, normal sintering temperatures will also usually allow significant amounts of diffusion, both in the bulk and in the surface. By such mechanisms, scratches on some metal surfaces will be "healed" if the sample is heated to a temperature well below the melting point.

"HISTORY" AND THE CHARACTERISTICS
OF SOLID SURFACES

As indicated above, the relative immobility of atoms and molecules in the surface of solids well below their melting point carries with it the result that the surface energy and other related characteristics of the surface will depend to a great extent on the formational and environmental history of the sample. For example, a clean cleaved crystal surface will almost certainly have a different surface energy than a surface of the same material that has been ground or polished. In this case, the cleaved surface will probably be of lower energy (assuming the absence of surface contamination) since the cleaving process will tend to occur along the crystal face of lowest energy. Grinding and polishing, on the other hand, are rather indescriminant in their action and tend to leave a significant number of small but energetically significant crystal defects increase the surface energy.

Other factors that can affect the apparent energy of a surface include the action of friction (see Chapter 18), corrosive action (which actually changes the chemical nature of the surface), and adsorption. All of these items are pointed out because they should always be considered and taken into account before one undertakes efforts to determine the surface energy of a solid.

Of equal or greater importance to the nature of a surface is the question of the history of its formation (as opposed to the treatments noted above). For crystalline materials especially, it must be kept in mind that commonly a crystal surface contains some or all of the defects already mentioned in various contexts, as well as missing layers, and screw and spiral dislocations. All such defects will alter the surface

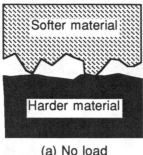
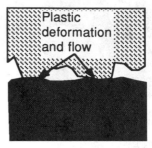

(a) No load   (b) Load pressure, P

Figure 7.4. Schematic illustration of plastic deformation and flow under load at points of contact between asperities of a soft material and a relatively hard surface.

energy of the crystal and complicate the analysis of phenomena related to it. Obviously, then, solid surfaces are more difficult to analyze and understand not only because of their inherent anisotropic nature, but also because of the potential role of history in determining the exact nature of the surface exposed at formation.

Equally important to the energetic nature of a surface is the presence (or absence) of adsorbed species and surface contamination. For low-energy, amorphous solids such as most polymers, surface contamination due to adsorption is normally not a major problem and can be controlled relatively easily. Higher energy crystalline, metallic, and inorganic glassy materials, however, pose significant experimental problems. "Clean" surfaces of such materials will have surface energies of 10^2 --10^3 mJ m^{-2}. They will literally "do anything" in order to lower their surface energies, including especially the adsorption of almost any available molecule --- nitrogen, oxygen, water, or any other material present in the environment. For that reason, it is difficult or impossible to prepare a truly clean surface of many solids without the use of exceedingly stringent environmental controls. Almost any low-energy material present in the environment will tend to be adsorbed leading to contamination of the surface and incorrect analysis of surface energy. As a result, most exact surface studies on solids are carried out in high vacuum and under ultraclean conditions. That is not to say, however, that meaningful and very useful results related to solid surface energies cannot be obtained by much easier techniques.

SOLID SURFACE FREE ENERGY vs SURFACE TENSION

As can be surmised from the above comments, the surface tension or surface free energy of solids, unlike liquids, cannot be equated with the total energy in the surface layer --- the "native" specific surface free energy plus "stress" energy due to factors mentioned above. By definition, the former is the work performed in *forming* a unit area of new

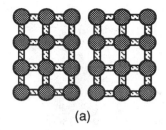 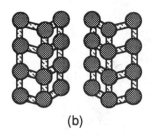

(a)   (b)

Figure 7.5. Schematic illustration of a stepwise mechanism for the formation of new surface: (a) initial cleavage; (b) rearrangement of surface units due to "excess" attraction by bulk underlying units.

surface while the latter pertains to the work spent in *stretching* the surface. In order to see the difference between the two concepts it is useful to visualize the process of new surface formation as a two-step process. First, the condensed phase is divided to produce the new surface, but the molecules in the new surface are held in the exact locations (relative to the remaining bulk phases) they occupied in the bulk, as illustrated schematically in Figure 7.5a. In the second step, the atoms in the newly formed surface are allowed to relocate into their most stable configuration. What this means, in effect, is that some of the units in the original (new) surface are "pulled" into the bulk by the unbalanced forces acting on them (Figure 7.5b). In a liquid system these two steps will occur simultaneously because of the mobility of the units.

In solids, on the other hand, the greatly reduced mobility means that the rearrangement will occur much more slowly, or perhaps not at all on a reasonable time scale. The density of units in the new solid surface will therefore be something other than the "equilibrium" value. The surface may therefore be compressed or stretched with no coincident change in the surface unit density. What changes will be the distance between units, which, as we have seen from Chapters 4 and 5, means a change in their interaction forces and therefore their free energy.

A mechanical model is useful for understanding exactly what is meant by *surface stress* in a solid, as opposed to its surface tension. Suppose that a solid surface is cleaved in a direction perpendicular to the surface. As pointed out above, the solid units in the new surface will not be able to relocate to attain their equilibrium positions relative to the bulk. In order to "make" their positions into "equilibrium" positions, one can think of applying some external force or lateral pressure on the surface units to hold them in place. The force per unit length of new surface needed for this equilibrium situation is the *surface stress* . If one takes the average surface stress for two mutually perpendicular cuts, one will obtain the surface tension of the solid. For the cases of liquids or *isotropic* solids, the two stresses are equal and the surface stress and surface tension are equivalent, as already stated. For an anisotropic solid, however, the two will not be equal (except perhaps by coincidence), so that the differentiation must be made.

THE FORMATION OF SOLID SURFACES

Because of the obvious importance of history to the nature of a solid surface, it is useful to understand some of the basic principles underlying the formation of such surfaces. The following section will briefly address the subject for two important classes of solids --- crystals and amorphous solids. Because of the complexity of describing metallic surfaces, they will not be treated here.

Crystalline Surfaces

Crystallization is a process in which an ordered solid phase is precipitated from a gaseous, liquid, or solid phase. The liquid phase may be either a melt or a solution. For most, but certainly not all, of the most important crystallization processes, that from solution is most important and will be emphasized here. A solid phase is precipitated from a solution if the chemical potential of the solid phase is less than that of the material in solution. A solution in which the chemical potential of the dissolved component is the same as that of the solid phase is said to be in equilibrium under the given set of conditions and is termed a *saturated* solution. The equilibrium state is defined by the concentration of the saturated component at a given temperature and concentration of other components, that is, by its *solubility* under those conditions.

In order for crystallization to occur, the equilibrium concentration of the component of interest must be exceeded by some *supersaturation* method, among which are included:

1. *Cooling* a solution in which the solubility of the component increases with increasing temperature or *heating* a solution in which the solubility of the material decreases with increasing temperature

2. *Evaporating* the solvent under heating

3. *Adiabatic evaporation* of the solvent, where removal of the heat of vaporization of the solvent is reflected in a decrease in the temperature of the solution

4. Adding to the solution another solvent that is miscible with the primary solvent, but is a *poorer* solvent for the material being crystallized

5. *Salting-out* through the addition of substances that may contain a common ion with the crystallized substance and thereby reduces its solubility, or by changing other factors affecting the ability of the solvent to solvate the material

6. Chemical reaction in the solution changing a soluble substance into an insoluble one

7. Various other specialized processes

In any of these ways the supersaturated solution necessary for crystallization can be obtained. If the supersaturation is attained by cooling, then the difference in temperature between that exactly corresponding to saturation and the actual temperature of the solution is termed *supercooling* . Provided that the supersaturation is not too great, the rate of formation of new crystals (*nucleation*) will be small and the solution will be in what is termed the *metastable* region; new crystals are formed to a limited extent and crystals already present grow. This, of course, corresponds to an ideal situation for the growth of a few very large (or single) crystals. If the state of supersaturation is increased, then the *maximum possible supersaturation* is attained, which is the *upper boundary* of the metastable region. When this boundary is exceeded, the rate of nucleation rapidly increases and the crystallization process becomes essentially uncontrolled. Thus, in order to have a controlled crystallization process, it is necessary to maintain the solution within the metastable region --- bounded on one side by the saturation concentration and on the other by the upper boundary of the metastable region.

The kinetics of crystallization can be usefully divided into two stages: *formation of new crystal nuclei* or *nucleation* and *crystal growth* proper. Both stages occur simultaneously, but they can to a greater or lesser extent be separated and considered independently in discussions of the crystallization process.

Nucleation Processes

Depending on the the crystallization process being employed, it is usual to divide nucleation into two types: *primary nucleation* , in which crystals begin to form in the absence of solid particles of the crystallized substance --- that is, spontaneously --- and *secondary nucleation* , which requires the presence of preformed or *seed* crystals of the substance of interest. Primary nucleation can be further subdivided into *homogeneous* and *heterogeneous* nucleation. In the former, spontaneous nucleation occurs without the intervention of any solid phase, while for the latter, the presence of a foreign surface such as dust, colloids, or vessel walls acts as a catalyst to initiate crystal formation. Secondary nucleation can also be subdivided into *true* , *apparent* , and *contact* nucleation. However, those topics will be left for the interested reader to pursue.

While there are a number of theories that attempt to predict crystal growth patterns or *habits* and growth rates based on thermodynamic and kinetic principles, only one will be mentioned here. The *theory of limiting faces* is of particular interest because it is relatively simple, it is directly related to the question of the surface energy of a surface, and it provides a way of estimating the energy of a given crystal face based on the shape of the crystal.

The Theory of LImiting Faces

The theory of limiting faces is the oldest of the thermodynamic approaches to explaining crystal growth and habit. It is based on the the derivation of the relationship between the crystal habit and the surface

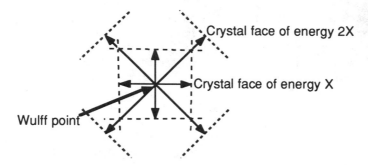

Figure 7.6. A Wulff construction for a hypothetical, two-dimensional crystal with surface energies X and 2X, from which the "ideal" geometric shape of the crystal can be predicted. The arrows iminating from the common point are proportional to the surface free energy of the intersecting crystal faces.

energy of the various crystal faces that go to produce a particular shape. If one assumes that a crystal is in equilibrium with its mother liquor (or vapor phase), then one can argue that the Gibbs condition for the most favorable crystal shape has been achieved; that is, the overall surface energy is a minimum for a given temperature and volume. If the surface energy of the i th face is defined as the product of the specific surface energy, σ_i, and the area of that face, A_i, then the Gibbs equilibrium condition can be expressed as

$$\Delta \sum_i^n \sigma_i A_i = 0 \quad [V, T = \text{constant}] \qquad (7.2)$$

where the summation is carried out over all n planes. Equation 7.2 leads directly to the *Wulff theorem*, which is given as

$$\sigma_1/h_1 = \sigma_2/h_2 = \ldots = \sigma_i/h_i = \ldots = \sigma_n/h_n \qquad (7.3)$$

where h_i is the distance of the i th plane from the center of the crystal, the *Wulff point* (Figure 7.6). What the theorem says, in summary, is that a crystal in equilibrium with its mother liquor contains a point whose distance from the various crystal faces (not in contact with a foreign surface) is proportional to the specific surface free energy of that face.

If equilibrium is disturbed, then the system will attempt to reestablish its condition of minimum energy by precipitating (or dissolving) more substance on (from) those crystal faces from which the greatest amount of energy will be released. Those faces will then correspond to areas of greater linear growth rate (or shrinkage). Thus faces with smaller linear growth rates will become larger and those with greater growth rates will become smaller until they eventually disappear.

The relationships among the growth rates of the individual faces can be affected in various ways (temperature, pressure, additives, etc) to achieve, in principle at least, various desired results.

Crystal Growth Modification

As a practical example of crystal habit modification, one can consider the growth of ice crystals in ice cream. If the crystals grow too large or attain certain shapes, the perceived quality of the product will be reduced significantly. In practice, the crystallization phenomenon is controlled by the addition of various natural gums (eg, locust bean gum) which presumably adsorb on specific crystal faces and retard or prevent further deposition of water molecules on "undesirable" crystal faces.

Another less tasty but potentially more important, application of such a phenomenon is the control of the growth of ice crystals in biological systems. In cold Antarctic seas, for example, where water temperatures may fall to - 2°C, fish swim through water thick with ice. The fish themselves, however, are protected from freezing by a natural antifreeze: compounds of protein and sugar that keep the liquids in a fish's body from freezing.

Just how the "antifreeze" works is something of a mystery, although it is assumed that the protein compounds act as a vigilant defense force, homing in on ice crystals as they begin to form and quickly adsorbing onto specific crystal faces. They apparently do not actually lower the freezing point of the fluids significantly, but inhibit the growth of crystals by adsorption onto the preferred crystal face, thereby slowing growth at temperatures where it would normally be very rapid.

Medical researchers are interested in retarding ice crystal growth in organs destined for transplantation. Normally, temperatures that halt organic decay in the organs also cause intercellular fluids to freeze, a process which tears the tissues. If the "fish antifreeze" or related substances could be modified to keep human organs free from ice crystals at, say, - 10°C, the viability of potential transplant organs might be extended many times over what is now possible.

Amorphous Solid Surfaces

From a practical standpoint, the "plastic" nature of our modern existence carries with it important questions concerning the surface characteristics and interactions of primarily amorphous (ie, noncrystalline) polymeric surfaces. Because of the molecular size, polydispersity, and generally random nature of polymeric solids (and their surfaces), many of the principles applied to studying and modeling ordered crystalline surfaces are no longer valid. Like crystals, polymer surface energies can be history dependent; however, since polymers in general have relatively low surface energies, complications due to rapid adsorption and contamination are somewhat reduced.

Dynamics in Polymer Surfaces

The concepts of solid surfaces discussed above assumed that the surfaces in question were effectively rigid and immobile. Such assumptions allow one to develop certain models and mathematical relationships useful for estimating and understanding surface energies, surface stresses, and specific interactions, such as adsorption, wetting, and contact angles. It is assumed that the surfaces themselves do not change or respond in any specific way to the presence of a contacting liquid phases, thereby altering their specific surface energy. Although such assumptions are (or may be) valid for truly rigid crystalline or amorphous solids, they more often than not do not apply strictly to polymeric surfaces. Glass, for that matter, has been shown to undergo interactions with liquids such as water which lead to specific alterations in its surface properties. And glass is popularly considered to be an inert, rigid solid.

The structure and characteristics of polymeric surfaces, like other solids, is generally time and environment dependent. The reason, of course, is that polymers are composed of long, very anisotropic molecules of a variety of molecular weights (*polydispersity*), all of which will seldom have the opportunity to achieve their equilibrium condition. Solid polymer surfaces, therefore, are inherently nonequilibrium structures and exhibit a variety of time- and condition-dependent properties that may change dramatically with those variables. Although often recognized in bulk polymer problems, such character changes are often ignored in the context of surface properties, sometimes much to the dismay and detriment of researchers, manufacturers, and users.

Polymer Motions in Surfaces

Because of their large molecular size, complex bonding patterns, the presence of side chains, etc, polymers exhibit a number of characteristics in the solid state that are much less common in crystalline solids. In the study of bulk polymers, the time, temperature, and other variable-related characteristics have come to be classed as either *relaxations* or *transitions* . As a general definition, a *relaxation* can be considered a time-dependent motion in a polymer system in which the molecules return to an equilibrium from which it has been displaced by the action of some external force. For example, if a polymer sample is compressed under some external load that forces the molecules to rearrange to attain a new equlibrium state and the force is then removed, the material will, with time, *relax* or return to its original state (before compression).

A *transition* in a polymer system is considered to be a temperature-dependent process. In a crystalline material we are familiar with the transition from the solid to the liquid phase --- melting --- which will nornally occur at some relatively sharp, well defined temperature. Similar process occur in polymers, but because of their nature, they are seldom sharply defined, but rather occur over a broad temperature range. The melting process is ideally an equilibrium process and occurs essentially independent of time. In amorphous polymers and glasses, there exists a temperature range over which the system undergoes a dramatic change in its physical properties. It changes from a rigid, possibly brittle system to a viscous liquid. Its tensile strength, elasticity, and other properties change dramatically. It undergoes a *transition* to an essentially distinct class of material.

Not surprisingly, polymer molecules in or near a surface are also found to undergo relaxations and transitions similar to those found in the bulk. However, those motions in a surface are somewhat different because of the different environment encountered there. They no longer interact with other polymer molecules only, but also with the surrounding phase. Like liquids and solids, surface polymer molecules will, given sufficient time, reorient themselves at the surface so as to attain the configuration of minimum surface energy. The reorientations will, of course, be time/temperature dependent and correspond to related bulk phenomena. At low temperatures, the transitions may require long equilibration times to become evident, while at higher temperatures the effects may become apparent in short order.

In contact with condensed phases, especially liquids, surface relaxations and transitions can become quite important. Even very hydrophobic, rigid polymers such as poly(methylmethacrylate) which contains somewhat hydrophilic ester side chains will, in contact with water, undergo surface molecular reorientation due to the interaction of water with the ester groups. The interfacial region becomes *plasticized* (roughly put, softened) because the water--ester interaction liberates to some extent the side chains and increases their mobility. The important point is that these surface interactions can dramatically change the interfacial characteristics of a polymer with possibly important consequences in a particular application. And since the processes are time dependent, the changes may not be evident over the short span of a normal experiment. For critical applications in which a polymer surface will be in contact with a liquid phase, it is important to know not only the surface characteristics (eg, coefficient of friction, adhesion, adsorption) under normal experimental conditions, but also to determine the effects of prolonged (equilibrium) exposure to the liquid medium of interest.

As a practical example, take the use of a polymer in some biomedical application such as an implant device, in which the polymer surface will continually contact blood or other body fluids. Classical surface studies using contact angle measurements, wetting phenomena, x-ray photoelectron spectroscopy, or other analytical techniques may indicate that the material should be biocompatible and not cause problems of blood platelet deposition and clot formation. Those analyses, however, are not normally or cannot be carried out under conditions of use. Under such conditions, surface transitions and relaxations may occur with time which will transform the polymer surface into one that is no longer biocompatible from the standpoint of blood interactions. The result could be catastrophic for the recipient of the transplant made of such material.

It is therefore important for biomedical as well as many other applications that the surface characteristics of a material of interest be determined under conditions that closely mimic the conditions of use and over extended periods of exposure to those conditions, in addition to the usual characterizations.

ADSORPTION AT THE SOLID--VAPOR INTERFACE

As has been repeatedly emphasized, the surface energy of a solid may be very history dependent, so that it is not surprising to find that the adsorption characteristics of solids may be as well. There are several underlying principles of adsorption which represent the bedrock for understanding all adsorption phenomena, especially the Gibbs equation, the Kelvin equation, and the Young-Laplace equation. Those equations have already been introduced, and they will arise again and again, so that it is important to have them well in mind.

A freshly formed, clean solid surface will often be of quite high surface energy (except for most polymer surfaces) so that there will exist a strong driving force for the reduction of the excess surface energy by whatever process may be available. In a liquid, some of that excess energy can be dissipated by spontaneously reducing the total interfacial area --- ie, the liquid forms a spherical drop (or as close to it as gravity and physical restraints allow). A solid does not have that option so that solid surfaces tend to adsorb materials that will not adsorb appreciably at liquid interfaces, namely gases such as nitrogen, oxygen, or carbon monoxide. Adsorption occurs because it reduces the imbalance of forces acting on the surface molecules of the solid or liquid. The energetics of adsorption will be essentially the same for any type of interface. However, a solid surface will almost certainly be heterogeneous in terms of the distribution of its excess surface energy, meaning that adsorption will not be a uniform process, while for liquids it is assumed to be so.

Energetic Considerations: Physical Adsorption vs Chemisorption

The forces involved in adsorption processes are the same as those encountered in every other interfacial process, and for that matter every chemical or physical phenomenon above the atomic level. Those include the nonspecific van der Waals forces, ionic or electrostatic forces, and specific forces involved in the formation of chemical bonds. Because the nonspecific interactions are orders of magnitude smaller than the specific forces, adsorption processes that involve only nonspecific interactions are referred to as *physical adsorption* while those in which stronger interactions occur are termed *chemisorption* .

Since adsorption onto a solid surface by a vapor is a spontaneous process, the overall free energy change for the process must be negative. However, in the process, the adsorbing molecules lose a degree of freedom; that is, they become restricted to two instead of three degrees of freedom and their entropy drcreases. From the thermodynamic relationship

$$\Delta G = \Delta H - T \Delta S \qquad (7.4)$$

it is clear that for ΔG to be negative, ΔH must be negative; that is, adsorption must be an exothermic process. The situation may be different in the case of adsorption from solution due to the effects of changes in solvation, etc. Because of its exothermic nature, the amount of gas adsorbed onto a solid will decrease as the temperature of the system is increased. That is why clean solid surfaces are more easily prepared by heating to high temperatures. In addition, high vacuum reduces the vapor pressure of the adsorbed gas and reduces its tendency to adsorb.

The heats of adsorption, ΔH, for gases onto a given solid can, in principle, be measured in a variety of ways and will, in reversible systems, adhere to the Clausius-Clapeyron equation

$$(\partial \ln p /\partial T)_v = - \Delta H_{ads} / RT^2 \qquad (7.5)$$

In systems where adsorption is stricktly of the physical type, it is found that ΔH_{ads} is generally of the same magnitude as the heat of condensation for the vapor. In the case of nitrogen, for example, the heat of condensation is approximately -6 kJ mol^{-1}. The heat of physical adsorption of nitrogen is found to be in the range of -10 kJ mol^{-1} on iron, -12 kJ mol^{-1} on graphite, and -14 kJ mol^{-1} on TiO_2. In the case of chemisorption of nitrogen on iron the heat of adsorption rises to about -150 kJ mol^{-1}, comparable in magnitude to the strength of chemical bonds. Obviously, chemisorption involves much stronger interactions.

In the absence of complicating factors such as capillary condensation (see below), the process of physical adsorption has no activation energy; that is, it is diffusion controlled and occurs essentially as rapidly as vapor molecules can arrive at the surface. The process will be reversible and equilibrium will be attained rapidly. Because the forces involved are the same as those involved in condensation, physical adsorption will generally be a multilayer process --- that is, the amount of vapor which can be adsorbed onto a surface will not be limited simply by the available solid surface area, but molecules can "stack up" to several molecular thicknesses in a pseudo-liquid assembly (Figure 7.7). If the vapor pressure of the gas reaches saturation level, in fact, the condensation and adsorption processes overlap and become indistinguishable. The fact that physical adsorption can be a multilayer process is very important to the mathematical modeling and analysis of the process, as will be seen below.

Unlike physical adsorption, chemisorption involves very specific interactions between the solid surface and the adsorbing molecules, as

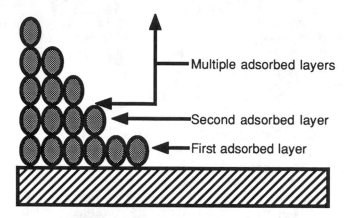

Figure 7.7. Multilayer adsorption on a solid surface. The first layer may be physically adsorbed or chemisorbed. Subsequent layers will be physically adsorbed only.

illustrated by the much higher heats of adsorption. Another important result of that specificity is that chemisorption is by nature limited to the formation of a monomolecular adsorbed layer. In addition, chemisorption processes will generally have some activation energy and may therefore be much slower than physical adsorption, and they may exhibit hysteresis --- that is, they may not be readily reversible.

The energetic relationship between the two processes can be illustrated by analysis of the schematic energy diagrams shown in Figure 7.8. Curve 1 in the figure represents the energy diagram for physical adsorption. At large distances, there is essentially no attraction between the surface and the vapor molecule. As the vapor approaches the surface there develops an attraction due to van der Waals interactions leading to an energy minimum representing the heat of adsorption, ΔH_{ads}. At some distance, in this case the molecular radius of the vapor, there begins to develop some overlap between electron clouds, leading to the development of a repulsive interaction (Born repulsion).

Curve 2 represents the process of chemisorption. Because chemisorption involves specific interactions between the adsorbent and the adsorbate, the process must also involve some specific change in the molecular structure of the adsorbate molecule. For example, for a diatomic molecule A_2 that involves π bonding, chemisorption may involve the rupture of a π bond between the two atoms. The starting point for the isolated process (at large distances) must be the energy of the bond being broken, ΔH_π, which would represent the activation energy of the process. (One may think in terms of an activated molecule, A_2^*.) Because of the magnitude and nature of the specific interactions involved, the energy minimum for chemisorption (ΔH_{chem}) will be much deeper thatn ΔH_{ads} and will occur at closer distances.

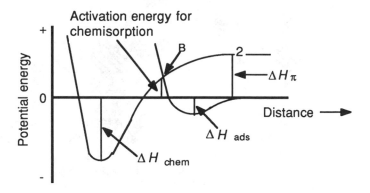

Figure 7.8. Schematic illustration of the energetic differences between physical adsorption and chemisorption.

If the two adsorption processes were mutually exclusive, the activation energy of chemisorption would be so high that it would occur only under rather vigorous conditions. However, if the two are seen to occur in a cooperative way, it becomes clear that physical adsorption is an important component of the overall chemisorption process. Referring again to Figure 7.8, one can see that if physical adsorption of the gas molecule occurs first, it can approach the solid surface along a pathway of much lower energy than that of chemisorption alone. At the point where the two curves intersect (B), the energy difference between the physically adsorbed molecule and the activated molecule is greatly reduced. That difference is the activation energy for chemisorption. The activation energy clearly depends on the shapes of the two energy curves and will vary greatly from one system to another.

Both physical adsorption and chemisorption have very important practical applications and implications. Because physical adsorption is a rapid and reversible process, it can be considered to occur to some extent on almost all solid surfaces (with the possible exception of some perfluorinated materials) except under extreme conditions of high temperature and high vacuum. If the physically adsorbed molecules are such things as oxygen or nitrogen from the air, their presence on the surface may cause no practical problems. However, if the atmosphere contains materials such as oil from fans and motors, which can be just as easily adsorbed, the nature of the solid surface may be altered to the extent that it no longer functions in a desired way.

For example, a freshly cleaned silicon surface intended for use in the manufacture of microelectronic devices must be coated with a so-called *microresist* , a photosensitive polymer that allows the manufacturer to transfer the microcircuit design from a large mask or negative to the surface of the wafer. Since the size of the circuit components may be in the micron range, the resist coating must be essentially free of defects. For a clean surface (or at least one with only atmospheric gases adsorbed), the microresist solution will uniformly wet

Figure 7.9. Illustration of the effect of adsorbed contaminants on coating processes: (a) the uncontaminated surface gives a smooth, uniform coated layer; (b) contamination by a lower energy material on the surface will often result in the formation of "repellency spots" or other defects, producing a less than optimum coating.

and spread to give the quality of surface necessary for the production of usable microcircuits (Figure 7.9). If the surface has organic materials adsorbed (eg, spots of oily material) the coating will not spread uniformly, and an irregular, useless layer will result. More details on the effects of adsorbed materials on wetting and spreading are given in Chapter 17. It may be said, then, that random physical adsorption (which is practically unavoidable in any case) may be innocuous or perhaps detrimental in practical situations, depending on the nature of the adsorbed species.

Chemisorption and Heterogeneous Catalysis

Chemisorption, while generally slower, less ubiquitous, and more easily avoided in most cases, has some very important practical aspects. The most important of those is its role in heterogeneous catalytic processes. As pointed out above, chemisorption involves such strong interactions between adsorbent and adsorbate that one may assume in some cases the formation of chemical bonds between the two. Even if that is not the case, it has been shown repeatedly that just the "forceful presence" of the solid surface can alter the electronic structure of an adsorbed molecule sufficiently to alter its electronic and vibrational spectra, and therefore its chemical reactivity. In addition, if two species are chemisorbed on a surface, their close proximity in the "activated" chemisorbed state may lead to chemical reaction between the two.

The number of important chemical processes that involve the chemisorption of gases at solid surfaces is quite large and cannot be even partially covered here. However, a few general examples of reaction types include: combination (hydrogen + alkene $\rightarrow$ alkane), decomposition (ethanol $\rightarrow$ ethylene + water), isomerization (n-alkane $\rightarrow$ branched alkane), polymerization (ethylene $\rightarrow$ polyethylene), and various other mixed reactions, including photolytic and photosynthetic processes.

The variations possible in the details of heterogeneous catalytic reactions are such that each specific case must be considered individually, and few if any are really fully understood. There are, however, a few fundamental aspects that can be considered in order to get an overall picture of what individual molecular processes may be involved. Those fundamentals include: (1) the initial physical adsorption process; (2) possible surface diffusion of the adsorbed species; (3) chemisorption processes (eg bond breaking, if it is involved); (4) chemical reaction between adsorbed species; and (5) desorption of the product. All or only some of those steps (or perhaps others not mentioned) may be involved in a given catalytic process. Any one of them may be the rate-determining step.

To illustrate the processes involved, consider a hypothetical heterogeneous oxidation with molecular oxygen

$$O_2 + 2\,A + catalyst \rightarrow 2\,AO + catalyst$$

for which it is known that the actual oxidizing species is atomic oxygen. The first step in the process must be the physical adsorption of the reactants onto the catalyst surface, followed, perhaps, by chemisorption of O_2 and dissociation to form two oxygen atoms.

$$A + catalyst \rightarrow A\ (adsorbed) \qquad rate\ constant = k_1$$

$$O_2 + catalyst \rightarrow O_2\ (adsorbed) \qquad rate\ constant = k_2$$

$$O_2\ (adsorbed) \rightarrow O_2\ (chemisorbed) \rightarrow 2O\cdot \qquad rate\ constant = k_3$$

Once atomic oxygen has been formed, it must encounter an adsorbed molecule of A in order for reaction to occur. That process may require surface diffusion with an accompanying rate constant k_4. The reaction between O· and A will obviously have its own rate constant k_5 to be thrown into the soup. Finally, the overall rate of the process will depend on the availability of surface sites for adsorption, so that as product AO is formed, it must be desorbed to free up space for further desired reaction (k_6). Clearly, in order to understand such a catalytic process fully, one must understand a variety of independent but interrelated processes. It is easy to see, therefore, why the subject of heterogeneous catalysis is so complex and in many cases poorly understood.

Chemisorption and heterogeneous catalysis also occurs at the solid--liquid interface. The basic concepts mentioned above remain valid in such systems, although the situation will naturally be complicated by such factors at solvation and solvent adsorption.

Catalytic Promoters and Poisons

It is well known in the science and art of heterogeneous catalysis that the presence of small amounts of certain materials can greatly improve or disastrously ruin a catalytic reaction. Where improvement is found the additive is referred to as a *promoter*. When disaster results, the additive is termed a *poison*.

The exact role of promoters is not very well understood in most cases, but it is now generally accepted that it is related to the formation of certain specific electronic surface states necessary for the given catalytic reaction. It apparently does not matter how that electronic state is produced --- in the "native" catalyst surface or by the presence of some other component which "induces" the state. For example, the activity of iron-iron oxide catalyst in the Haber ammonia synthesis is found to be greatly improved by the presence of small amounts of aluminum oxide *and* potassium oxide. Either alone appears to have no significant effect.

Catalyst "poisons" are materials that significantly reduce or completely destroy the activity of a given catalyst. Such materials generally function by binding strongly and (effectively) irreversibly to the specific surface sites necessary for the functioning of the desired process. Particularly troublesome materials in that sense are sulfur-containing compounds, especially thiols and thioethers. For example, the catalytic converters used to oxidize hydrocarbon residues in automobile exhausts will rapidly lose their effectiveness if exposed to such materials.

On the other hand, "poisoned" catalysts can have their uses. In some hydrogenations of organic molecules, for example, it may be desirable to produce a reaction between hydrogen and one functional group in the molecule, while leaving untouched another functionality that would normally react as well. By selectively poisoning the catalyst, surface states necessary for the desired reaction may be left untouched while those for the unwanted reaction are blocked.

Clearly, chemisorption and related catalytic processes are quite complex and remain a relatively poorly understood area of surface science. Modern surface analytical techniques have added much to our understanding of the molecular processes involved, but much remains in the realm of art (or perhaps black magic).

Solid--Vapor Adsorption Isotherms

Study of the adsorption of gases onto solid surfaces has a long and illustrious history, with some of the fundamental aspects of physical adsorption being recognized early in the 19 [th] century. It was known as early as 1814, for example, that the amount of a gas adsorbed by a given amount of a particular solid under "standard" conditions (room temperature and atmospheric pressure) was directly related to its "condensibility" under those conditions ($NH_3 > H_2S > CO_2 > N_2 > H_2$). It was also recognized that heat was evolved when gas was adsorbed onto a solid. With the more detailed and quantitative analysis of adsorption process in modern times, it became obvious that adsorption processes were far from simple; that they were, in fact, quite varied and sometimes distinct, depending on the specific system under consideration.

Some systems, for example, were found to be completely reversible while others exhibited hysteresis, depending on from which side of the equilibrium an approach was made. The effects of temperature and gas pressure on the amount of gas adsorbed was noted. And in some cases it was noted that the gases given off as vacuum was applied were chemically different from those that were originally adsorbed. It was a confusing situation that required some innovative conceptual and mathematical modeling in order to satisfy the need to have a clear theoretical picture of the process.

The general relationship between the amount of gas (volume, V) adsorbed by a solid at a constant temperature (T) and as a function of the gas pressure (P) is defined as its *adsorption isotherm.* It is also possible to study adsorption in terms of V and T at constant pressure, termed *isobars*, and in terms of T and P at constant volume, termed *isosteres*. The experimentally most accessible quantity is the isotherm, although the isosteres are sometimes used to determine heats of adsorption using the Clausius-Clapeyron equation. In addition to the observations on adsorption phenomena noted above, it was also noted that the shape of the adsorption isotherm changed with temperature. The problem for the physical chemist early in this century was to correlate experimental facts with molecular models for the processes involved and relate them all mathematically.

Classification of Adsorption Isotherms

A useful model for a simple adsorption isotherm should take into consideration all phenomena involved in the process, including the initial monomolecular adsorption process at both low and high coverages, multilayer adsorption, if present, and accompanying (and complicating) phenomena such as chemisorption and capillary condensation. In a given situation, some or all of those factors may be important. As a result, there exists a wide variety of isotherm types, five of which are generally considered to be important in solid--vapor adsorption processes. The theoretical details behind each of the five major isotherm types will not be given here; however, each will be described qualitatively in terms of the relationship between shape and the (presumed) molecular processes involved. Because of their practical experimental utility, the mathematical foundations of three classical adsorption isotherms --- the Langmuir, Freundlich, and Brunauer-Emmett-Teller (BET) isotherms --- will be given in the following sections.

The five isotherms to be considered are shown schematically in Figure 7.10. The type I isotherm, usually termed the Langmuir type, is characterized by a fairly rapid initial rise in the amount of gas adsorbed with increasing gas pressure until some limiting value is reached. That limiting value is usually identified with the attainment of a complete monolayer coverage. Such an isotherm would be expected, for example, in chemisorption where the system is naturally limited to a monolayer. It may also be found for systems in which there is a strong nonspecific attractive interaction between the adsorbate and adsorbent, but weak attraction between the absorbate molecules themselves. Such isotherms may also be encountered in systems in which the solid has a very fine

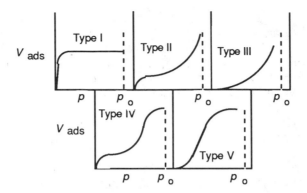

Figure 7.10. Schematic illustration of the shapes of the five principle types of adsorption isotherms.

microporous structure.

Type II isotherms are typical of physical adsorption on nonporous solids. In contrast to type I, the adsorbate molecules in these cases also have relatively strong mutual interactions, which leads to the tendency for multilayer formation. The initial rapidly rising part of the isotherm corresponds to the equivalent type I adsorption. Point B on the curve is identified with complete monolayer coverage. Multilayer formation then begins which may lead to surface condensation. Type II isotherms are sometimes encountered for microporous solids in which case point B would correspond to both completition of monolayer coverage and filling of the micropores by capillary condensation. The rest of the curve would then correspond to normal multilayer formation.

Types III and V isotherms are relatively rare and correspond to systems in which the interaction between adsorbate molecules is stronger than that between adsorbate and adsorbent. In these cases, the uptake of gas molecules is initially slow until surface coverage is sufficient so that interactions between adsorbed and free molecules begins to dominate the process. One might say that the processes are autocatalytic in terms of the adsorption process.

Type IV isotherms are obviously similar to type II and usually correspond to systems involving capillary condensation in porous solids. In this case, however, once the pores have become filled, further adsorption to form multilayers does not occur and the terminating plateau region results. This would indicate a relatively weak interaction between the adsorbate molecules. While more complex isotherm classifications are available, they generally represent combinations and extensions of the five basic types described above.

The Langmuir Isotherm

Because of its simplicity and wide utility, the Langmuir isotherm has found wide applicability in a number of useful situations. Like many such "classical" approaches, it has its fundamental weaknesses, but its utility generally outweighs its shortcomings. The Langmuir isotherm model is based on the assumptions that adsorption is restricted to monolayer coverage, that adsorption is localized (ie, that specific adsorption sites exist and interactions are between the site and a specific molecule), and that the heat of adsorption is independent of the amount of material adsorbed. The Langmuir approach is based on a molecular kinetic model of the adsorption--desorption process in which the rate of adsorption (rate constant k_A) is assumed to be proportional to the pressure of the adsorbate (p) and the number of unoccupied adsorption sites ($N - n$), where N is the total number of adsorption sites on the surface and n is the number of occupied sites, and the rate of desorption (rate constant k_D) is proportional to n.

At equilibrium, the rates of adsorption and desorption will be equal so that

$$k_A p (N - n) = k_D n \qquad (7.5)$$

Ignoring entropy effects, the equilibrium constant for the process will be $K_{eq} = k_A / k_D$, so that

$$K_{eq} = n / p(N - n) = \exp(-\Delta H^\circ / RT) \qquad (7.6)$$

where ΔH° is the heat of adsorption per mole at temperature T and standard pressure P_{st}.

The fraction of the adsorption sites occupied at a given time, θ, is given by

$$\theta = n / N \qquad (7.7)$$

so that eq. 7.6 can be written

$$\theta = K_{eq} p / (1 + K_{eq} p) \qquad (7.8)$$

or

$$\theta = p / (p + K_{eq}^{-1}) = p / [p + P_{st} \exp(\Delta H^\circ / RT)] \qquad (7.9)$$

A useful characteristic of the Langmuir isotherm is that it can be rearranged to the linearized form

$$n^{-1} = N^{-1} + (K_{eq} N p)^{-1} \qquad (7.10)$$

so that a plot of n^{-1} vs p^{-1} should be linear and yield values of K_{eq} and N from the slope and intercept. If the plot is not linear, then the langmuir model does not fit the adsorption process in question.

The Freundlich Adsorption Isotherm

It is commonly found that the simple Langmuir isotherm does not adequately describe many adsorption systems. Another classical isotherm that has found application in describing adsorption, especially at moderate pressures, is the so-called *Freundlich adsorption isotherm*

$$V = k p^{1/a} \qquad (7.11)$$

where V is the volume of adsorbed gas and k and a are constants, a usually being greater than 1. Equation 7.11 can be linearized by taking the logarithm of each side to give

$$\ln V = \ln k + 1/a \ln p \qquad (7.12)$$

Obviously, a plot of $\ln V$ vs $\ln p$ should give a straight line. Although originally derived empirically, the Freundlich equation can also be derived theoretically using a model in which it is assumed that the heat of adsorption is not constant but varies exponentially with the extent of surface coverage, which is probably more near to the truth than the Langmuir assumption in most cases.

The Brunauer-Emmett-Teller (BET) Isotherm

A major assumption of the Langmuir isotherm model is that adsorption stops at monolayer coverage. However, since the van der Waals forces leading to physical adsorption are the same as those involved in the formation of the liquid state, it should not be surprising to find that the limitation of adsorption to a single monolayer is unrealistic in most cases. As mentioned previously, such a case is expected only when the interactions bewteen adsorbate molecules is very much weaker than that between adsorbate and adsorbent.

Asuming the formation of a multilayer of adsorbed molecules, Brunauer, Emmett, and Teller (BET) modified the Langmuir approach of balancing the rates of adsorption and desorption for the various molecular layers. The BET model works from the assumption that the adsorption of the first monolayer has a characteristic heat of adsorption ΔH_A, but that subsequent layers are controlled by the heat of condensation of the vapor in question, ΔH_L.

The BET equation will not be derived here, but the most common form of the final equation is

$$p / [V(p_0 - p)] = 1 / V_m c + [(c-1)p / V_m c p_0] \qquad (7.13)$$

where V is the volume of adsorbed vapor at STP, V_m is the monolayer capacity at STP, p is the partial pressure of the gas, p_0 is the saturation vapor pressure, and

$$c \approx \exp[(\Delta H_A - \Delta H_L) / RT]$$ (7.14)

The BET isotherm was developed primarily to describe the commonly encountered type II isotherm, such as for the adsorption of relatively inert gases (N_2, Ar, He, etc) on polar surfaces ($c \approx 100$). However, it reduces to the Langmuir isotherm when restricted to monolayer coverage ($\Delta H_A \gg \Delta H_L$), and describes type III isotherms in the unusual situation where the adsorption of the first monolayer is less exothermic that that of the subsequent layers (eg, $c < 1$).

Surface Areas from the BET Isotherm

One of the most common uses of the BET isotherm is for determining the surface area of finely divided solids by physical adsorption. Such information can be of great importance in a number of areas including, heterogeneous catalysis and various sorption applications. While the BET model for multilayer adsorption contains several potential pitfalls due to the assumptions of the absence of lateral interactions between adsorbed molecules, the constancy of the heat of adsorption (after the first monolayer), and solid surface homogeneity, it generally produces useful results at pressures, p, of between 0.05 p_o and 0.35 p_o. It must be used with great caution, however, for porous solids that show adsorption hysteresis, or when point B on the isotherm corresponding to V_m cannot be accurately determined.

The monolayer capacity V_m is of primary interest because it allows for the calculation of the surface area based upon the area occupied by each adsorbed gas molecule. According to eq. 7.13, a plot of $p / [V(p_o - p)]$ versus p/p_o should be linear over the pressure region noted above. From the slope of the line, $S = (c - 1) / V_m c$ and the intercept $I = 1/V_m c$ one can calculate the monolayer capacity V_m of the solid and thereby its specific surface area A_s.

$$V_m = 1 / (S + I)$$ (7.15)

$$A_s = V_m k / \text{sample weight}$$ (7.16)

$$k = NA / M_v$$ (7.17)

where N is Avagadro's number, A is the area per molecule of the adsorbed gas, and M_v is the gram molecular volume of gas (22.400 L at STP).

The adsorbate most commonly employed for BET surface area determinations is nitrogen, which has an effective area per molecule of

0.162 nm^2 at liquid nitrogen temperature (77 K). Nitrogen produces good results because it generally gives a well-defined value of B (ie, $-\Delta H_A \gg -\Delta H_L$), while not having c so large that adsorption becomes localized. Other useful gases include argon (A = 0.138 nm^2) and krypton (A = 0.195 nm^2). Other vapors can be used so long as their effective molecular area is known. Such can be determined by using a solid of known specific surface area, as determined by nitrogen adsorption, for example, for calibration. The potential problem with many other adsorbed gases is that, due to specific interactions with the surface sites, adsorption may become localized in the first monolayer. It is not uncommon, therefore, to find that A for a given molecule will vary significantly from one solid to another.

Capliiary Condensation

Reference has been made several times to adsorption isotherm hysteresis and problems with microporous solids related to capillary condensation. The apparently anomolous adsorption behavior of porous solids can be explained in terms of capillary phenomena and the Kelvin equation (see Chapter 6).

A liquid that wets the walls of a capillary will exhibit a concave liquid--vapor interface and therefore a lower vapor pressure than it does in the bulk vapor phase. According to the Kelvin equation

$$RT \ln p/p_o = -\sigma v (1/r_1 + 1/r_2) \qquad (7.18)$$

where v is the molar volume of the condensed adsorbate and the other terms are as previously defined. Because the pressure in the capillary is lower than the normal saturation vapor pressure, condensation can take place at lower partial pressures than would normally occur. As adsorption proceeds and the adsorbate begins to fill the capillaries, r_1 = the pore radius and $1/r_2 = 0$. If the contact angle of the condensed liquid on the solid is small or zero, at a given pressure, capillary rise will be so large that pores of a certain size and smaller will normally be filled

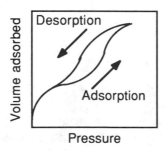

Figure 7.11. Typical shape of an adsorption isotherm showing a hysteresis loop due to capillary condensation in pores.

(a)   (b)

Figure 7.12. Pore shapes and their effect on capillary condensation and hysteresis: (a) for V-shaped pores hysteresis will be minimal since the capillary pressure will always be such that the pores fill and empty at essentially the same rates; (b) for "ink-bottle" pores in which the radius of the mouth, r_1, is less than that of the main body of the pore, r_2, the pores will fill up faster than the general rate of adsorption and empty more slowly, thus giving rise to the observed hysteresis.

completely before condensation begins in the larger ones.

It is probably realistic to assume that although condensation in larger pores occurs last in the overall process, monolayer adsorption will occur relatively uniformly. In order to analyze the effects of capillary condensation quantitatively, then, an adjustment to the pore radii corresponding to the adsorbed layer thickness should be made. For example, it would be an oversimplification to assume that the radius of curvature of the meniscus of the condensed liquid was equal to the "dry" pore radius. In addition, the assumption that the meniscus is hemispherical as the pore fills and empties must be strongly questioned, so that the application of eq. 7.18 must be done with some "intelligent" modification of the variables.

Once a pore is filled and the *desorption* cycle begins, the situation described above changes significantly. In that case, the radii of curvature become (approximately) $r_1 = r_2 = r$, the pore radius. The equilibrium vapor pressure of the gas over the condensed liquid is thus smaller during desorption than that during adsorption. As a result the smaller pores, which were filled first, will be emptied last. In the adsorption isotherm of a porous solid, which will undoubtedly have pores of a variety of sizes and shapes, the effect of capillary condensation will be to produce the commonly observed hysteresis in the isotherm (Figure 7.11). Capillary condensation may not be able to account for all of the observed hysteresis, but it certainly is an important feature to be taken into account.

The Kelvin equation represents a handy way to evaluate effective pore radii with the assumption of uniform, cylindrical pore shapes. In reality, of course, one can expect pores to be rather irregular in shape, with complex interconnections in some cases, and complete "dead ends" in others. That makes complete analysis of a particular situation difficult or impossible. For example, it is often suggested that two primary types of pores will be present in a given sample --- V-shaped pores in which the "mouth" is larger than the interior radius and "ink-bottle" pores in which the interior radius is larger (Figure 7.12). In the case of the V-shaped pores, one assumes that the adsorption and condensation processes occur more or less reversibly, introducing little hysteresis. In the "ink bottles," however, the pores will fill at a rate controlled primarily by the interior radius of the pore, while the emptying process will be controlled by the relatively smaller radius of the mouth. The net result will be a significant difference in the two rates, ie, hysteresis.

There is, of course, much more that could be said about solid surfaces and solid--vapor interfaces --- their formation, energetics, theoretical models, adsorption characteristics, etc. However, space limitations dictate that the discussion move along to other topics. Because of the intimate and intricate relationships among the various areas of surface and colloid science, the concepts outlined above should serve as a good introduction to the topics to follow.

CHAPTER 8

LIQUID--FLUID INTERFACES

Liquids have several distinct characteristics that differentiate them from solid and gas phases. One of the more important ones (from the point of view of surface chemistry, at least) is that, unlike a gas, liquids have a relatively high density and fixed volume, while they possess a mobility at the molecular level that is many orders of magnitude greater than that in solids. As a result of that mobility, interfaces involving liquids and another fluid generally behave as though homogeneous and therefore lack many of the complications encountered when considering solid surfaces.

THE NATURE OF A LIQUID SURFACE:
SURFACE TENSION

It is common practice to describe a liquid surface as having an elastic "skin" that causes the liquid to assume a shape of minimum surface area, its final shape being determined by the "strength" of that skin relative to other external factors such as gravity. In the absence of gravity, or when suspended in another immiscible liquid of equal density, a liquid spontaneously assumes the shape of a sphere. In order to distort the sphere, work must be done on the liquid surface, increasing the total surface area and therefore the free energy of the system. When the external force is removed, the contractile skin then forces the drop to return to its equilibrium shape.

While the picture of a skin like a balloon on the surface of a liquid is easy to visualize and serves a useful educational purpose, it can be quite misleading, since there is no skin or tangential force as such at the surface of a pure liquid. It is actually an imbalance of forces on surface molecules pulling into the bulk liquid and out into the adjoining vapor phase that produces the apparent contractile skin effect. The forces involved are, of course, the same van der Waals interactions that account for the liquid state and for most physical interactions between atoms and molecules. Because the liquid state is of higher density than the vapor,

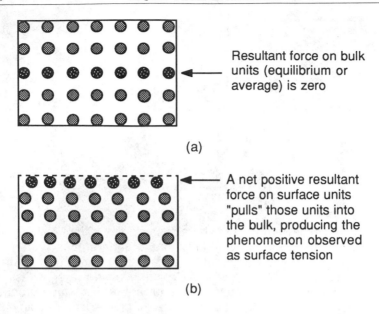

Resultant force on bulk
units (equilibrium or
average) is zero

(a)

A net positive resultant
force on surface units
"pulls" those units into
the bulk, producing the
phenomenon observed
as surface tension

(b)

Figure 8.1. A schematic illustration of intermolecular (or atomic) forces acting on surface units to produce the effect of surface tension in liquids.

surface molecules are pulled away from the surface and into the bulk liquid, causing the surface to contract spontaneously (Figure 8.1). For that reason, it is more accurate to think of surface tension (or surface energy) in terms of the amount of work required to increase the surface area of the liquid isothermally and reversibly by a unit amount, rather than in terms of some tangential contractile force.

As will be seen later in the chapter, the same basic ideas that are used to describe the liquid--vapor interface apply to the liquid--liquid interface. However, since a second liquid phase is much more dense than a vapor phase, the various attractive interactions among units of the two phases *across* the interface, which depend on the number density of interacting units (see Chapter 4), are significantly greater. As a result, for a given increase in interfacial area, the excess surface energy of each unit (and therefore the total energy) will be lower. In other words, the net work required to increase the interfacial area, the interfacial tension, will be reduced. Table 8.1 lists the surface tensions of several typical liquids and their corresponding interfacial tensions against water and mercury.

There are two quick observations that one may note from the data in Table 8.1: (1) The interfacial tension between a given liquid and water is always less than the surface tension of water; and (2) for an homologous series of materials such as the normal alkanes, the interfacial tension between the members of the series and water (or any other immiscible liquid) will change only slightly as a function of the molecular weight of the material. Those characteristics are a direct consequence of the nature of the interactions at the interface. Where the

Table 8.1. Typical liquid surface and interfacial tensions at 20°C (mN m^{-1}).

Liquid	Surface tension	Interfacial tension vs water
Water	72.8	---
Ethanol	22.3	---
n-Octanol	27.5	8.5
Acetic Acid	27.6	---
Oleic Acid	32.5	7.0
Acetone	23.7	---
Carbon Tetrachloride	26.8	45.1
Benzene	28.9	35.0 (357 vs Hg)
n-Hexane	18.4	51.1 (378 vs Hg)
n-Octane	21.8	50.8
Mercury	485	375

two liquids are highly immiscible, the interfacial tension will lie between the two surface tensions (eg, water--benzene); if significant miscibility exists, the interfacial tension will be lower than the lower of the two surface tensions (eg, water--octanol). The difference stems from the *surface activity* of the molecules of the miscible liquid (in water), a topic introduced in Chapter 3 which will be addressed again in Chapter 15.

Most commonly encountered room temperature liquids have surface tensions against air or their vapors that lie in the range of 10--80 mN m^{-1}. Liquid metals and other inorganic materials in the molten state exhibit significantly higher values as a result of the much greater and more diverse interactions occurring in such systems. Water, the most important liquid commonly encountered in both laboratory and practical situations, lies at the upper scale of what are considered normal surface tensions, with a value in the range of 72--73 mN m^{-1} at room temperature, while hydrocarbons reside at the lower end, falling in the lower to middle 20s.

Modern treatments of van der Waals forces have made it possible for one to calculate with good accuracy the expected interfacial tensions of many systems that do not involve strong specific interactions, such as dipolar and hydrogen bonding.

Surface Mobility

The common concept of interfacial tensions is simplistic in the sense that it implies that the surface or interface is a static entity. There is, in reality, a constant and for liquids and gases, rapid interchange of molecules between the bulk and interfacial region, and between the liquid and vapor phases. If it is assumed that molecules leave the interfacial region at the same rate that they arrive, it is possible to estimate the exchange rate, β, of an individual molecule from the relationship

$$\beta = \alpha(2\pi mkT)^{1/2} p_o \qquad (8.1)$$

where α is a so-called *sticking coefficient* (ie, the fraction of molecules striking the surface that actually becomes part of it), p_o is the equilibrium vapor pressure of the liquid, m the mass of the molecule, and k Boltzman's constant. Assuming α to lie in the range of 0.03--1.0, a water molecule at 25°C will have an average residence time of 3 μs or less at the air--water interface. The corresponding residence time for a mercury atom would be roughly 5 ms, while that for a tungsten atom would be 10^{37} s at room temperature.

With such molecular mobility, it is clear that the surface of a pure liquid offers little resistance to forces that may act to change its shape. That is, there will be very little viscous or elastic resistance to the deformation of the surface. An important consequence of that fact is that a pure liquid does not support a foam for more than a small fraction of a second (see Chapter 12). A similar situation exists at the liquid--liquid interface. As we shall see in later chapters, the highly mobile nature of liquid interfaces has significant implications for many technological applications such as emulsions and foams, and forms the basis for many of the most important applications of surface-active materials or surfactants.

Temperature Effects on Surface Tension

Because of the mobility of molecules at fluid interfaces, it is not surprising to find that temperature can have a large effect on the surface tension of a liquid (or the interfacial tension between two liquids). An increase in surface mobility due to an increase in temperature will clearly increase the total entropy of the surface and thereby reduce its free energy, ΔG. Since the surface tension has been thermodynamically defined as

$$\sigma = \Delta G / \Delta A \qquad (8.2)$$

one would expect to encounter a negative temperature coefficient for σ. While that is the case for most normal liquids, including most molten metals and their oxides, positive coefficients have been encountered. While the reason for that phenomenon is not entirely clear, it probably results from some change in the actual atomic composition of the surface as the temperature is increased.

At temperatures near the critical temperature of a liquid, the cohesive forces acting between molecules in the liquid become very small and the surface tension approaches zero. That is, since the vapor cannot be condensed at the critical temperature, there will be no surface tension. A number of empirical equations that attempt to predict the temperature coefficient of surface tension have been proposed, with one of the most useful being that of Ramsey and Shields,[1]

$$\sigma(Mx/\rho)^{2/3} = k_s(T_c - T - 6 \tag{8.3}$$

where M is the molar mass of the liquid, ρ its density, x the degree of association, T_c the critical temperature, and k_s a constant.

The Effect of Surface Curvature

Because many practical situations involve surfaces and interfaces with high degrees of curvature, it is important to understand the effect of curvature on interfacial properties. As pointed out in Chapters 2 and 6, there will develop a pressure differential across any curved surface, with the pressure being greater on the concave side of the interface. That is, the pressure inside a bubble will always be greater than that in the continuous phase. The Young-Laplace equation,

$$\Delta p = \sigma(1/r_1 + 1/r_2) \tag{8.4}$$

in which Δp is the drop in pressure across a curved interface, r_1 and r_2 are the principal radii of curvature, and σ is the surface (or interfacial) tension, relates the quantities of interest in this situation. For a spherical surface where $r_1 = r_2$, the equation reduces to

$$\Delta p = 2\sigma/r \tag{8.5}$$

For a very small drop of liquid in which there is a large surface to volume ratio, the vapor pressure is higher than that over a flat surface of equal area. The movement of liquid from a flat interface into a volume with a curved interface requires the input of energy into the system since the surface free energy of the curved volume increases. If the radius of a drop is increased by dr, the surface area increases from $4\pi r^2$ to $4\pi(r + dr)^2$, or by a factor of $8\pi r\, dr$. The free energy increase is $8\pi\sigma r\, dr$. If during the process ∂n moles of liquid are transferred from the flat phase with a vapor pressure of p_o to the drop with vapor pressure p_r, the free energy increase also given by

$$\Delta G = \delta n RT \ln(p_r/p_o) \tag{8.6}$$

Equating the two relationships leads to the expression

(a) (b)

Figure 8.2. Dynamic surface tension in pure liquids: (a) for a liquid of isotropic molecular shape, dynamic surface tension effects are controlled by the rate of diffusion of molecules from the bulk to the new surface; (b) in polar or anisotropic liquids, the situation may be further complicated by the question of molecular orientation at the surface.

$$RT \ln (p_r / p_o) = 2 \sigma M / \rho r = 2\sigma V_m / r \qquad (8.7)$$

known as the Kelvin equation. In eq. 8.7, ρ is the density, M the molar mass, and V_m the molar volume of the liquid. It can be shown that extremely small radii of curvature can lead to the development of significant pressure differences in drops. For a drop of water with a radius of 1 nm, the partial pressure ratio from eq. 8.7 is about 3. Obviously, the condensation of liquid molecules to form very small drops will be retarded by a relatively high energy barrier due to curvature. Understanding the consequences of the Kelvin equation helps in explaining the ability of many liquid--vapor systems to become supersaturated, when logic says that condensation should readily occur. It is the input of energy by scratching, agitation, etc, or the provision of a heterogeneous nucleation site, thereby reducing the *effective* radius of curvature of the incipient drops, that brings about the rapid condensation or crystallization of a supersaturated system.

Dynamic Surface Tension

For a pure liquid in equilibrium with its vapor, the number density and orientation of molecules at the surface will be different from that of bulk molecules (Figure 8.2). When new surface is created, it is reasonable to assume a finite amount of time to be required for new molecules forced to the surface by the increased area to attain their equilibrium values again. In that interum, as short as it may be, the measured surface tension of the system should be different from that of the system in equilibrium. The surface tension of such "new surface" is referred to as the *dynamic surface tension* .

Qualitatively, it is assumed that the time required to attain equilibrium after formation of new surface is related to the time for diffusion of new liquid molecules to the surface --- that is, to the self-diffusion constant. Diffusion times are usually on the order of 10^{-6} cm^2 s^{-1}, which translates to times of miliseconds for the attainment of equilibrium. The accurate measurement of surface tensions over such short time frames is difficult at best, so that a great deal is still in question

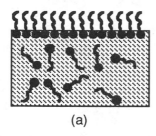 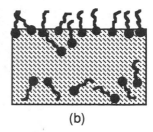

(a)   (b)

Figure 8.3. Dynamic surface tension effects in surfactant solutions: (a) an equilibrium surface; (b) as the surface is enlarged, additional surfactant molecules must diffuse to the new surface, temporarily depleting the concentration of free surfactant just below the surface.

concerning the thermodynamics and kinetics of such fresh surfaces.

If a surface-active solute is present, it becomes rather easy to demonstrate the dynamic surface tension effect experimentally, although the problem is complicated by the questions of solute concentration and orientation (among others). While a number of theories exist concerning dynamic processes of adsorption at freshly formed surfaces, the uncertainties involved make them somewhat problematical for most practical purposes. As a general concept, however, it is usually assumed that the initial rate of adsorption at the new surface approximately equals the adsorption rate for molecules of the pure liquid. However, as surface adsorption occurs, the solution region just below the surface becomes depleted of solute, and diffussion is slowed until more solute diffuses into the region from the bulk (Figure 8.3). Obviously, the rate of such movement is related to the solute diffusion constant; the smaller the constant, the longer the time required to attain equilibrium. For relatively large solute molecules, reliable data have been obtained which indicate that, in agreement with intuition, the attainment of equilibrium surface tension values takes longer for larger solute molecules and for lower bulk concentrations.

An additional complication in evaluating dynamic surface tensions may arise in terms of molecular orientation at the surface. For a symmetrical molecule, orientation will not be a problem; however, for many systems, especially those involving asymmetric (in both shape and chemical nature) molecules such as alcohols and other surface- active organic materials, the surface tension will be a function of the orientation of the molecule at the surface. For an aqueous solution of a long-chain alcohol, the equilibrium surface tension results when the adsorbed molecules are oriented with the alkyl chain pointing out toward the vapor phase and the hydroxyl group "buried" in the water (Figure 8.4). Some finite time is required for such orientation to occur, so that two materials with essentially identical bulk diffusion coefficients will exhibit distinct dynamic surface tension characteristics due to differences in orientation rates. That effect is especially apparent in systems in which the solute is

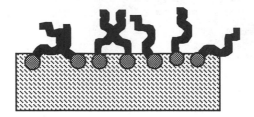

Figure 8.4. Polar molecules in solution will, when possible, orient themselves at surfaces and interfaces in order to minimize the overall interfacial energy of the system.

a macromolecule. It has been found that polymer solutions may take minutes, hours, or even days to attain their equilibrium surface tension, primarily due to the long times required for the chains to orient and accommodate themselves at the surface.

A great deal of practical work has been done on the effects of dynamic surface tension in processes such as high-speed coating operations. Unfortunately, much of it has been in terms of specific industrial systems and the results remain buried in the never-never land of "proprietary information."

SURFACE TENSIONS OF SOLUTIONS

Because of differences between the shape, size, and/or chemical nature of a solute relative to a given solvent, the presence of the solute often results in the alteration of the surface tension of the solution relative to that of the pure liquid. Most commonly, such effect is to lower the surface tension, although the opposite effect is also found.

Intuitively, the surface tension of a solution may be expected to be some mathematical average of that of the two pure components. The simplest such combination for a binary mixture would be an additive combination related to the quantity of each component in the mixture, such as mole fraction. Such a relationship may be written

$$\sigma_{mix} = \sigma_1 X + \sigma_2 (1 - X) \tag{8.8}$$

where σ_{mix} is the surface tension of the solution, σ_1 and σ_2 are the surface tensions of the respective components, and X is the mole fraction of component 1 in the mixture. In ideal systems where the vapor pressure of the solution is a linear function of the composition, such relationships are found. Normally, however, there will be some positive or negative deviation from linearity, with the latter being most commonly encountered. Some examples of the variation of the surface tension of mixtures with composition are shown in Figure 8.5.

Taking water as an example, when the second component is an inorganic electrolyte that requires significant solvation, the relationship

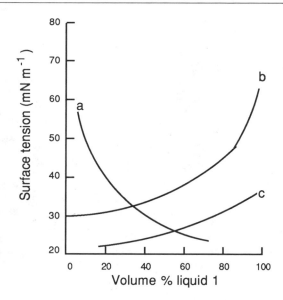

Figure 8.5. Variations in the surface tension of liquid mixtures: curve a, acetone--water; curve b, glycerol--2-ethoxyethanol; curve c, aniline--cyclohexane.

between surface tension and composition may be quite varied, depending on the exact nature of the interaction. It is generally found, for example, that the addition of inorganic electrolyte to water results in an increase in the surface tension of the solution, although the effect is not dramatic and requires rather high salt concentrations to become significant (Figure 8.6). The relative effectiveness of ions at increasing the surface tension of water generally follows the Hofmeister series: Li^+ > Na^+ > K^+, and F^- > Cl^- > Br^- > I^-.

Unlike the inorganic electrolytes, the presence of an organic material in aqueous solution will result in a decrease in the surface tension of the system. The extent of such lowering depends upon a number of factors, including the relative miscibility of the system (or the solubility of the organic solute) and the tendency of the organic material to adsorb preferentially at the water--air interface. Liquids such as ethanol or acetic acid produce gradual decreases in the surface tension of their aqueous solutions (as in Figure 8.5, curve a), while longer chain organics such as butanol can produce more dramatic effects (Figure 8.7).

When the organic solute has a limited solubility in water, the effect on surface tension becomes characteristic of solutions of surface-active materials or surfactants. For solutions of such materials, one encounters a steady decrease in surface tension with increased solute concentration. At some point, a minimum value of σ will be obtained as the solute concentration increases before surface saturation or some form of solute behavior change (precipitation, micelle formation, etc)

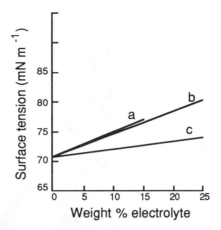

Figure 8.6. The effect of common electrolytes on the surface tension of water: curve a, LiCl; curve b, NaCl; curve c, NaBr.

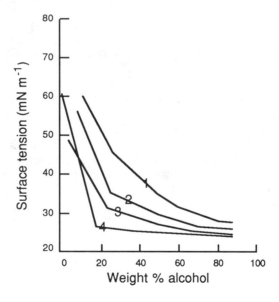

Figure 8.7. The effect of short-chain alcohols on the surface tension of water: 1 = methyl, 2 = ethyl, 3 = n-propyl, and 4 = n-butyl alcohol.

prevents further change in the surface tension. A typical surface tension--concentration curve is shown in Figure 8.8. Surface-active agents or surfactants form a large class of materials whose importance to our existence (both literally and figuratively) cannot be overstated. Such materials, including classical soaps, detergents, and hundreds of other types of materials, and their production is part of a multibillion dollar industry worldwide. But that is only part of the story. Surfactants,

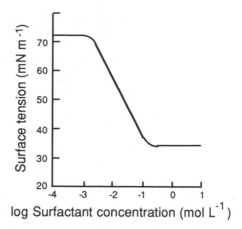

log Surfactant concentration (mol L^{-1})

Figure 8.8 Schematic illustration of a "typical" surface tension-- concentration curve for an aqueous surfactant solution.

because of their properties for lowering surface tension and altering related phenomena, are practically indispensible in an even wider range of processes and products, so that their actual economic value is many times that of their direct production value.

Of greater importance from an existential point of view, however, is the vital importance of surface-active molecules in our most important biological functions. The walls of the cell which make up our bodies are assemblies of surface active molecules, plus other components. Vital functions such as respiration, blood transport phenomena, nutrition, and the functioning of our antibody systems all involve surface activity and surface interactions. In a nutshell, we are creatures of surface activity.

Surfactants and the Reduction of Surface Tension

The fundamental principle underlying our present understanding of surface activity is that old reliable, the Gibbs adsorption equation. Since the surface tension of a liquid is determined by the excess energy of the molecules in the interfacial region, the displacement of surface liquid molecules by adsorbed solute directly affects the measured value of σ. It is the relationship between the chemical structure of an adsorbing molecule and the rate and extent of adsorption under given circumstances that differentiates the various surfactant types and determines their utility in applications where surface tension lowering is of importance.

In aqueous solutions, the interface between the liquid and vapor phases involves interactions between relatively densely packed, highly polar water molecules, and relatively sparse, nonpolar gases. The result is a large imbalance of forces acting on the surface molecules and the observed high surface tension of water (72.8 mN m^{-1}). If the surface solvent molecules are replaced by adsorbed surfactant molecules with lower specific excess surface energy, the surface tension of the solution is decreased accordingly, with the amount of reduction being related to the surface excess concentration of solute and the nature of the adsorbed molecule.

A large body of literature has grown up around the question of how one can relate the chemical structure of a surfactant to its surface activity in various situations. At the present time, most such structure--property relationships are semiquantitative at best, but they can serve one well as a guide to the choice of the best surfactant for a given situation. Several works reviewing the literature in that area are cited in the Bibliography, and some specific examples will be noted in the appropriate context of later chapters.

If the vapor phase is replaced by a condensed phase that has a higher molecular density and more opportunity for attractive interaction between molecules in the interfacial region, the interfacial tension will be reduced significantly. In the case of water, the presence of a liquid such as octane, which interacts only by relatively weak dispersion forces, lowers the interfacial free energy to 52 mN m^{-1}. If the extent of molecular interaction between phases can be increased by the introduction of polar groups that interact more specifically with the water, as, for instance, in octanol, the interfacial energy reduction will be even greater (to 8.5 mN m^{-1}). Clearly, any alteration in the nature of the molecules composing the liquid surface would be expected to result in a lowering of the interfacial energy of the system. And therein lies the basic explanation for the action of surfactants in lowering the surface and interfacial tension of aqueous solutions.

The same qualitative reasoning also explains why most surfactants do not affect the surface tension of organic liquids --- the molecular nature of the liquid and the surfactant are not sufficiently different to make adsorption a particularly favorable process. Moreover, if adsorption occurs, the energy gain is not sufficient to produce a significant change in the surface energy to be useful. The actions of fluorocarbon and siloxane surfactants are exceptions since the specific surface free energy of such materials may be significantly lower than that of most hydrocarbons. They will therefore be positively adsorbed at hydrocarbon surfaces and lower the surface tension of their solutions.

SURFACTANT ADSORPTION AND GIBBS MONOLAYERS

The basic concepts behind the factors governing the adsorption of surface-active molecules at interfaces has already been mentioned several times in terms of the Gibbs adsorption isotherm, which relates the surfaces excess concentration of the adsorbed species to the surface or interfacial tension of the system. Because of the broad range of areas in which such adsorption occurs and produces significant alterations in the surface chemistry of the systens involved, specific details of the phenomenon and resultant effects will be covered in chapters related to specific topics, such as emulsions, foams, solid-liquid interfaces, and wetting phenomena.

For now, the discussion will be limited to some general concepts related to adsorption at liquid--fluid interfaces, such as some general relationships between surfactant structure and the rate and effect of adsorption.

Efficiency, Effectiveness, and Surfactant Structure

When discussing the performance of a surfactant in lowering the surface tension of a solution it is necessary to consider two aspects of the process: (1) the concentration of surfactant in the bulk phase required to produce a given surface tension reduction, and (2) the maximum reduction in surface tension that can be obtained, regardless of the concentration of surfactant present. The two effects may be somewhat arbitrarily defined as follows: the surfactant *efficiency* is the bulk phase concentration necessary to reduce the surface tension by a predetermined amount, say 20 mN m^{-1}. Its *effectiveness* is the maximum reduction in σ that can be obtained by the addition of any quantity of surfactant.

Because the extent of reduction of the surface tension of a solution depends upon the substitution of surfactant for solvent molecules at the interface, the relative concentration of surfactant in the bulk and interfacial phases should serve as an indicator of the adsorption efficiency of a given surfactant and, therefore, as a quantitative measure of the activity of the material at the solution--vapor interface. For a given homologous series of straight-chain surfactants in water, $CH_3(CH_2)_n$-S, where S is the hydrophilic head group and n is the number of methylene units in the chain, an analysis based on the thermodynamics of transfer of a surfactant molecule from the bulk phase to the interface leads to the conclusion that the above-defined efficiency of adsorption is directly related to the length of the hydrophobic chain. The efficiency can be defined mathematically by the expression

$$-\log (C)_{20} = pC_{20} = n\,(-A/2.3RT) + (-B/2.3RT) + K \qquad (8.9)$$

where A, B, and K are terms for the free energies of transfer of methylene, terminal methyl, and head groups, respectively, from the bulk solution to the interface, and C is the concentration of surfactant required to lower σ by 20 mN m^{-1}. For a given head group at constant temperature, solvent composition, etc, the equation reduces to a direct dependence of efficiency on the length of the hydrocarbon chain n, as expected.

Since the surfactant efficiency is directly related to the thermodynamics of chain transfer from bulk to interface, it is reasonable to expect chain modifications that alter that characteristic, that is, changes in the hydrophobic character of the surfactant, to produce parallel changes in adsorption efficiency. For example, branching in the hydrophobic group results in a reduction in the hydrophobicity of a surfactant chain relative to that of a related straight-chain material with the same total carbon content. Carbon atoms located on branch sites contribute approximately two-thirds as much to the character of a surfactant molecule as one located in the main chain. Similar results are observed for surfactants with two or more short-chain hydrophobes of equal total carbon content (eg, internal substitution of the hydrophilic group) and for the presence of unsaturation in the chain. A benzene ring usually contributes an effect equivalent to approximately 3.5 methylene groups.

If a surfactant possesses two polar groups, the methylene groups lying between the two polar groups contribute an effect equivalent to approximately one-half that found for such groups located in the main body of the hydrophobe.

In cationic surfactants, the presence of short-chain alkyl groups (fewer than four carbon atoms) attached to the nitrogen seem to have little effect on the efficiency of adsorption of the molecule. The dominant factor will always be the length of the primary hydrophobic chain. That effect is true whether the alkyl groups are attached to a quaternary ammonium group, an amine oxide, or a heterocyclic nucleus such as pyridine.

Within limits, the nature of the charge on an ionic surfactant has little effect on the efficiency of surfactant adsorption. Again, it is the nature of the hydrophobic group that predominates. Some increase in efficiency, however, is seen if the counterion is one that is highly ion paired, that is, one that is not highly solvated in the system and therefore produces a lower net electrical charge as the molecules are adsorbed at the interface, facilitating the movement of molecules into the interface. The addition of neutral electrolyte to an ionic surfactant solution produces a similar result in increasing the efficiency of adsorption by compression of the electrical double layer associated with the ionic head group.

Polyoxyethylene (POE) nonionic surfactants with the same hydrophobic group and an average of 7--30 OE units, exhibit adsorption efficiencies that follow an approximately linear relationship of the form

$$pC_{20} = A_{tr} + nB_{tr} \qquad (8.10)$$

where A_{tr} and B_{tr} are constants related to the free energy of transfer of ---CH_2--- and OE groups, respectively, from the bulk phase to the interface and n is the number of OE units in the POE chain. As is usually the case for POE nonionics, most data reported have been obtained using nonhomogeneous POE chains. The available data indicate that the efficiency of adsorption decreases slightly as the number of OE units on the surfactant increases.

To this point, we have seen that the efficiency of surfactant adsorption at the solution--vapor interface is dominated by the nature of the hydrophobic group and is relatively little affected by the hydrophilic head group. It is often found that the second characteristic of the adsorption process, the so-called adsorption effectiveness, is much more sensitive to other factors and quite often does not parallel the trends found for adsorption efficiency.

Adsorption Effectiveness

The choice of 20 mN m^{-1} as a standard value of surface tension lowering for the definition of adsorption efficiency is convenient, but, as mentioned, somewhat arbitrary. When one discusses the effectiveness of adsorption, as defined as the maximum lowering of surface tension regardless of surfactant concentration, the value of σ_{min} is determined only by the system itself and represents a more firmly fixed point of reference. The value of σ_{min} for a given surfactant will be determined by one of two factors: (1) the solubility limit or Krafft temperature (T_k) of the compound, or (2) the critical micelle concentration (CMC). In either case, the maximum amount of surfactant adsorbed is reached, for all practical purposes, at the maximum bulk concentration of free surfactant.

Because the activity of surfactants used below T_k cannot reach their theoretical maximum as determined by the thermodynamics of surfactant aggregation (see Chapter 15), they will also be unable to achieve their maximum degree of adsorption at the solution--vapor interface. It is therefore important to know the value of T_k for a given system before considering its application. Most surfactants, however, are employed well above their Krafft temperature, so that the controlling factor for the determination of their effectiveness will be the CMC.

When one examines the shape of the surface tension--ln C curve for a surfactant, it can be seen that the curve becomes approximately linear at some concentration below the CMC. It can be shown that the effectiveness of the adsorption of a surfactant, $\Delta\sigma_{CMC}$, can be quantitatively related to the concentration of surfactant at which the Gibbs equation becomes linear, C_1, the surface tension attained at C_1, σ_1, and the CMC. The relationship has the general form

$$- \Delta\sigma_{CMC} = (\sigma_0 - \sigma_1) + 2.3\Omega RT\, \Gamma_m \log (C_{CMC}/C_1) \qquad (8.11)$$

where σ_0 is the surface tension of the pure solvent and Γ_m is the maximum in surface excess of adsorbed surfactant at the interface. The factor Ω in eq. 8.11 is related to the number of molecular or atomic units that become adsorbed at the interface with the adsorption of each surfactant molecule; for nonionic surfactants or ionic materials in the

Table 8.2. Experimental values of CMC/C_{20}, Γ_{20} (x 10^{10} mol cm^{-2}) , and σ_{min} (mN m^{-1}) for some typical surfactants in aqueous solution.

Surfactant	Temperature (°C)	CMC/C_{20}	Γ_{20}	σ_{min}
$C_{12}H_{25}SO_4^-Na^+$	25	2.0	3.3	40.3
$C_{12}H_{25}SO_3^-Na^+$	25	2.3	2.9	40.8
$C_{16}H_{33}SO_4^-Na^+$	60	2.5	3.3	37.8
$C_{12}H_{25}C_6H_4SO_3^-Na^+$	70	1.3	3.7	47.0
$C_{12}H_{25}C_5H_5N^+Br^-$	30	2.1	2.8	42.8
$C_{14}H_{29}C_5H_5N^+Br^-$	30	2.2	2.8	41.8
$C_{12}H_{25}N(CH_3)_3^+Br^-$	30	2.1	2.7	41.8
$C_{10}H_{21}(POE)_6OH$	25	17.0	3.0	30.8
$C_{12}H_{25}(POE)_6OH$	25	9.6	3.7	31.8
$C_{16}H_{33}(POE)_6OH$	25	6.3	4.4	32.8
$C_{12}H_{25}(POE)_9OH$	23	17.0	2.3	36.8
$C_{16}H_{33}(POE)_9OH$	25	7.8	3.1	36.8
$C_{12}H_{25}(POE)_{12}OH$	23	11.8	1.9	40.8
$C_{16}H_{33}(POE)_{12}OH$	25	8.5	2.3	39.8
$C_{16}H_{33}(POE)_{15}OH$	25	8.9	2.1	40.8

presence of a large excess of neutral electrolyte, $\Omega = 1$; for ionic surfactants $\Omega = 2$, since one counterion must be adsorbed for each surfactant molecule giving a total of two species.

The effectiveness of a surfactant can be conveniently quantified by using a value of C_1 at which the surface tension has been reduced by 20 mN m^{-1}, assuming $\Gamma_{20} \approx \Gamma_m$. Application of eq. 8.11 allows the calculation of a standard quantity, CMC /C_{20}, which serves as a useful measure for evaluating surfactant effectiveness. For several surfactants of comparable CMC, for example, a larger value for CMC /C_{20} indicates that at a given concentration, the available free surfactant is being used more effectively at the surface. Some representative values that illustrate the effects of well-controlled changes in surfactant structure are given in Table 8.2.

It is often found that the efficiency and effectiveness of surfactants do not run parallel; in fact it is commonly observed that materials that produce significant lowering of the surface tension at low concentrations (ie, are more efficient) are less effective (ie, have a smaller Γ_m). This follows from the complex relationship between adsorption at the interface and micelle formation in the solution.

On a molecular basis, the conflicting factors can be seen conceptually as arising from the different roles of the molecular structure in the adsorption process. Surfactant efficiency is related to the extent of adsorption at the interface as a function of bulk surfactant concentration. At a concentration well below that at which micellization becomes a factor, efficiency can be structurally related to the hydrophobicity of the surfactant tail and the nature of the head group. For a given homologous series of surfactants it will be a functionof the thermodynamics of transfer of the hydrophobic tail from the bulk to the surface phase. A plot of σ vs ln C for such a series will exhibit a relatively regular shift in the linear portion of the curve to lower concentrations (for a given σ) as methylene groups are added to the chain.

While the role of molecular structure in determining surfactant efficiency is primarily thermodynamic, its role in effectiveness is more directly related to the size of the hydrophobic and hydrophilic portions of the adsorbing molecules. When one considers the adsorption of molecules at an interface, it can be seen that the maximum number of molecules that can be fitted into a given area depends upon the area occupied by each molecule. That area will, to a good approximation, be determined by either the cross-sectional area of the hydrophobic chain or the area required for the arrangement for closest packing of the head groups (Figure 8.9), whichever is greater. For straight-chain 1:1 ionic surfactants, it is usually found that the head group requirement predominates, so that for a given homologous series, the surface tension minimum obtained varies only slightly with the length of the hydrocarbon chain.

Since the decrease in surface tension obtained is directly related to the surface excess adsorption of the surfactant by the Gibbs equation, a reduction in the amount of material that can be adsorbed in a given surface area reduces the ultimate surface tension lowering attained. The efficiency will, of course, change more or less regularly with the chain length. The sign of the charge on the ionic surfactant has only a minor effect on the ultimate surface tension attained, indicating that the geometric requirements (including electrostatic effects) are fairly constant from one head group to the next. In the presence of neutral electrolyte, of course, electrostatic repulsions between adjacent molecules are reduced, so that their effective areas are smaller. The net result is a slight increase in surfactant effectiveness.

While an increase in the hydrocarbon chain length in a series of normal alkyl surfactants between C_8 and C_{20} carbon chains will have a minor effect on the effectiveness of a surfactant, other structural changes can produce much more dramatic effects. As will be seen in more detail in Chapter 15, structural features such as branching and multiple-chain hydrophobes will generally result in increases in the CMC of surfactants with the same total carbon content. Those changes seem to have a much smaller effect on the efficiency of the surfactant (C_{20}) than on its effectiveness.

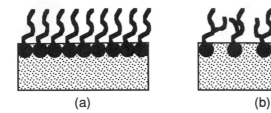

(a)   (b)

Figure 8.9. Schematic illustration of the role of surfactant tail and head group in determining packing efficiency at an interface: (a) straight chains favor close, efficient packing, limited by the size of the head group; (b) branched or bulky chians hinder efficient packing.

The introduction of polar groups such as ethylenic unsaturation; ether, ester, or amide linkages; or hydroxyls located well away from the head group, usually results in a significant lowering of both the efficiency and effectiveness of the surfactant as compared to a similar material with no polar units. Such a result has generally been attributed to changes in orientation of the adsorbed molecule with respect to the surface due to interactions between the polar group and the water (Figure 8.10). If the polar group is situated very near to the primary hydrophilic group, its orientational effect will be much less dramatic, although it may still have a significant effect on the CMC of the material.

Changes in the hydrophobic group in which fluorine atoms are substituted for hydrogen usually result in significant increases in the efficiency and effectiveness of the surfactant. The substitution of fluorine for hydrogen in a straight-chain surfactant results in a relatively small increase in chain cross-sectional area, as compared to a methyl branch, for example, so that the changes must be related to the chemical nature of the substitution. As has already been pointed out, fluorinated organic materials have a relatively low cohesive energy density and therefore little interaction with adjacent phases, or themselves, for that matter. They therefore have very favorable thermodynamic driving forces for adsorption (leading to high efficiency), as well as low surface energies. Their effectiveness is reflected in the very low surface tension values produced (as low as 20 mN m^{-1} in some instances).

There appear to be only relatively minor variations in effect from one head group to another in anionic surfactants. The difference in cross-sectional area between sulfate and sulfonate groups does not appear to influence greatly the activity of surfactants in lowering surface tensions, although some difference can be noted when differences in CMC are taken into consideration. The role of the counterion can be important when changes result in significant alterations in the ion binding properties of the molecule. Tight ion binding, for instance, reduces the extent of electrostatic repulsion between adsorbed molecules, allowing for tighter packing of surfactant at the interface and, in general, increases in both the efficiency and the effectiveness of the surfactant. A similar

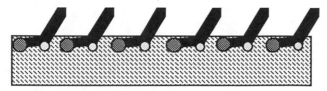

Figure 8.10. Illustration of the effect of a second polar group in the surfactant on its orientation at the interface.

result is obtained by the addition of neutral electrolyte.

While the the head group may be of minor importance for hydrophiles closely related in size and charge character, alterations in those factors can produce significant changes in their activity at the solution--fluid interface. A class of surfactants well suited to the study of such effects is that of the quaternary ammonium salts in which three of the alkyl groups are short-chain units such as methyl, ethyl, and propyl. The substitution of the larger alkyl groups for methyl surfactants of the type $RN^+(CH_3)_3X^-$ results in a significant reduction in the efficiency of adsorption, while not affecting σ_{min} significantly. Presumably, the presence of bulkier alkyl substituents on the head group greatly increases its area and therefore reduces its adsorption efficiency.

The practical effects of surface tension lowering have not been addressed here because they are generally more meaningful when presented in the context of related phenomena such as emulsification, foaming, wetting, and detergency. For further details on the subject of surface tension lowering and surfactant adsorption at fluid interfaces, the reader is referred to the works cited in the Bibliography.

INSOLUBLE MONOMOLECULAR FILMS

The above discussion of liquid surface tensions and the effects of adsorbed molecules on them was aimed primarily at systems in which the adsorbing species has a sufficiently large solubility in the solvent that ,when the surface is saturated, there would remains in solution a reservoir of dissolved surface-active molecules to "fill in the blanks" if more surface is created, or to participate in other phenomena such as micelle formation. Surface films of materials that do not have sufficient solubility to fit this model, but that also exhibit interesting and useful properties have also been recognized for centuries and have added much to our understanding of adsorption and surface phenomena in general. Such materials form the so-called *insoluble monolayers* --- that is, monomolecular layers or films of adsorbed molecules which have very low solubility in the supporting liquid phase, so that they are essesntially isolated on the surface.

A schematic comparison of the two situations is shown in Figure 8.11. On the left , a reservoir that has a movable barrier (B) is filled with pure water. The surface tension of the water on each side of the barrier

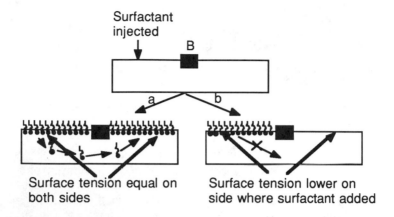

Figure 8.11. A schematic comparison of Gibbs (a) and insoluble (b) monolayers.

will be the same (σ_0). If a quantity of surfactant is added to one side of the reservoir, time is allowed for the system to reach equilibrium, and the surface tension of each side is measured (σ_L and σ_R), one finds that $\sigma_L = \sigma_R$. Surfactant molecules placed on one side will dissolve in the supporting liquid and be adsorbed on the other side of the barrier. Addition can be continued until a saturated monolayer (a Gibbs monolayer) is produced; the important point is that the monolayer formed on *both sides* of the barrier due to dissolution and diffusion processes.

Consider now the situation on the right in which an essentially insoluble, but still surface-active, material such as stearic acid is placed on one side of the barrier. Assuming that the barrier does not "leak," the system may be left for any practical period of time and when the surface tension of each side is measured, it will be found that σ for the side to which stearic acid has been added has been lowered, while the other side remains that of the pure water, σ_0. Stearic acid may be added until a monolayer is formed with no change in σ for the "clean" side. The added stearic acid molecules, being insoluble in water, cannot be dissolved and transported through the water to be adsorbed on the other side of the barrier. The monolayer ultimately formed is an *insoluble monolayer* .

If the barrier B is made movable and connected to some indicating device (a torsion wire or some electronic transducer, for example) one will see that for the experiment on the left, at equilibrium, there is no change in the location of the barrier --- the readout is zero. For that on the right, the barrier will be seen to move away from the side to which the stearic acid has been added, effectively increasing the area of surface containing stearic acid relative to clean surface (arrow). By analogy to a three- dimensional system, the system on the right behaves as if some pressure has been applied to the barrier, pushing it away from the added

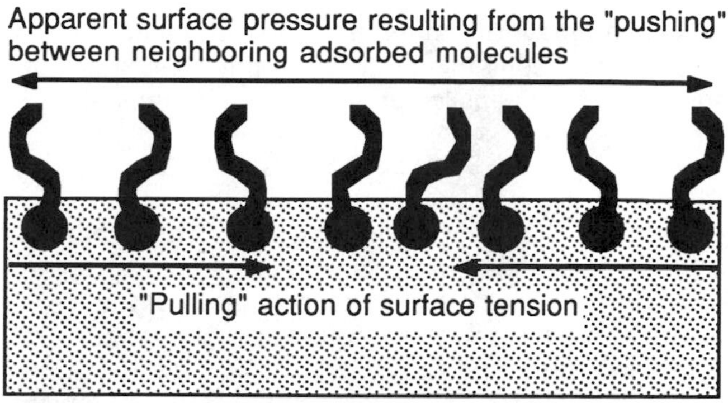

Apparent surface pressure resulting from the "pushing"
between neighboring adsorbed molecules

Figure 8.12. An illustration of the pulling action of surface tension
working against the "pushing" of adjacent adsorbed molecules.

material.

Surface Pressure

The *surface pressure* of a monolayer film, π, is defined as the
difference between the surface tension of the pure supporting liquid, σ_0,
and that of the liquid with an adsorbed film, σ.

$$\pi = \sigma_0 - \sigma \qquad (12)$$

The phenomenon of surface pressure has been studied since the
late 19th century. Some consequences of insoluble monolayer formation
were known (but not understood) as early as Biblical times and were of
interest to the likes of Benjamin Franklin. For example, the pouring of oil
on stormy waters was recognized as an effective measure to protect
fragile ships in a storm.

The surface pressure, as defined by eq. 8.12, represents an
expanding pressure exerted by the monolayer acting against the
(contracting) surface tension of the pure liquid substrate (Figure 8.12).
Analogous to the pressure--volume (P--V) curve of a three-dimensional
bulk material, one can construct a pressure--area (π-A) curve for a
monolayer. Experimentally, one can either work with a fixed surface area
and increase the pressure by incrementally adding more of the adsorbed
material, or a known amount of material can be added to a surface and
the pressure increased by slowly decreasing the available area. In
practice, the latter approach, as represented by the classical Langmuir
trough, is much preferred, since it is easier to measure area reproducibly
than to measure the addition of small amounts of dilute solutions.
(Typically, the adsorbed material is added as a dilute solution in a volitile
solvent.)

With a known amount of material on the surface, the π-A curve allows one to determine something about the physical nature of the film and some molecular characteristics of the adsorbed material.

Surface Potential

It is a general fact that at an interface or phase boundary between two dissimilar materials, there exists a surface electrical potential that reflects differences in the electronic makeup of the two phases. Because almost all surface-active materials (for aqueous systems, at least) have a polar head group, when the molecules adsorb at the surface, the dipole moments of those groups become at least partially oriented with respect to the interface. As a result of the orientation of the dipoles (or charges), the potential difference across the interface will be altered. The *surface film potential* due to the monolayer, ΔV, is the change in the interfacial potential due to the presence of the monolayer.

If the monolayer is treated as a parallel plate condenser, the measured surface film potential can be used to deduce information about the orientation of the adsorbed molecules. While the calculation is only approximate, if n is the number of molecules in the adsorbed film (a known quantity), μ is the dipole moment of the head group (also known or accessible), and ε is the permittivity of the film (its dielectric constant x the permittivity of a vacuum), the approximate relationship

$$\Delta V = n \; \mu \cos \theta \, / \, \varepsilon \qquad (8.13)$$

allows one to estimate the angle of inclination of the dipole to the normal to the surface. Using known bond angles and distances, one can then deduce the orientation of the entire molecule with respect to the surface.

If a mixed monolayer is present (see below), the surface film potential can be used to estimate the homogeneity of the film or, for a homogeneous mixture the film composition (assuming that the values for the pure films are known). The technique is particularly useful for studying the *penetration* of insoluble monolayer films by surface-active molecules injected just below the surface. For example, if a monolayer of a film analogous to a biological membrane is formed and another material of interest is injected below the film (a drug, for instance), surface potential measurements may indicate whether the injected material can penetrate the membrane, and if so, how fast. While the results of such experiments cannot be considered to be absolute in the sense of mimicking an *in vivo* system, they may be very useful as an aid in interpreting other data.

Surface Rheology

Because of the mobility of molecules in the surface of a pure

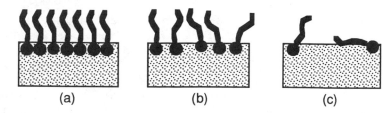

(a)   (b)   (c)

Figure 8.13. Schematic illustration of the primary "states" of adsorbed monolayer films: (a) the condensed state---tight packing of head groups and limited mobility of tails; (b) liquid expanded---relatively tight head group packing, but significant mobility in the tails; (c) gaseous---wide separation of molecules with little interaction between neighbors.

liquid, such surfaces have very little elasticity. For that reason, pure liquids cannot support a foam. In the presence of an adsorbed monolayer film, however, the rheological properties of the surface can change dramatically. By analogy with bulk phases, the physical state of a surface film can be distinguished by its viscosity.

If talc powder is gently placed on a liquid surface, gently blowing on the surface causes the particles to move relatively freely. If the surface is covered with a low density monolayer film movement becomes more restricted, but relatively free movement is still evident. If the film pressure is increased (ie, more molecules per unit area of surface), at some point the particles become fixed in place --- the surface viscosity has increased substantially and the film behaves as if it is in a condensed state (liquid or solid).

It is the increase in surface viscosity produced by adsorbed films (insoluble and Gibbs monolayers, adsorbed polymers, etc) that leads to the production of persistent foams, helps stabilize emulsions, and explains the role of spread monolayers in dampening surface waves, among other important interfacial phenomena.

The Physical States of Monolayer Films

Like bulk materials, monolayer films exhibit characteristics that can (sometimes with a bit of imagination) be equated to the solid, liquid, and gaseous states of matter. For films, the equivalent states are roughly defined as:

a. *Condensed* (solid) films, which are coherent, rigid (essentially incompressible), and densely packed, with high surface viscosity. The molecules have little mobility and are oriented perpendicular (or almost so) to the surface (Figure 8.13a).

b. *Expanded* films, roughly equivalent to the liquid state, in which the monolayer is still coherent and relatively densely packed but is much more compressible than condensed films. Molecular orientation is still approximately perpendicular to the surface, but the tails are less rigidly packed (Figure 8.13b).

     c. Gaseous films, in which the molecules are relatively far apart and have significant surface mobility. The molecules act essentially independently, much as a bulk phase gas and molecular orientation will be random (Figure 8.13c).

     Each type of monolayer film exhibits its own special characteristics analogous to the corresponding bulk phase, as well as distinct "phase transitions" which are useful in characterizing the nature of the film in terms if its equation of state, molecular orientation, interfacial interactions, etc.

Gaseous Films

     An ideal (bulk) gas will have an equation of state given by the ideal gas law

$$PV = nRT \qquad (8.14)$$

Similarly, an "ideal" gaseous monolayer should follow the corresponding law

$$\pi A = kT \qquad (8.15)$$

which means that the π-A curve should be a rectangular parabola such as that in Figure 8.14. In fact, such ideal behavior is rare due to the finite size of the adsorbed molecules.

     If molecular interactions are taken into consideration, eq. 8.15 can be modified to

$$\pi A = xkT \qquad (8.16)$$

where x is a constant (usually between 1 and 2) which adjusts the system to allow for lateral interaction between molecules. Because the value of x must be determined independently for each material, eq. 8.16 is sometimes inconvenient to use. A more generally useful modification is to employ a constant A_0, which is essentially the area excluded (or occupied) by the adsorbed molecule.

$$\pi (A - A_0) = kT \qquad (8.17)$$

It is tempting to equate A_0 directly with the cross-sectional area (in the plane of the surface) of the vertically adsorbed molecule. However, lateral interactions tend to reduce (or in cases of repulsion increase) its value, so that for many films, A_0 may be significantly less than or greater than the expected cross-section of the adsorbate molecule.

     Of course, even more complicated equations of state such as a combination of eqs. 8.16 and 8.17 can be devised, as for example

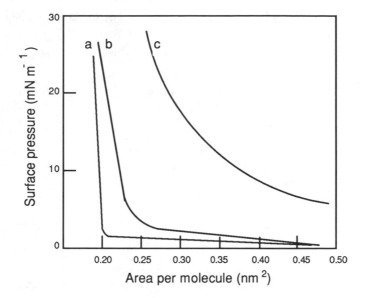

Figure 8.14. Schematic illustration of the π--A curves for the three principle states of monolayer films: (a) condensed; (b) liquid expanded; (c) gaseous.

$$\pi(A - A_0) = xkT \qquad\qquad (8.18)$$

However, since fitting experimental data (which may likely contain some uncertainty) to isotherms and equations of state is sometimes quite subjective, the use of overly complicated relationships involving fitted parameters such as A_0 and x are seldom truly justified.

Gaseous films are common for soluble surfactants solutions (Gibbs monolayers) since solvent--adsorbed solute interactions tend to keep the adsorbed molecules "independent" of neighboring molecules. While they are also encountered in insoluble monolayers, many materials of interest are not so "well behaved" in that they do not exhibit the parabolic π--A curve of Figure 8.14.

Liquid Films

Liquid films are coherent in that they appear to involve some degree of cooperative interaction between portions of the adsorbed molecules, either head groups or tails. They exhibit characteristics of a fluid in that they appear to have no yield point, yet their π--A curve extrapolates to zero at molecular areas significantly larger than than that corresponding to the "theoretical" cross-sectional area (Figure 8.14b). This indicates the presence of molecular interactions at relatively long distances --- a coherent structure, albeit loose or disorganized.

It is sometimes found convenient to designate two subclasses of liquid films, *liquid expanded* (L_1) and *liquid condensed* (L_2), based on subtle differences in respective π--A curves. The L_1 curve is one which typically extrapolates to a limiting value of π (sometimes zero) at a molecular area of about 0.5 nm^2. In contrast to the bulk liquid analogue, such films exhibit a significant degree of compressibility but show no signs of "island" or *hemimicelle* formation; that is, it appears to maintain the characteristics of a uniform phase. In many cases, L_1 films show a transition to a gaseous film at low pressures and perhaps to an L_2 film as the available area per molecule is decreased.

The L_2 films are characterized by the fact that they have considerably lower compressibility than L_1 films and that their π--A curve undergoes a gradual transition to linearity, reminiscent of the solid films. Such films are commonly viewed as having head groups that are close-packed, but that can, under pressure, be rearranged somewhat to give a still tighter packing arrangement. For example, in Figure 8.15, the head groups are shown schematically in a square (or cubic, in three dimensions) lattice at π_1. If π is increased ($\pi_2 > \pi_1$), the arrangement may change to a hexagonal structure, with tighter packing, at which point further significant changes in "crystal structure" are precluded and the π--A curve becomes linear and steep.

While L_2 films ultimately extrapolate to some limiting area at high pressures, that area is usually found to be some 20% larger than the cross-sectional area of a hydrocarbon chain taken from x-ray data, or 10% greater than the respective condensed film (0.22 versus 0.205 nm^2).

Pressure increase

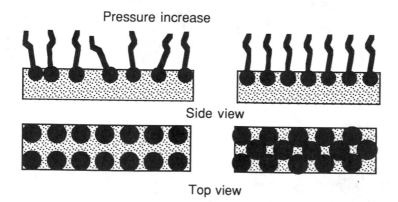

Side view

Top view

Figure 8.15. Schematic illustration of the effect of increased surface pressure on the packing of adsorbed molecules.

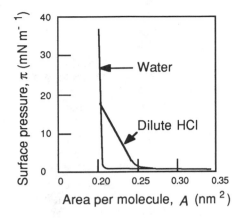

Figure 8.16. The typical shapes of the π--A curves for long-chain saturated carboxylic acids on water and dilute acid.

Condensed Films

Condensed films are composed of densely packed, highly oriented molecules with little mobility and low compressibility. Unlike the gaseous films, the π--A curve for such materials as the saturated, straight-chain carboxylic acids (eg, stearic acid) exhibit minimal change is pressure as the surface area is decreased until a critical area is reached, at which point the pressure will increase rapidly (Figure 8.16). The interpretation of the form of the π--A curve is that in such cases, the cohesive interaction between adsorbed molecules is sufficient to cause the formation of clusters of molecules or hemimicelles on the surface. Because of the strong cohesion, as the available area is decreased, the clusters grow in size and/or number, while intercluster interaction remains small, and it is the interaction between floating clusters that is measured as changes in π. When the area is reduced to the point that the clusters are forced to interact (by physical contact), the pressure increases rapidly.

A typical π--A curve for stearic acid on water at 20°C, plotted as area per molecule, is shown in Figure 8.16. The curve is found to become very steep at an area per molecule of 0.205 nm^2. If the area is decreased further, the pressure suddenly falls, indicating a buckling or collapse of the film. That point is commonly referred to as the *yield point*. The critical molecular area of 0.205 nm^2 found for stearic acid is the same as that for palmitic, myristic, and the other members of the series with more than 12 carbons in the chain. X-Ray diffraction data indicate that the cross-sectional area of the stearic acid molecule is 0.185 nm^2, suggesting that the critical limiting area represents the point at which the molecules become more efficiently packed, approximating the packing in the solid crystal.

With very careful experimental work, it is sometimes possible to identify various intermediate phase transitions occurring before the formation of the condensed film. For example, myristic acid spread on $0.1N$ HCl at 14°C can (with extreme care) produce a curve similar to that in Figure 8.16, in which an intermediate liquid expanded phase is found between the gaseous and condensed phases.

Some Factors Affecting the Type of Film Formed

The type of monolayer film formed by a given material will depend on a number of factors, both intrinsic and external, including the natures of the tails and head groups, the degree and nature of solvation of the head group, the nature of the substrate or supporting liquid phase, and temperature.

The Nature of the Tail

The first factor to consider is the nature of the hydrocarbon tail. For simple, straight-chain materials such as the saturated carboxylic acids, $CH_3(CH_2)_n COOH$, solid or L_2 films will be favored. At a given temperature, long chain lengths (eg, $n \geq 14$ carbons at 20°C) tend to favor solid films, while shorter chains ($10 < n < 14$) tend to produce L_2. For $n < 8$, the acids begin to have significant water solubility, so that gaseous films may result. Similar trends will be found for other classes of n-alkyl materials (alcohols, amines, etc), although such factors as solvation of the head group may become important. Cetyl alcohol ($C_{18}OH$), for example, can be compressed to an L_2 film at 20°, but not to a condensed film, presumably due to the solvation of the ---OH group by hydrogen-bonded water. The carboxyl group, on the other hand, is much more strongly associated with "its own kind," and produces a condensed film.

If the tail is branched, the larger cross-sectional area of the molecule precludes the close packing and lateral cohesion required for the formation of condensed films and expanded films result. Similarly, molecules having two (or more) hydrocarbon groups such as esters of polyhydric alcohols, are limited in their lateral interaction and generally produce expanded films, although this depends somewhat on the length of the chains and the temperature.

Molecules containing two hydrophilic groups will exhibit characteristics curves reflecting the interaction of the second (and usually weaker) hydrophile with the substrate. Unsaturated and hydroxycarboxylic acids, some esters and amides, etc, tend to lie more or less flat on the water surface at low pressures as a result of interactions between the water and the second hydrophilic group (Figure 8.17). As the pressure is increased, energy is required to force those groups away

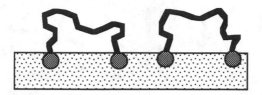

Figure 8.17. Schematic illustration of the mode of adsorption of molecules having two widely separated hydrophilic groups.

from the surface so that for a given area, π will be greater than that for a normal chain material such as stearic acid. The process of standing up the tails in such materials is gradual, so that a curve characteristic of an expanded film will result.

For unsaturated carboxylic acids, the nature of the monolayer film will depend of the configuration of the double bond. For a trans double bond, the hydrocarbon chain will be more or less straight, so that lateral interactions and good packing efficiency may lead to the formation of a solid or an L_2 film. The corresponding cis isomer has a forced bent structure, reducing its ability to pack tightly and leading to expanded (probably L_1) film formation (Figure 8.18).

If hydrogen atoms on, for example, a long, straight-chain acid are substituted by fluorine or other halides, the film type gradualy changes from solid to L_2, presumably due to packing difficulties imparted by the bulkier halogen atoms as well as the weaker cohesive interactions present in such materials. Irregular and complex molecules such as steroids, dyes, and polymers usually exhibit complex phase behavior in monolayer films and often defy clear classification. In such cases, one may say that a monolayer is "essentially" solid, fluid, etc, with the understanding that some allowance is being made for the nature of the beast.

The Effect of the Head Group

A second important factor in determining the type of film formed is the nature of the head group. A bulky head group that requires more surface area to accomodate it tends to keep the tails farther apart so that lateral cohesive interactions, and therefore efficient packing, are prevented. Such systems tend to form expanded rather than solid films. Charged head groups act similarly relative to uncharged species, in that the electrostatic repulsion between adjacent molecules will force them apart and reduce the interactions neccessary for the formation of a tightly packed solid film.

Finally, the degree of solvation of the head group will affect its effective size; groups that are in fact relatively small, such as the hydroxyl group, but which require a significant amount of solvation, form more expanded films than a similar molecule with a larger, but less solvated group (eg, ---COOH).

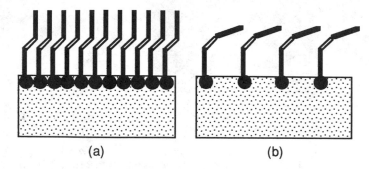

(a)   (b)

Figure 8.18. The effect of unsaturation in the tail on packing at the interface: (a) trans unsaturation allows for a reasonably uniform close arrangement of molecules and a more condensed layer; (b) cis isomers, due to the inherent curve of the molecule, prevent close packing and produce much more expanded films.

The Effect of Temperature

The discussion to this point has been couched more or less in terms of a fixed temperature (eg, 20°C). Just like three-dimensional phases, however, monolayer phases and phase changes are sensitive to temperature. In general, as the temperature is lowered, the behavior of a given film goes from expanded to condensed or solid. The temperature at which the transition occurs depends on the specific molecule; however, for a homologous series of materials, it is usually found that the addition of one ---CH_2--- group to the chain corresponds to an increase of 5°C in the temperature at which the transition from condensed to expanded film occurs.

For ionized or bulky head groups, the temperature at which the expanded to condensed fiilm transition occurs will be lower than that for the corresponding unionized material. Slightly ionized salts of polyvalent cations, on the other hand, will have higher transition temperatures.

Effects of Changes in the Nature of the Substrate

It is often found that the nature of the film formed by a given type of molecule will depend greatly on the pH and other characteristics of the aqueous substrate (ie, concentration and valence of solute ions). This is especially true for ionizable materials such as carboxylic acids and amines. In the first case, at low pH, the unionized acid will tend to form solid or condensed films, depending on other factors mentioned above. As the pH is raised, however, the degree of ionization of the head group will increase, leading to the expansion of the film for reasons already mentioned. For amines, the effect will be the opposite - lower pH leading to greater ionization and film expansion.

In the presence of polyvalent ions such as Ca^{2+} carboxylic acids

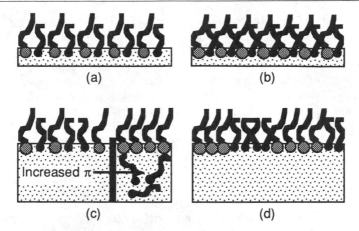

(a)   (b)

Increased π

(c)   (d)

Figure 8.19. Schematic illustration of various mixed-film possibilities: (a) an "ideal" film; (b) a synergistic film involving specific interactions between components; (c) immiscible components, with expulsion of the more soluble component at high surface pressures; (d) heterogeneous mixed films with "islands" of each component.

tend to form metal soaps, which have significantly lower solubility in water than the corresponding acid or alkali salt. Because in such cases each cation becomes associated with two molecules of the adsorbate, the result is a tighter packing of the molecules and a transition to a solid or condensed film at higher temperatures. For ionized species, a similar result may be obtained with a substrate containing a relatively high concentration of neutral electrolyte (eg, NaCl), which can reduce the electrostatic repulsion between head groups and thereby enhance packing efficiency. The result is the formation of a condensed or solid film at temperatures that produced expanded films on water alone.

Mixed-Film Formation

Brief reference has already been made to the formation of mixed monolayer films and some of their probable chacteristics. For completeness, the basic ideas will be summarized again. In studies of mixed films of materials that form monolayers alone, the type of film formed may vary from an "ideal solution" film, through films involving specific compound formation, to essentially "immiscible" systems, all depending on the specifics of molecular interactions between the components. The simplest possibilities are illustrated schematically in Figure 8.19.

If two film components are structurally similar (eg, two normal-chain carboxylic acids) the characteristics of the film produced by the mixture will lie between that formed by each separately (Figure 8.19a). For example, if the two each form expanded films alone, the mixed film will also be of the expanded type. If, on the other hand, one is a condensed film and the other expanded, the mixture will be more condensed than the expanded film or more expanded that the condensed film.

If dissimilar materials are mixed that can undergo specific interactions (eg, alcohols with carboxylic acids), interesting effects can be observed (Figure 8.19b). For example, if an alcohol is added to an acid layer of the same chain length, the layer becomes more condensed than if the same amount of the acid is added to the film. The explanation is that the alcohol can undergo a strong specific interaction (hydrogen bonding) with the acid, reducing the effective molecular area of each molecule and essentially shrinking the area of the film (or reducing the film pressure at constant area). If the two materials are sufficiently different and no specific interactions occur (Figure 8.19c), increasing the surface pressure may cause the complete expulsion of one component from the surface. In the absence of the effects mentioned above, two materials may form a heterogeneous film, with islands of one film "floating" in a sea of the other. One might think of it as a two-dimensional emulsion or dispersion (Figure 8.19d).

A mixed-film phenomenon of particluar interest in the biological and medical areas is that referred to as film *penetration* , in which a soluble surface-active material in the substrate enters into the surface film in sufficient quantity to alter its nature significantly, or to undergo some alternative physical or chemical process related to the surface (Figure 8.20). Such penetration studies using films of biological materials have been used to mimic phenomena in biological systems (cell walls and membranes, for example) that cannot readily be studied directly. Of particular interest are such topics as cell surface reactions, catalysis, transport across membranes, etc.

A typical penetration experiment might involve the formation of an insoluble monolayer at a surface pressure π, after which a soluble surface-active material is injected below the monolayer and changes in surface pressure (at constant area) due to penetration or inclusion of the new material in the monolayer are monitored. Alternatively, one can study changes in surface area at constant π, changes in surface potential, or a combination of any or all.

Surface Films of Polymers and Proteins

High molecular weight polymers, including proteins, also form surface monolayer films. However, due to the length of the polymer molecules and the complex interactions involved in intra- and interchain interactions, the properties of such films are less distinct and more difficult to determine with any degree of confidence (quantitatively, at least). In order for a macromlecular film to reach its "true" equilibrium pressure, every unit of every chain must orient itself to provide the optimum thermodynamic situation. Such a process may involve hours (or more likely, days), which means that the problems associated with

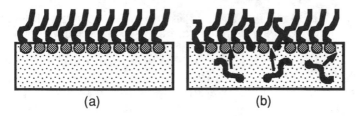

(a)   (b)

Figure 8.20. Schematic illustration of film penetration: (a) a monolayer film before subsurface injection of second component; (b) the second component penetrates the original monolayer producing a mixed monolayer film with new properties.

maintaining a stable, clean system are magnified enormously.

When carried out carefully, however, a great deal of interesting information about the macromolecule can be gained. In the case of proteins, for example, it has been shown that the secondary and tertiary structure of the spread protein is significantly different from that of the same material in its "native" solution configuration --- it is *denatured* . In addition, at low pressures, it is generally found that the amino acid side chains are lying flat on the surface of the water, while higher pressures will cause them to stand up and point out into the air. Given time, the protien monolayer may even form a substantially rigid and strong gel or "skin" which can be physically removed from the surface as a unit (or at least in large units).

Nonprotein polymer films generally behave similarly to the protein films in terms of their π--A curves, orientation, compressibility, etc, depending on the nature of the sidechains and the possibilities of nonspecific and specific interactions between neighboring units.

Monolayer Films at Liquid--Liquid Interfaces and on Liquids Other Than Water

With improvements in apparatus for doing monolayer film studies has come more interest in studying the nature of monolayer films at liquid--liquid interfaces, with one liquid usually being water or an aqueous solution. For monolayer films deposited at a water--nonpolar liquid interface, it is normally found that the area per molecule for alcohols and carboxylic acids is larger than the same material spread at the water--air interface. The accepted explanation is that the presence of the nonpolar liquid reduces the lateral cohesive interactions between adjacent tails, causing what may be termed a swelling of the monolayer film. A similar "swelling" effect can be noted for protein and other polymer films at the water--liquid interface. Since biologically important monolayer film models generally involve aqueous--oil type systems, the behavior of monolayer films at such interfaces can be of particular interest.

Studies of monolayer films on liquids other than water have been somewhat limited for various reasons, including experimental difficulties and possibly lack of obvious practical relevance. However, some work has been done using mercury, long-chain hydrocarbons, mineral oil, etc. Mercury, due to its high surface tension, adsorbs almost anything. That broadens the choice of monolayer material one might use and facilitates the formation of the film. However, it also creates the significant problem of assuring the presence of a clean surface before deposition of the film of interest. As a result little quantitatively reliable information is available.

Silicone polymers and fully fluorinated surface-active materials have been found to be the best candidates for spread monolayer film studies on nonpolar liquids. Because nonpolar liquids are more difficult to manipulate in terms of their solvent properties (eg, by changing pH, electrolyte content, etc) it is often necessary to talk in terms of adsorbed Gibbs monolayers, rather than true insoluble monolayers. However, sometimes we must take what we can get from nature and make the most of it.

Deposited Monolayer and Multilayer Films

It was discovered early in the studies of insoluble monolayer films that the adsorbed monolayer could, with carefull attention to detail, be transferred from the liquid surface to a solid substrate which was passed through the surface. The technique, commonly referred to as the *Langmuir-Blodgett technique* , is illustrated schematically in Figure 8.21. The solid surface on which the layer is deposited is usually glass or metal, although any material that has a relatively strong affinity for one part of the monolayer material will serve.

If an insoluble polyvalent salt of a carboxylic acid is deposited on a glass slide (which would normally be completely wetted by water), the resulting surface exhibits a water contact angle equal to or greater than that of water on pure paraffin. The explanation is that the monolayer is transferred intact to the solid surface, producing a densely packed layer with the hydrocarbon tails oriented out toward the air (Figure 8.21a). If the deposited monolayer is passed back down through a new monolayer, adsorbtion will occur with the reverse orientation ("back-to-back" or Y films, Figure 8.21b), producing a bilayer film but this time one that is completely wetted again. The process can be repeated many times, always producing films of alternate wetting and nonwetting character.

With modifications in the dipping technique, it is possible to produce multilayer films with each layer having the same orientation (head-to-tail or X films, Figure 8.21c). It is usually found that the Y films are more stable than their X film counterparts, a result that is intuitively satisfying considering the interfilm interactions involved.

The adsorption of deposited monolayer films can be very tenacious, especially when the solid substrate can react with the deposited material. Carboxylic acid monolayers deposited on metal or metal oxide surfaces, for example, almost certainly form metal soaps that

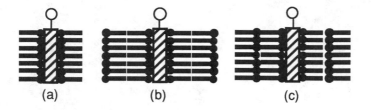

Figure 8.21. Illustration of orientations in the deposition of Langmuir-Blodgett films: (a) first layer deposition; (b) second layer with back-to-back or Y-film orientation; (c) second layer with head-to-tail or X-film orientation.

are extremely difficult to remove by any means other than direct chemical etching. For example, such strongly adsorbed films are very useful in friction and wear studies of boundary lubrication (see Chapter 18).

Monolayer films on solid substrates need not necessarily be deposited from a monolayer system. They can also be deposited directly from solution or a melt of the material to be deposited, or by a vapor deposition process. Such films are generally referred to as *self-assembled monolayers*. There are several possible advantages to such monolayers, mostly associated with simplification of the process (ie, no need to be concerned with depositing the monolayer on a liquid surface, maintaining a constant surface pressure, problems of surface contamination, etc.)

Deposited monolayer films have received a great deal of attention in recent years because of the possibility of using surface-active materials that can be reacted after deposition to produce very thin, strong films. If a multilayer (50 molecules thick, for example) of photosensitive material is deposited on the surface of a semiconductor, the resulting "microphotoresist" coating can be used to produce a circuit with a definition and resolution several orders of magnitude better than that possible using a normal photoresist coating. The result is the possibility of significantly higher packing of circuit information in a given area of semiconductor --- ie, greater miniturization.

FINAL COMMENTS

The above discussion of liquid--fluid interfaces, adsorbtion, monolayer films, etc, was extremely limited and a great deal of interesting and useful information was excluded. Although the theories and techniques discussed may be considered to be old in comparison to much of modern science, one finds repeatedly that a great many modern technlogical problems can be understood and solved on the basis of a little understanding of the classical ideas of such interfaces. In the study of *belles lettres* we are constantly reminded to remember the "classics." The same holds for surface science.

CHAPTER 9

ADSORPTION AT SOLID--LIQUID INTERFACES

Adsorption phenomena at solid--vapor, liquid--vapor, and liquid--liquid interfaces hve been discussed in the general context of those interfaces. This chapter will close out the general discussion of adsorption with a discussion of adsorption at the solid--liquid interface. While being treated last, it is far from the least in terms of functional importance in nature or technology. Interactions between solid surfaces and solutions are of fundamental importance in many biological systems (joint lubrication and movement, implant rejection, etc), in mechanics (lubrication and adhesion), in agriculture (soil wetting and conditioning and pesticide application), in communications (ink and pigment dispersions), in electronics (microcircuit fabrication), in energy production (secondary and tertiary oil recovery techniques), in foods (starch--water interactions in bakery doughs), in paint production and application (latex polymer and pigment dispersion stabilization), etc, etc, etc.

THE ADSORPTION MODEL

The adsorption of molecules at an solid--liquid interface creates a transition region on the order of molecular dimensions in which the composition of the system changes from that of the bulk solid to that of the bulk liquid. A "typical" interfacial region is shown in Figure 9.1. It will be noted that the concentration of one component of the liquid phase (the black circles) is apparently higher near the surface than in the bulk of the liquid phase. In a pure liquid, that component will be nothing more than more liquid molecules. If specific interactions occur between the liquid and the solid the liquid molecules near the interface may undergo a specific orientation (solvation of the surface, if you like) that may change the density, dielectric constant, or other physical (or even chemical) characteristics of the liquid near the surface. Except for the most delicate experimental work or in the context of catalytic processes, such effects are of little concern. It is the adsorption of solute molecules at the solid--solution interface that is generally of most concern.

For a solution, the higher concentration of the solute (black circles) near the interface will reflect the specific adsorption of the solute

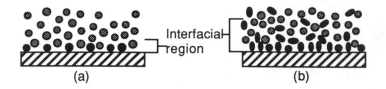

(a)   (b)

Figure 9.1. Illustration of a solid--liquid interface: (a) in the liquid, the molecular distribution is approximately uniform, although some molecules may be adsorbed to produce a small surface excess (black circles); (b) for a solution of surface-active solute (black ovals) extensive adsorption will occur, producing a significant interfacial region of excess solute concentration.

molecules. A typical concentration profile of the adsorbate is shown in Figure 9.1b. From both a theoretical and practical standpoint, it is of interest to know the characteristics of such adsorption profiles for a given system in order to understand the mechanism of the adsorption process, as well as its consequences.

We have already seen (and will continue to see) the Gibbs adsorption equation

$$\Gamma_2^{(1)} = -1/RT \, (\delta\sigma / \delta \ln c) \qquad (9.1)$$

in terms of adsorption at solid--vapor, liquid--vapor, and liquid--liquid interfaces. The relationship is every bit as applicable and useful for solid--liquid systems as for the others. While for liquid--fluid systems, the equation is normally employed to determine the amount of adsorbed material at an interface as a function of interfacial tension, σ, in the case of solid surfaces, it is difficult or impossible to determine σ directly. It is, however, relatively easy to determine the amount of adsorbed material directly and use that information to calculate a value of the interfacial tension. Such exercises are of great theoretical importance in understanding why and how molecules are adsorbed at an interface, and of even greater practical importance for understanding how such adsorption affects the characteristics of the interface and its interaction with its surroundings, especially in the context of colloidal stability.

Some degree of adsorption will occur at any solid--liquid interface, although it may be so small as to be effectively negligable. In fact, the adsorption may even be negative; that is, the concentration of the "adsorbed" component may be lower near the interface than in the bulk. Such situations, however, are rather rare. Important exceptions being in the context of electrical doulble-layer theory and some polymer solutions. Of more interest are systems in which one or more components of the liquid phase are strongly (and positively) adsorbed at the interface, bringing about a significant lowering of the interfacial tension and, in some cases, a significant change in the nature of the interface alltogether. The effects of such strong adsorption are of great practical importance and allow us to manipulate solid–liquid interfaces to our own best advantage.

When the adsorption of a surfactant onto a solid surface is considered, there are several quantitative and qualitative points that are of interest. They include (1) the amount of surfactant adsorbed per unit mass or area of solid, (2) the solution surfactant concentration required to produce a given surface coverage or degree of adsorption, (3) the surfactant concentration at which surface saturation occurs, (4) the orientation of the adsorbed molecules relative to the surface and solution, and (5) the effect of adsorption on the properties of the solid relative to the rest of the system. In all of the above, it is assumed that such factors as temperature and pressure are held constant.

QUANTIFICATION OF SURFACTANT ADSORPTION

The classical method for determining the above quantities in a given system is by way of the adsorption isotherm, introduced in Chapter 7. The basic quantitative equation describing the adsorption of one component of a binary solution onto a solid substrate can be written as

$$n_0 \Delta x_1 / m = n_1^s x_2 - n_2^s x_1 \qquad (9.2)$$

where n_0 is the total number of moles of solution before adsorption, $x_{1,0}$ the mole fraction of the adsorbing component 2 in solution before adsorption, x_1 and x_2 the mole fractions of components 1 and 2 at adsorption equilibrium ($\Delta x_1 = x_{1,0} - x_1$), m is the mass of solid adsorbent present, in grams, and n_1, n_2 the number of moles of components 1 and 2 adsorbed per gram of solid at equilibrium.

In the case of a dilute surfactant solution where the surfactant (2) is much more strongly adsorbed than the solvent (1), the equation simplifies to

$$n_2^s = \Delta n_2 / m = \Delta C_2 V / m \qquad (9.3)$$

where $C_{2,0}$ is the molar concentration of 2 before adsorption, C_2 the molar concentration of 2 at equilibrium, $\Delta C_2 = C_{2,0} - C_2$, and V is the volume of the liquid phase in liters.

For surfactant systems, the concentration of adsorbed material can be calculated from the known amount of material present before adsorption and that present in solution after adsorption equilibrium has been reached. A wide variety of analytical methods for determining the solution concentration of surfactants are available and almost all have been used at one time or other, the use of the Gibbs equation and measurements of σ being the simplest. The utility of a specific method will depend ultimately on the exact nature of the system involved and the resources available to the investigator.

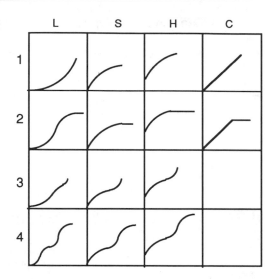

Figure 9.2. General shapes of typical solid--liquid adsorption isotherms.

Adsorption Isotherms In Solid--Liquid Systems

The experimental evaluation of the adsorption from solution of surface-active agents at the solid--liquid interface usually involves the measurement of changes in the concentration of the active solute in the solution after adsorption has occurred. The usual method for evaluating the adsorption mechanism is through the adsorption isotherm. The important factors to be considered are: (1) the nature of the interaction(s) between the adsorbate and the adsorbent, (2) the rate of adsorption, (3) the shape of the adsorption isotherm and the significance of plateaus, points of inflection, etc, (4) the extent of adsorption (ie, monolayer or multilayer formation), (5) the interaction of solvent with the solid surface (solvation effects), (6) the orientation of the adsorbed molecules at the interface, and (7) the effect of environmental factors such as temperature, solvent composition, and pH.

Just as in the case of solid--vapor adsorption, interactions between the adsorbent and adsorbate may fall into two categories --- relatively weak and reversible physical adsorption, and stronger and sometimes irreversible chemisorption. Because of the varied possibilities of adsorption mechanisms, a variety of isotherm shapes have been determined experimentally. Although most are found to fall into two main categories, a satisfying theory of adsorption must encompass the complete range of forms. A general classification of isotherms with various shapes that have been justified theoretically is shown in Figure 9.2. The classification system[1] gives four fundamental isotherm shapes based on the form of the isotherm at low concentrations; the subgroups are then determined by their behavior at higher concentrations.

The L-class (Langmuir) isotherm is the most common and is

identified by having its initial region (L1) concave to the concentration axis. As the concentration of adsorbate increases, the isotherm may reach a plateau (L2), followed by a section convex to the concentration axis (L3). If the L3 region attains a second plateau, the region is designated L4.

In the S class of isotherms, the initial slope is convex to the concentration axis (S1) and is often broken by a point of inflection leading to the characteristic S shape (S2). Further concentration increases may then parallel those of the L class. The H or high-affinity class of isotherm occurs as a result of very strong adsorption at low adsorbate concentrations. The result is that the isotherm appears to have a positive intercept on the ordinate. Higher concentrations lead to similar changes to those found in the L and S classes.

The final type of isotherm is the C class. Such systems exhibit an initial linear portion of the isotherm, indicating a constant partitioning of the adsorbate between the solution and the solid. Such isotherms are not found for homogeneous solid surfaces but occur in systems in which the solid is microporous. The above classification system has proved very useful in providing information about the mechanism of adsorption.

ADSORPTION AND THE MODIFICATION OF THE SOLID--LIQUID INTERFACE

The adsorption of surface-active materials onto a solid surface from solution is an important process in many situations, including those in which we may want to remove unwanted materials from a system (detergency), change the wetting characteristics of a surface (coating and waterproofing) or stabilize a finely divided solid system in a liquid where stability may otherwise be absent (dispersion stabilization). In these and related applications, the ability of surface-active materials to adsorb at the solid--liquid interface with a specific orientation and produce a desired effect is controlled by the chemical natures of the components of the system: the solid, the surfactant, and the solvent. The following discussions will summarize some of the factors related to chemical structures that significantly affect the mechanisms of surfactant adsorption and the orientation with which adsorption occurs.

The generally heterogeneous nature of solid surfaces (in both a physical and a chemical sense) has already been discussed in various contexts. As in the case of adsorption at solid--vapor interfaces, the exact nature of the adsorption process will depend to a great extent on the nature of the surface and its potential for interaction with the contacting solvent and dissolved species. Those interactions are classically studied and interpreted in terms of adsorption isotherms, related to, but in some ways distinct from, those already discussed.

Adsorption and the Natureof the Adsorbent Surface

As indicated above, the nature of the solid surface involved in the

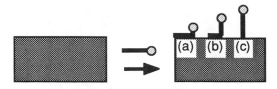

Figure 9.3. Modes of adsorption of surfactants on nonpolar surfaces: (a) flat trains; (b) L's; (c) perpendicular.

adsorption process is a major factor in determining the manner and extent of surfactant adsorption. When one considers the possible nature of an adsorbent surface, three principal groups readily come to mind: (1) surfaces that are essentially nonpolar and hydrophobic, such as polyethylene; (2) those that are polar but do not possess discrete surface charges, such as polyesters and natural fibers such as cotton; and (3) those that possess strongly charged surface sites. Each of these surface types will be discussed, beginning with what is probably the simplest, type 1.

Nonpolar, Hydrophobic Surfaces

Adsorption of surfactants onto nonpolar surfaces is primarily by dispersion force interactions. From aqueous solution, it is obvious that the orientation of the adsorbed molecules will be such that the hydrophobic groups are associated with the solid surface with the hydrophilic group directed toward the aqueous phase. In the early stages of adsorption it is likely that the hydrophobe will be lying on the surface much like trains or L's (Figure 9.3a,b). As the degree of adsorption increases, however, the molecules will gradually become oriented more perpendicular to the surface until, at saturation, an approximately close-packed assembly will result (Figure 9.3c). It is generally found that surface saturation is attained at or near the CMC for the surfactant. In many cases the isotherm is continuous, while in others an inflection point may be found. The existence of the inflection point is usually attributed to a relatively sudden change in surfactant orientation --- from train or L shaped to a more perpendicular arrangement. Because the orientation of the adsorbed molecules is with the hydrophilic group directed outward from the solid surface, there will normally be no inclination for the formation of a second adsorbed layer. That is, the process will usually be limited to monolayer formation.

The adsorption of surface-active agents onto nonpolar surfaces from nonaqueous solvents has been much less intensively studied than aqueous systems. Such studies have generally been limited to various carbon black dispersions in hydrocarbon solvents. The orientation of the adsorbed molecules in that case appears to remain more or less parallel to the surface, although the exact details will depend on the history of the carbon surface (eg, the presence of oxide layers, or charges).

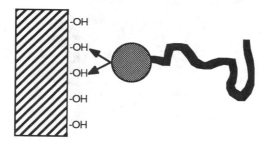

Figure 9.4. Schematic illustration of the orientation of adsorbed surfactant in the presence of specific interactions at polar surfaces.

Polar, Uncharged Surfaces

Polar, uncharged surfaces include many of the synthetic polymeric materials such as polyesters, polyamides, and polyacrylates, as well as many natural materials such as cotton and silk. As a result of their surface makeup, the mechanism and extent of surfactant adsorption onto such materials has rather great potential technological importance. The mechanism of adsorption onto these surfaces will be much more complex than that of the nonpolar case discussed above, since such factors as orientation will be determined by a balance of several forces.

The potential forces operating at a polar surface include the ever-present dispersion forces, dipolar interactions, and hydrogen bonding and other acid--base interactions. The relative balance between the dispersion forces and the uniquely polar interactions is of supreme importance in determining the mode of surfactant adsorption. If the dispersion forces predominate, adsorption will occur in a manner essentially equivalent to that for the nonpolar surfaces (Figure 9.3). If, on the other hand, polar interactions dominate, adsorption may occur in a reverse mode; that is, the surfactant molecules will orient in such a way that the hydrophilic head group will be at the solid surface and the hydrophobic group oriented toward the aqueous phase (Figure 9.4).

Obviously, the net result of the two adsorption modes will be drastically different. In aqueous systems, the final orientation will also be affected by the relative strength of solven--adsorbent and solvent--adsorbate interactions. In marginal cases, the mode of adsorption by be reversed by small, subtle changes in the nature of the solvent (eg, pH, electrolyte content, presence of a cosolvent.)

Surfaces Having Discrete Electrical Charges

The final class of adsorbent surfaces is the most complex of the three for several reasons. From the standpoint of the nature of the surface, these materials are capable of undergoing adsorption by all of the previously mentioned mechanisms. Possibly more important, however, is the fact that adsorption involving charge--charge interactions is significantly more sensitive to external conditions such as pH, neutral

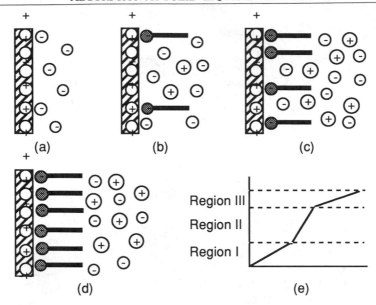

Figure 9.5. Schematic illustration of the stages of adsorption of a surfactant onto a surface of opposite charge: (a) native surface; (b) ion exchange; (c) ion pairing; (d) complete charge neutralization; (e) representative isotherms.

electrolyte, and the presence of non-surface-active cosolutes than are the other mechanisms.

Materials possessing charged surfaces include almost all of the inorganic oxides and salts of technological importance (silica, alumina, titania, etc), the silver halides, latex polymers containing ionic comonomers, many natural surfaces such as proteins, and cellulose. It is very important, therefore, to be able to understand the interactions of such surfaces with surfactants in order to optimize their effects in such applications as paint and pigment dispersions, paper making, textiles, pharmaceuticals, biomedical implants, etc.

Because of the large number of possible interactions in systems containing charged surfaces and ionic surfactants, it is very important to control closely all of the variables in the system. As adsorption proceeds, the dominant mechanism may go from ion exchange through ion bonding to dispersion or hydrophobic interactions. As a result, adsorption isotherms may be much more complex than those for the simpler systems.

Studies of the adsorption of surfactants onto surfaces of opposite charge have resulted in the identification of three principle regions of adsorption in which the rates vary due to changes in the mechanism of adsorption. It is generally assumed that such adsorption patterns involve three mechanisms as illustrated in Figure 9.5. In the early stages (region 1), adsorption occurs primarily as a result of ion exchange in which adsorbed counterions are displaced by surfactant molecules. During that stage the electrical characteristics (ie, the surface charge or surface potential) of the surface may remain essentially unchanged. As adsorption continues, ion pairing may become important (region 2), resulting in a net decrease in surface charge. Such electrical properties as the surface and zeta potentials will tend toward zero during this process. It is often found that in region 2 the rate of adsorption will increase significantly. The observed increase may be due to the cooperative effects of electrostatic attraction and lateral interaction between hydrophobic groups of adsorbed surfactants as packing density increases.

As the adsorption process approaches the level of complete neutralization of the native surface charge by adsorbed surfactant, the system will go through its zero point of charge (ZPC), where all of the surface charges have been paired with adsorbed surfactant molecules (region 3). In that region, lateral interactions between adjacent surfactant tails may become significant, often leading to the formation of aggregate structures or hemimicelles. If the interaction between surfactant tails is weak (due to short or bulky structures) or if electrostatic repulsion between head groups cannot be overcome (due to the presence of more than one charge of the same sign or low ionic strength), the enhanced adsorption rate of region 2 may not occur and hemimicelle formation may be absent. An additional result of the onset of dispersion-force-dominated adsorption may be the occurrence of charge reversal as adsorption proceeds, which will be covered below.

ENVIRONMENTAL EFFECTS ON ADSORPTION

Surfaces possessing charged groups in aqueous solvents are especially sensitive to environmental conditions such as electrolyte content and the pH of the aqueous phase. In the presence of high electrolyte concentrations, the surface of the solid may possess such a high degree of bound counterions that ion exchange is the only mechanism of adsorption available other than dispersion or hydrophobic interactions. Not only will the electrical double layer of the surface be collapsed to a few angstroms thickness, but attraction between unlike charge groups on the surface and the surfactant, and repulsion between the like charges of the surfactant molecules, will be suppressed. The result is often be almost linear adsorption isotherm, lacking any of the mechanisms described above.

An increase in the electrolyte content will generally cause a decrease in adsorption of surfactants onto oppositely charged surfaces and an increase in adsorption of like charged molecules. The presence in the solution of polyvalent cations such as Ca^{2+} or Al^{3+} generally increase the adsorption of anionic surfactants. Such ions are characteristically tightly bound to a negatively charged surface, effectively neutralizing charge repulsions. They also can serve as an efficient bridging ion by association with both the negative surface and the anionic surfactant head group.

An increase in the temperature usually results in a decrease in the adsorption of ionic surfactants, although the change may be small when compared to those due to pH and electrolyte changes. Nonionic surfactants solubilized by hydrogen bonding, which usually have an inverse temperature--solubility relationship in aqueous solution, generally exhibit the opposite effect. That is, adsorption will increase as the temperature increases, often having a maximum near the Krafft point of the particular surfactant.

Adsorption onto solid surfaces having weak acid or basic groups such as proteins, cellulosics, and many polyacrylates can be especially sensitive to variations in solution pH. As the pH of the aqueous phase is reduced, the net charge on the solid surface will tend to become more positive. That is not to say that actual positive charges will necessarily develop; instead, ionization of the weak acid groups (eg, carboxylic acids) will be suppressed. The net result will be that the surface may become more favorable for the adsorption of surfactants of like charge (eg, anionic surfactants onto carboxyl surfaces) and less favorable for adsorption of oppositely charged surfactants. For surfaces containing weak basic groups such as amines, the opposite would be true. That is, lowering the pH will lead to ionization of surface basic groups, increased adsorption of oppositely charged (negative) molecules, and decreased interaction with materials of the same charge.

THE EFFECTS OF ADSORPTION ON THE NATURE OF THE SOLID SURFACE

When a surfactant is adsorbed onto a solid surface, the resultant effect on the character of that surface will depend largely upon the dominant mechanism of adsorption. For a highly charged surface, if adsorption is a result of ion exchange, the electrical nature of the surface will not be altered significantly. If, on the other hand, ion pairing becomes important, the potential at the Stern layer will decrease until it is completely neutralized. In a dispersed system stabilized by electrostatic repulsion, such a reduction in surface potential will result in a loss of stability and eventual coagulation or flocculation of the particles (see below).

In addition to the electrostatic consequences of specific charge--charge interactions, surfactant adsorption by ion exchange or ion pairing results in the orientation of the molecules with their hydrophobic groups toward the aqueous phase; therefore the surface becomes hydrophobic and less easily wetted by that phase. Once the solid surface has become hydrophobic, it is possible for adsorption to continue by dispersion force interactions. When that occurs, the charge on the surface will be reversed, acquiring a charge opposite in sign to that of the original surface, because the hydrophilic group will now be oriented toward the aqueous phase (Figure 9.6). In a system normally wetted by water, the adsorption process reduces the wettability of the solid surface making its interaction with other less polar phases (eg, air) more favorable.

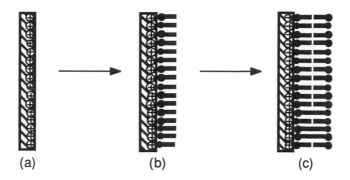

(a)   (b)   (c)

Figure 9.6. Illustration of surface charge reversal by surfactant adsorption: (a) native surface (counterions omitted); (b) complete charge neutralization; (c) in large excess of surfactant, charge reversal by bilayer adssorption.

Industrially, the production of a hydrophobic surface by the adsorption of surfactant lies at the heart of the froth flotation process for mineral ore separation. Because different minerals possess varied surface charge characteristics, leading to differences in adsorption effectiveness and efficiency, it becomes possible to obtain good separation by the proper choice of surfactant type and concentration.

Although surfactant adsorption and its effect on solid surface properties is often discussed in terms of colloidal systems, the same results can be of technological importance for macrosurfaces, especially in the control of the wetting or nonwetting properties of materials (in waterproofing), detergency, lubrication (with cutting oils and other lubricants), the control of fluid flow through porous media (crude oil production), and corrosion control. Almost any process or product that involves the interaction of a solid and a liquid phase will be affected by the process of surfactant adsorption; thus the area represents a major segment of the technological application of surfactants. More detail on the effects of surfactant adsorption of wetting phenomena will be given in Chapter 17.

CHAPTER 10

COLLOIDS AND COLLOIDAL STABILITY

In Chapter 1 the importance of the various classes of colloidal systems to modern science and technology was indicated in a general way. Because of the wide variety of colloidal systems one encounters, each having certain unique features that distinguish it from the others, it is convenient to discuss each major classification separately. For that reason, specific chapters have been devoted to specific systems such as solids, aerosols, emulsions, foams, lyophilic colloids, and association colloids. There is a great deal of overlap in many aspects of the formation, stabilization, and destruction of those systems, and an effort will be made not to repeat more than is necessary. However, for purposes of clarity, some repetition is unavoidable.

The following discussion, while being general in nature, is intended to apply primarily to "classical" colloidal systems --- dispersions of small solid particles in a liquid medium or *sols*. A bare minimum of information that may impart a general conceptual understanding of the phenomena is presented below. It is intended to serve as a useful lead into more detailed information as needed.

THE IMPORTANCE OF COLLOIDS AND COLLOIDAL PHENOMENA

The importance of colloids and colloidal phenomena to our modern technological society cannot be overstated, even though it is quite commonly overlooked. An abbreviated list of some important products and processes involving colloids is given in Table 10.1. Considering the limited nature of that compilation, it should be obvious that the principles involved are basic to an immense variety of practical areas. The listing also omits "natural" colloidal systems, which are of vital importance in biology, medicine, meteorology, and to some extent even cosmology. Unfortunately, experience indicates that few individuals working in those areas have more than a very cursory understanding of the subject and therefore may overlook important concepts related to their specific area of interest . While ignorance may be bliss in some

Table 10.1. Illustrative examples of the practical application of colloids and colloidal phenomena.

Application	Principles involved
Pharmaceuticals, cosmetics, inks, paints, foods, lubrication, food products, dyestuffs, foams, colloids chemicals	Formation and stabilization of agricultural for end-use products
Photographic products, ceramics, paper coatings, magnetic media, catalysts, chromatographic adsorbents, membrane and latex film production electrophotographic toners	Formation of colloids for use in subsequent manufacturing processes
Wetting of powders, enhanced petroleum recovery detergency, mineral ore flotation, purification by adsorption, electrolytic coatings, industrial crystallization, chemical waste control, electro-photography, lithography	Direct application of colloidal phenomena to processing
Pumping of slurries, coating technology, caking, powder flow, filtration	Handling properties of colloids, rheology, sintering
Water purification, sewage disposal, dispersal of aerosols, pollution control, fining of wines and beers, radioactive waste disposal, breaking of unwanted emulsions and foams	Destruction of unwanted colloidal systems

situations, in this case, a little bit of knowledge properly applied may pay practical dividends many times the value of the effort expended to gain that knowledge. Previous chapters have covered the main concepts behind what is generally considered "surface chemistry." We now turn our attention to the basics concepts of "colloid chemistry" --- what colloids are and how they are formed, stabilized, used, and destroyed.

COLLOIDS: A WORKING DEFINITION

Any attempt to define the term "colloid" rigourously will usually be found to be unsatisfactory in that it will restrict the range of systems excessively, especially in that systems which operate according to the "rules" of colloidal behavior may be excluded by some rather arbitrary factor, such as size. A working definition may reasonably be arrived at if one understands the "nature of the beast" in the context of the "normal" bulk states of matter --- solids, liquids, and gases. We normally have few conceptual problems with respect to those three states since they possess certain characteristics that can readily be associated with each, including, for example, rigidity, fixed volumes under specified conditions, and characteristic phase transformations such as melting, boiling, and sublimation. We also understand a great deal about atoms, molecules, and solutions, and the forces controlling interactions at that level. Yet understanding of the "twilight zone" between bulk phases and the molecular level continues to be a mystery to most.

In homogeneous solutions, there exists a mixture of distinct species that are intermixed or dispersed as individual molecules (where the molecular size of the two materials are comparable in size). However, between pure bulk materials and molecularly dispersed solutions lies a wide variety of important systems in which one phase is dispersed in a second, but in units which are much larger than the molecular unit (eg, a classical sol) or in which the molecular size of the dispersed material is significantly greater than that of the solvent or continuous phase (a macromolecular or polymer solution). Such systems are generally defined as colloids, although there are generally accepted limitation on the unit size of the dispersed phase, beyond which other terminology may be used. The importance of many systems falling into the "colloidal" category has been recognized and recorded for thousands of years, although a reasonable understanding of the forces and concepts involved has lagged behind other areas of chemistry, physics, biology, medicine, etc.

To define a colloid more or less adequately one must consider two aspects of the system: structure --- how the components of the system are put together or mixed --- and size --- what the dimensions are of the dispersed units in the system.

Colloid Structure

In general, a colloid is a system consisting of one substance (the *dispersed phase* --- a solid, liquid, or gas finely divided and distributed evenly (relatively speaking) throughout a second substance (the *dispersion medium* or *continuous phase* --- a solid, liquid, or gas). Commonly encountered examples of colloids are milk (liquid fat dispersed as fine drops in an aqueous phase), smoke)solid particles dispersed in air), fog (small liquid droplets dispersed in air), paints (small solid particles dispersed in liquid), jelly (large protein molecules dispersed in water), and bone (small particles of calcium phosphate dispersed in a solid matrix of collagen).

The different types of dispersed systems are classified depending on the nature of the dispersed phase and the continuous phase. A solid or liquid dispersed in a gas is termed an *aerosol* , with *smoke* being a common example of the solid-in-air system. A liquid-in-air system is a *mist* or *fog* . Milk is an *emulsion* , in which a liquid is dispersed in another liquid. Paints and inks are *sols* or *colloidal dispersions* and consist of solid particles dispersed in a liquid.

A second class of colloids of particular importance in the general context of surface chemistry, but less familiar to most people, is that of the so-called *association colloids* . Association colloids consist of aggregates or units of a number (sometimes hundreds or thousands) of molecules that associate in a dynamic and thermodynamically driven process leading to a system that may be simultaneously a molecular solution and a true colloidal system. As we shall see, the formation of association colloids involving specific substances will often depend on various factors such as concentration, temperature, solvent composition, and specific chemical structure. Many biological systems, including cell membrane formation, certain digestive processes, and blood transport phenomena, involve various forms of associated colloidal structures. This class of colloidal materials will be discussed in more detail in Chapter15.

In addition to the colloids composed of insoluble or immiscible components, there are the *lyophilic* colloids which are in reality solutions, but in which the solute molecules (ie polymers) are much larger than those of the solvent. Lyophilic colloids are somewhat unique in that they have been able to "cross over" into another major area of science --- polymer science --- and thereby gain a great deal more general attention than more classical colloidal systems.

A fourth class of colloids often encountered is that of the *network colloids* . Such systems are difficult to define exactly because they consist of two interpenetrating networks, which make it hard to specify exactly which is the dispersed phase and which is the continuous phase. Classic examples of network colloids would be porous glass (air--glass), opal glass (solid--solid), and many gels and jellies. Practical examples of many of the colloids mentioned above are given in Table 10.2.

The above examples of colloids may be considered "simple" colloids because they involve one fairly distinct type of dispersed and continuous phase. In practice, many colloidal systems are much more complex in that they contain a variety of colloidal types, eg, a sol, an emulsion (or multiple emulsions), an association colloid, macromolecular species, plus the continuous phase. Such systems are often referred to as *complex* or *multiple colloids* . As we shall see, even the simplest colloids can be quite complex in their characteristics. It should be easy to understand, then, why the difficulty of understanding a multiple colloid increases dramatically with the number of components present.

Colloid Size

Working with the above general concept of what is meant by a "colloid" in structural terms, it is still necessary to specify what is meant in terms of the size of the dispersed units. The general definition refers to the dispersed phase as being "finely divided," but what exactly do we mean by that phrase? Experience over many years has shown that special "colloidal" properties are usually exhibited by systems in which the size of the dispersed phase falls in the range of 1--1000 nm, although those limits are far from rigid. It would be tempting, then, to limit the term to systems which fall into that size range. In practice, however,

Table 10.2. Examples of commonly encountered colloidal systems.

System	Type	Dispersed Phase	Continuous Phase
Fog, mist	Liquid aerosol	Liquid	Gas
Smoke	Solid aerosol	Solid	Gas
Shave cream	Foam	Gas	Liquid
Styrofoam	Solid foam	Gas	Solid
Milk	Emulsion	Liquid (fat)	Liquid (water)
Butter	Emulsion	Liquid (water)	Liquid (fat)
Toothpaste (nongel type)	Dispersion	Solid	Liquid
Paint	Dispersion	Solid	Liquid
Opal	Dispersion	Solid	Solid
Jello	Gel	Macromolecules	Liquid
Liquid soaps and detergents	Micellar solution	Micelles of detergent molecules	Liquid

many systems with dimensions beyond that range, in particular most emulsions, paints, and aerosols, must also be included under the colloid umbrella, since their characteristics allow no other realistic option. Other colloidal systems, such as fibers, clays, and thin films, many "qualify" as colloids because one or two dimensions fall into the designated range, and their properties adhere to the "rules" of colloidal behavior. That concept is illustrated schematically in Figure 10.1. Ultimately, the most useful definition to use is: if it *looks like* a colloid and *acts like* a colloid, it *is* a colloid, regardless of other more restrictive limitations.

Some Points of Nomenclature

In addition to the general definitions given above, there are a number of additional terms related to colloids --- their nature, characteristics, and stability --- which should be clearly defined (where possible) before beginning the discussion of the topic. Some of the more important include:

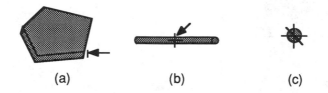

(a)   (b)   (c)

Figure 10.1. Illustration of the role of dimensions in the definition of a colloid: (a) a flat crystal or clay particle with one dimension in the "colloidal" range; (b) a needle crystal with two such dimensions; (c) a sol or polymer latex with three.

Coagulum : An aggregate of colloidal particles having a relatively tight, dense structure formed as a result of the inability of the colloidal system to maintain its dispersed state. Such aggregates are normally formed irreversibly; that is, they cannot be returned to the colloidal state without significant input of work.

Coagulation : The process of forming coagulum.

Creaming : The separation of coagulum or flocs from the continuous phase, where the aggregate is less dense than that phase. (Slightly different definitions of creaming and sedimentation are presented in Chapter 11 for application in the context of emulsions.)

Floc : An aggregate of individual colloidal particles related to a coagulum but generally having a rather loose, open structure. Flocs may sometimes be formed reversibly and returned to the dispersed state with minimal energy input.

Flocculation : The process of forming flocs.

Monodisperse : Having all particles in the colloidal system of approximately the same size (ie, a narrow size distribution).

Polydisperse : Having a rather broad range of particle sizes.

Sedimentation : As in creaming except that the aggregates are more dense than the liquid and settle to the bottom.

MECHANISMS OF COLLOID FORMATION

Having defined "colloids" (approximately, at least) another question of interest is how one may prepare systems of particles or drops of the proper size and stability. Since colloids represent a range of unit sizes intermediate between molecules and macroscopic bulk phases, it seems reasonable to expect that the problem can be attacked from two directions --- by breaking down large pieces to the size required, known generally as *comminution* , or by starting with a molecular dispersion and build up the size by aggregation, that is, by *condensation* . Special situations such as the formation of association colloids and lyophilic colloids will be addressed in other chapters.

Comminution or Dispersion Methods

It was shown in Chapter 2 that when a bulk material is divided into two pieces and separated to infinite distance there is a characteristic (theoretical) energy change per unit area that is termed the specific surface free energy. As the subdivision of the sample is continued, the total free energy change will be the product of the surface energy (or surface tension) of the material, σ, and the total new surface area produced by the process. If carried to its ultimate, that is, reduction to the molecular level, the theoretical work required would be equal to the energy of evaporation, sublimation, or dissolution, depending on the situation. In practice, of course, a great deal more than the theoretical energy will be required due to the various irreversible, heat-generating processes that naturally accompany any grinding or pulverizing process.

It is reasonable to assume that the work required to reduce a given material to colloidal size to vary directly with the surface energy of the material --- higher surface energy materials requiring more work input. In addition, as we have seen, the natural tendency of subdivided particles is to reduce the total surface area by some aggregation process, especially if produced in a vacuum or inert atmosphere. Such processes of "clumping" or sintering are very important and represent serious technological problems. As will be seen below, the attractive interaction between particles can be reduced by the introduction of an intervening medium, usually a liquid. The liquid medium will have two positive effects on the process: it will reduce the surface energy of the system by adsorption on the surface and it will usually reduce the van der Waals attraction between the particles by "averaging" its Hamaker constant with that of the particulate material (see below). For those reasons (plus a reduction in environmental problems such as dust formation), most industrial comminution processes are carried out in the presence of a liquid.

While the presence of the liquid normally facilitates the comminution process, the resulting dispersion still may not exhibit the stability necessary to make the process viable. That is, the dispersed particles may begin to flocculate or coagulate rapidly once the comminution process is halted. The solution for the second problem is normally the addition of new components (surfactant, polymer, small particles, etc) that adsorb at the solid--liquid interface and provide an electrostatic or steric barrier which retards or prevents "sticky" collisions between particles, thereby making the dispersion more stable. In the case of particulate colloids such additives are termed *dispersing* aids or agents .The mechanisms of action of such materials will be discussed in more detail in the context of specific colloidal phenomena.

The comminution of liquid phases is a special case of the above and is generally referred to as *emulsification* . Because of the nature of liquid systems, emusification has a number of additional variations not generally available for the formation of dispersions. These include what are termed *spontaneous emulsification* , *electroemulsification* , and *microemulsion* formation.

Two final important processes for the formation of sols involve first the formation of an emulsion or a liquid aerosol. In the first case, termed *suspension* or *dispersion polymerization* , a monomer or monomer mixture is emulsified to a drop size the same as that of the final desired particle. Polymerization is then initiated using an initiator soluble in the monomer, so that chain growth occurs within each individual drop. The result (with luck) is a dispersion of polymer particles with the same average size as the original monomer emulsion. Normally, some type of stabilizer system is employed in the emulsification stage --- a surfactant or very small particles of some material such as silica.

The second process involves the formation of a liquid aerosol or mist of the precursor to the solid. This is sprayed into a vapor atmosphere containing a reagent which induces the reaction leading to solid formation. In the case of polymer particles, for example, monomer drops are sprayed into a vapor containing an initiator producing spherical polymer particles. Particles of some inorganic materials can be prepared similarly by, for example, spraying a mist of a reactive precursor [eg, titanium (IV) ethoxide] into water vapor to produce spherical particles of the corresponding oxide.

Condensation Methods

Approaching the formation of colloids from the other end of the size range involves one of several growth mechanisms. Such processes are commonly employed for the production of dispersions and aerosols, and less commonly in the production of emulsions. Typical examples of important condensation processes include fog formation (both water and chemical), silver halide "emulsions" (really dispersions) for use in photographic products, crystallization processes, colloidal silica, latex polymers, etc. More details of some of those process are introduced in Chapters 11 and 13.

THE "ROOTS" OF COLLOIDAL BEHAVIOR

Earlier chapters have emphasized the fact that atoms or molecules at a surface have properties distinct from those located in a bulk phase or in solution. In normal bulk-phase chemistry and physics, the relative number of molecules located at a surface or interface is so small compared to that in the bulk that any surface effects will be easily overshadowed by bulk effects. However, as the bulk phase is subdivided into finer and finer particles, the relative ratio of surface to bulk molecules will increase until the effect of specific surface properties will begin to become significant, or even dominate the characteristics of the system.

For example, the amount of surface exposed by a given mass of material is inversely proportional to the linear dimension of the divided units. For unit mass of material, the "specific surface area" is given by $6/\rho p$, where ρ is the density and d particle dimension (diameter for a sphere or edge length of a cube). If the units (atoms or molecules) in the material have a linear dimension x and a molecular volume of x^3 (4/3 πx^3 for a sphere), the fraction of molecules at the surface will be approximately equal to 6(x/d). Thus, for a material with x = 0.4 nm a cube with d =1 cm will have roughly 0.000025% of its molecules at the surface, or about one in five million. If the cube is divided into small cubes with 1000 nm edges, the number of surface molecules increases to about one in 400 ($\approx$ 0.25%). If further divided to a dimension d = 10 nm, the system now contains approximately 25% of its units on the surface. Because of the higher energy, and therefore higher reactivity in many cases, of surface molecules, the observable effects of the surface begin to be significant as the colloidal size range is approached and usually becomes dominant very rapidly as the smaller end of the spectrum is approached.

Because the lower limit of the colloidal range is just larger than the size of some molecules and solvated species it is difficult to determine exactly where the distinction between "surface" and "bulk" ends and a *molecularly dispersed* system begins. For macromolecular systems, of course, the molecular size is such that even a "molecular dispersion" or solution easily falls into the size range of colloids. For that reason, primarily, such systems are referred to as lyophilic colloids, even though the properties of such systems are governed for the most part by phenomena distinct from the "classical" surface interactions considered in lyophobic colloids. It is no trivial matter, therefore, to decide just where surface effects end and the characteristics of the individual free, solvated units begin.

GROUND RULES FOR COLLOIDAL STABILITY

Keeping in mind what we mean by "colloid," the next step is to define what is meant by a "stable" colloid, or "stability" in general. Basic thermodynamics tells us that any system, left to its own devices, will spontaneously tend to alter its condition (chemical and/or physical) in an effort to attain the condition of minimum total free energy. It says nothing about how fast such a transformation will occur --- that is the province of kinetics --- nor necessarily whether the system will "stop along the way" in the form of some *metastable* configuration. We only know that given a viable mechanism, it will occur.

The idea of such changes in energy can be illustrated by analogy to a game of golf (Figure 10.2). The free energy of a golf ball lying on a flat putting green (in this case its gravitational potential energy) relative to the bottom of the cup is given by

$$\Delta G_g = mg\Delta h \qquad (10.1)$$

where m is the mass of the ball, g is the gravitational constant, and Δh is the distance of the green surface above the bottom of the cup. Let us assume, in order to enhance the analogy, that the "lip" of the cup in the green is slightly higher than the level of the green. In position (A) the ball has a higher free energy than in position (C); however, there is no available mechanism for it to reduce that energy spontaneously --- it is

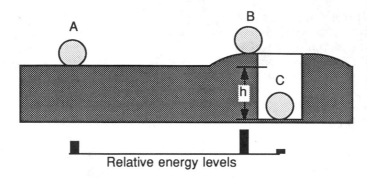

Relative energy levels

Figure 10.2. Schematic illustration of the "golf ball" analogy for energetic considerations of colloidal stability.

metastable . When the player strikes the ball (assuming a certain degree of skill) it will roll toward the cup and up the lip. Going up the lip will increase the energy of the ball, but if it is struck with the proper force, the excess kinetic energy will allow it to go over the lip and fall into the state of lower energy (for the ball and the player); it will be *stable* relative to the cup. If not struck with sufficient force the ball will not be able to surmount the lip and it will roll back down to some energy level still higher than that in the cup but lower than that at the top of the lip. (Under some circumstances, the ball may stop exactly at the top of the lip (B), in which case it will have an even higher energy than before, as will the player, in all likelihood. In such a situation, however, it will require only a very small force to cause the ball to roll in one direction back onto the green or the other into the cup.)

The same general scheme is used to describe the situation in a chemical or physicochemical system in the form of a reaction coordinate. In that case, the height of the lip above the level of the green corresponds to the activation energy for the system which must be overcome in order for a given reaction or transformation to occur. If elements of a physicochemical system have sufficient energy (kinetic, electronic, vibrational, etc) to overcome the barrier, reaction will occur. If not, the system will remain in the metastable state. In golf, of course, there is only one ball involved (we assume!) while chemical systems may involve on the order of 10^{13} (for typical "model" colloidal systems) or more. In such complicated systems, where there exists some statistical distribution of energies, it may be that a certain number of atoms or molecules may overcome the barrier, while the majority cannot. Likewise, there may be some which have sufficient energy to pass in the reverse direction (like a golf ball which falls into the cup with such force that it bounces back out again.) Such a distribution of energies leads to the establishment of an *equilibrium* in which individual units are constantly changing their energy "state," but in which the overall situation is constant. If the reverse reaction requires an energy that is essentially unavailable (ie, the barrier is very high), the system may slowly "bleed" off until all of the units are in the stable state. Such a process, however, may require such a long time that it is imperceptable, or insignificant in the time frame of the observations.

In terms of colloids, we can say the following: interfacial energy considerations dictate that the "position" of lowest energy for a given system (assuming a positive interfacial free energy) will be that in which there is a minimum in the interfacial area of contact between phases. Put another way, in the absence of other factors, colloids should be unstable and rapidly revert to a state of complete phase separation. However, Nature has designed things in such a way that we (or she) can impose barriers of various types between metastable and stable states so that useful (and vital) colloidal systems can be made to exist and persist for enough time so that they can serve a useful function (like make up a significant part of our biological systems!)

A Problem of Semantics

At this point a problem of semantics begins to rear its ugly head, specifically in the operational use of the terms "stable," "unstable," and "metastable." The definitions given above refer to energetic states, without regard to time (ie, kinetics). For practical purposes, however, it is often inconvenient and confusing to use the term "metastable," since its meaning can be somewhat ambiguous. As a general practice one commonly describes a colloid as "stable" if it remains in the energetically metastable state for some arbitrary or functionally determined length of time. It would be termed "unstable" if the system begins to loose its colloidal properties (eg, its degree of dispersion or size) before that predetermined time has passed.

Such designations of stable and unstable colloids are very relative and must be made in the context of the application in question. It may be, for example, that a colloid which maintains its characteristics for 2 days would be considered stable in one application, while another would require that a minimum of 2 years pass without change. Obviously, then, one must be careful when discussing colloidal stability and instability. From this point on in the discussion, unless otherwise indicated, the kinetic (rather than energetic) "definition" of stablity will be employed in its most general sense, it being assumed that all (or almost all) colloids are in reality metastable systems. Also, it must be kept in mind that stability in the present context is used in terms of lyophobic colloids, since lyophilic and association colloids will be inherently stable unless perturbed.

Mechanisms of Stabilization

Clearly, from what has been said so far, colloids must be considered to be metastable, in that surface forces "demand" that a state of fine dispersion represents a high-energy situation. Knowing that colloids exist and that some samples have been kept for over 100 years, tells us that energy barriers can be imposed which will make the transition from the metastable to the stable energy state difficult or impossible. On the other hand, if a colloid is metastable, and, for example, unwanted, changes in the system which lower existing energy barriers to a critical level will allow the system to spontaneously pass from metastable to stable.

In the analogy of the golf ball, the energy necessary for the transformation of the energy state of the ball was supplied by the player putting. In colloids, that energy is normally supplied by random collisions between the colloidal particles and molecules of the dispersion medium --- that is, by Brownian motion. The average translational energy imparted to a colloidal particle due to such a mechanism is $(3/2)kT$ per particle, where the terms k and T have their usual significance. At room temperature (298 K), then, the energy of an average particle will be on the order of 5×10^{-21} J. For two colliding particles, the energy of involved in the collision will be about 10^{-20} J. That will be for average particles; however, the actual energy of a given particle in a given collision may be much smaller or much larger, since the distribution of energies is in accordance with the Maxwell-Boltzman distribution.

If there exists an energy barier between particles of some mutiple, n, of kT, the probability of collisions with sufficient energy to overcome that barrier becomes smaller --- the larger the value of n, that is, the smaller is the probability of the two particles being able to make contact. Operationally, a value of $n = 10$ is usually cited as representing a condition in which a colloid can be considered stable. (That value did not, however, come down from the mountain cast in stone, so that for a specific situation, other conditions may prevail!) Conversely, if the barrier can be reduced to some value of n closer to one, instability will be induced. Several mechanisms exist for lowering the barrier, including changes in temperature, which may or may not be significant depending on the nature of the barrier (see below): solvent properties of the continuous phase; pressure; orelectrolyte content.

A Review of Basic Intermolecular Forces

According to the theories introduced in Chapter 4, the force of attraction between two molecules as a function of the distance of separation is given by

$$F^{att} = -Ar^{-7} \tag{10.2}$$

where A is a constant defined by eq. 4.31. The work required to separate two molecules from a distance d to infinity will be

$$\Delta W = -\int_{r}^{\infty} F^{att} dr = A \int_{r}^{\infty} (r^{-7}) dr = 6Ad^{-6} = A'd^{-6} \tag{10.3}$$

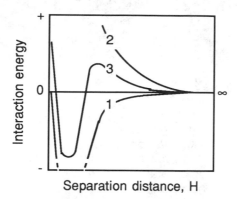

Separation distance, H

Figure 10.3. "Typical" interaction energy curves for two interacting units: (1) the attractive potential; (2) the repulsive contributions; (3) the total interaction potential.

Employing the convention that attractive energy is negative, at infinite separation the energy of the system will be zero, so that the free energy of attraction at distance d will be

$$\Delta G^{att} = -\Delta W = -A'd^{-6} \qquad (10.4)$$

The value of A' was previously defined as

$$A' = (3/4)\,h\,\nu\,\alpha^2 \qquad (10.5)$$

For two dissimilar materials, 1 and 2, the value of A' is given by

$$A'_{12} = (3/4)\,h\,(\nu_1\,\nu_2\,/\,\nu_1 + \nu_2)\,\alpha_1\alpha_2 \qquad (10.6)$$

According to eq. 10.4 the attractive forces between two molecules increases (ie, the energy becomes more negative) continuously as their distance of separation decreases. The relationship is shown graphically in Figure 10.3, curve 1. At some point in their approach, the electron clouds of the two molecules must begin to interact and ultimately overlap. If covalent bonding is not possible, that gives rise to a repulsion (the *Born repulsion term*) and consequent increase in free energy which becomes effectively infinite when interpenetration occurs (Figure 10.3, curve 2). As a good approximation, the Born repulsion term has the form

$$\Delta G^{rep} = B'd^{-12} \qquad (10.7)$$

The total free energy will be the sum of the attractive and repulsive terms

$$\Delta G = \Delta G^{rep} + \Delta G^{att} = B'd^{-12} - A'd^{-6} \qquad (10.8)$$

which is usually known as the Lennard-Jones 6/12 potential. The resulting energy relationship is shown as curve 3, Figure 10.3.

Fundamental Interparticle Forces

Given the above relationships for interactions between molecules, how can we convert that information into interactions between large groups of molecules, ie, colloidal particles? Mathematically, the simplest situation to analyze is that involving two hard, flat, nonpolar, effectively infinite surfaces separated by a distance H in a vacuum. Hamaker showed[1] that the free energy of attraction per unit area in such a case is given by

$$\Delta G^{att} = -A_H/(12\pi H^2) \qquad (10.9)$$

where A_H is termed the Hamaker constant. The value of A_H is related to A' by

$$A_H = (3/4)\, h\, v\, \alpha^2\, \pi^2\, q^2 = A'\, \pi^2\, q^2 \qquad (10.10)$$

In eq. 9.10, q is the number of atoms or molecules in a unit volume of the phase. For two identical spheres of radius a, where $H/a \ll 1$, a similar type of approximate equation is

$$\Delta G^{att} = -(A_H a/12H)[1 + (3/4)\, H/a + \text{higher terms}] \qquad (9.11)$$

In most practical instances, it is safe to neglect all of the higher terms.

A comparison of eqs. 10.4 and 10.10 shows that the free energy of attraction between two surfaces falls off much more slowly than that between individual molecules. This extended range of bulk interactions plays an important role in determining the properties of systems involving surfaces and interfaces.

Due to the fact that interactions between surfaces fall off much more slowly with distance than those for individual atoms or molecules, a significant complicating factor enters into the quantum mechanical evaluation of the attractive forces. The quantum mechanical effects leading to the London-van der Waals interactions occur close to the speed of light, yet even at the short distances involved in colloids, relativiistic effects can be significant.

At close distances, the interactions resulting from fluctuating dipoles is effectively instantaneous. However, as the distance between interacting units increases, the time required for the electromagnetic "signal" of one unit to travel to its neighbor (at the speed of light), polarize the local electron cloud, and receive the return "signal" of said polarization, is long relative to the "lifetime" of the original dipole. The result is that over "long" distances, the net attractive interaction is reduced from that which would be expected otherwise (ie, it is *retarded*). This effect is termed the "retardation effect" and can be generally ignored for atoms and molecules. For larger units such as colloidal particles, however, which interact over greater distances, the retardation effect becomes significant and causes the attractive interactions to fall off faster than they would otherwise (H^{-3} vs H^{-2}). While the retardation effect is important in quantitative theoretical discussions of surface interactions, from a practical standpoint, it is usually insignificant compared to other factors.

When both attractive and repulsive terms are taken into account, the interaction curve for particles resembles that shown in Figure 10.3, curve 3. In terms of colloidal stability, the key element in such a curve is the height of the so-called *primary maximum* indicated as ΔG_{max} on the curve. Later we will see whence comes that energy maximum.

One must keep in mind that the above discussion was given in terms of interactions in a vacuum or other inert environment, which is not a very practical situation for most applications. In order to understand "real" colloidal systems, one must take into consideration the effects of an intervening medium, the continuous phase, on the above interactions.

Fundamental Attractive Interactions in Media Other Than Vacuum

The equations for particle interactions given above were derived for the situation in which the interacting units were separated by a vacuum. "Real life" dictates that in all but a few situations, interacting units are separated by some medium which itself contains atoms or molecules that will impose their own effects on the system as a whole. How will the relevant equations be modified by the presence of the intervening medium?

Surfaces interacting through an intervening fluid medium will experience a reduced mutual attraction due to the presence of the units of the third component. The calculation of interactions through a vacuum involve certain simplifying assumptions; therefore, it should not be surprising to find that models for three-component systems (where each particle is considered a component) are even more theoretically complex.

Although a number of elegant approaches to the theoretical problem have been developed over the years, for most purposes a simple approximation of a composite Hamaker constant is found to be sufficient. Consider, for example, two particles of material 1 dispersed in medium 2 (Figure 10.4). The effective Hamaker constant (A_H^{eff}) for the two particles can be approximated by

$$A_H^{eff} = [A_{H(10)}^{1/2} - A_{H(20)}^{1/2}]^2 \qquad (10.12)$$

where $A_{H(10)}$ is the Hamaker constant for component 1 in a vacuum, and

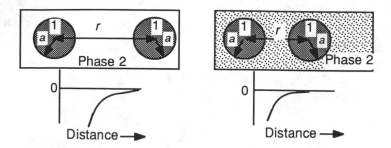

Figure 10.4. Two particles of phase 1 interacting in a medium of phase 2: as the nature of 2 approaches that of 1, the attractive van der Waals interaction between the two particles is reduced, as is the distance over which it acts.

the same for $A_{H(20)}$. An important result of the relationship in eq. 10.12 is that as the vacuum Hamaker constants for 1 and 2 become closer in value, A_H^{eff} tends toward zero, and the free energy of attraction between the two particles tends toward zero. That is, the force of attraction is decreased and any repulsive term necessary to maintain an adequate value of ΔG_{max} is also reduced. As we will see, such a reduction in the attractive forces due to an intervening medium gives one a handle on ways to manipulate the stability of colloids. Since eq. 10.12 involves the square of the difference between the Hamaker constants for components 1 and 2, the same effect applies for surfaces of component 2 separated by a continuous medium of 1. The form of the interaction curve for the above situation will be the same as that for the vacuum case, although the shape and values will differ because of the diferent value of the effective Hamaker constant.

SOURCES OF COLLOIDAL STABILITY

Knowing where the attractive interactions between colloidal particles stem from, we must now address the question of where the free energy maximum comes from, which gives a system what stability it may have. In general there are two mechanisms for stabilizing lyophobic colloids --- *electrostatic repulsion* between electrical double layers and *steric* or *entropic* stabilization. The earliest recognized, and in some ways the most easily understood is that arising from electrostatic interactions, so we will begin the discussion in that area.

Charged Surfaces and the Electrical Double Layer

Chapter 5 presented a capsule summary of the primary sources of electrical charges at surfaces and an introduction to the energetics of such systems, including the concept of the electrical double layer. We now turn our attention to the question of how such interactions tend to provide stability to a colloidal system. If we first consider the situation of

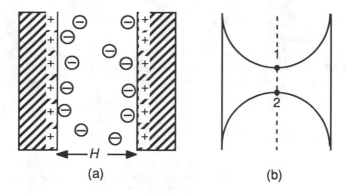

Figure 10.5. Schematic illustration of the interaction between two electrical double layers: (a) two positively charged surfaces with only counterions dissociated from the surface in between; (b) the counterion density profile (1) and the electrostatic potential profile (2).

two isolated, similarly charged particles (or flat surfaces, for that matter) with their associated electrical double layers, it is relatively easy to understand the basic concepts involved in the electrostatic stabilization of colloids. In the context of (kinetic) stability, one can say that a system is stable so long as the individual particles maintain their identities --- that is, so long as flocculation and/or coagulation do not occur. In order for aggregation to occur, of course, two particles must collide, and do so with sufficient force that the collision will be effective or "sticky." There are therefore two primary criteria which must be considered in the discussion of colloid stability: the number or frequency of particle collisions and the effectiveness of those collisions.

    With reference to Figure 10.5, as the two charged particles approach, the two charge clouds or electrical double layers (EDL) will begin to overlap. Since the EDL's are of the same sign, their interaction will be repulsive, leading to an increase in the electrical potential between the particles. As a first approximation, we can assume that the potentials of the two surfaces, represented by the solid lines in the curve, are additive so that the total electrostatic interaction potential will be similar to the dashed line. Since the increased electrical potential represents an increase in the total free energy of the system, such interactions will represent a barrier to the approach of the two particles, and if they cannot approach sufficiently closely to touch, they cannot form aggregates --- flocs or coagulum. The interaction of the respective electrical double layers, therefore, represents an energy barrier in the total interaction curve like the lip of the cup on the putting green.

    As the two particles approach, there will be two (at least) types of interaction: the repulsive interaction just described and the relentless van der Waals attractive interactions, which make most colloids inherently unstable. The total interaction energy for the system under consideration will be the sum of the two energies

$$\Delta G^{\text{total}} = -\Delta G^{\text{att}} + \Delta G^{\text{rep}} \tag{10.13}$$

A general expression for the repulsive interaction between the electrical double layers around two spherical particles is quite complex and does not warrant discussion here. A simple and relatively good approximate equation derived by Reerink and Overbeek[2] is

$$\Delta G^{\text{rep}} = (B e k^2 T^2 a \gamma^2 / z^2) \exp(-\kappa H) \tag{10.14}$$

where H is the distance between spheres of radius a, B is a constant equal to 3.93×10^{39} A^{-2} s^{-2}, z is the charge on the counter-ion, e in the unit electrical charge, and

$$\gamma = [\exp(ze\,\psi_s / 2kT) - 1] / [\exp(ze\,\psi_s / 2kT) + 1] \tag{10.15}$$

As pointed out in Chapter 5 the effective electrical potential of importance will be that at the Stern layer ψ_s rather than that actually at the surface, ψ_o.

The attractive interaction is given by eq. 10.11 so that the total interaction will be

$$\Delta G^{\text{total}} = [(B e k^2 T^2 a \gamma^2 / z^2) \exp(-\kappa H)] - [A_H / (12 \pi H^2)] \tag{10.16}$$

Curves illustrating the individual and summed interactions for various electrolyte concentrations are shown in Figure 10.6. It should be fairly clear that the key element in determining the height of the energy barrier imposed by the electrical double layers is the concentration and valency of electrolyte in the system. An increase in the electrolyte concentration reduces the repulsive electrostatic interaction, reducing the energy barrier and facilitating effective particle collisions --- ie, the system is less stable. A good approximation to the point at which the system will begin to undergo rapid coagulation (indicating a loss of stability) is that at which $\Delta G^{\text{total}} = 0$ and $d\Delta G^{\text{total}}/dH = 0$.

In general, the analysis according to eq. 10.16 overestimates the attractive interactions due to van der Waals forces, because it does not take into consideration the retardation effect mentioned above. The total energy of interaction has some interesting characteristics that warrant pointing out. For example, the repulsive potential energy function is an exponential function of the distance of separation of the particles H, usually a distance of the same order of magnitude as the thickness of the electrical double layer, while the attractive potential decreases as the inverse square of H. As a result, the total interaction will be controlled primarily by the attractive potential at small and large distances, while the double-layer repulsion potential may dominate at intermediate distances, depending on the relative values of the two.

In summary, then, if the total potential energy maximum is large

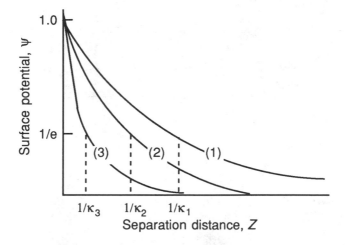

Figure 10.6. Repulsive interaction curves for electrostatically stabilized particles at various electrolyte concentrations; (1) low; (2) medium; (3) high.

compared to the average thermal energy of the particles, say 10 kT, then the system should be stable; if not, the system will flocculate or coagulate. The height of the barrier will be determined by the surface potential at the Stern layer ψ_s and the thickness of the double layer, $1/\kappa$.

COAGULATION KINETICS

The coagulation of emulsions and dispersions due to random Brownian motion has historically been the topic of most general interest to surface and colloid scientists because of the experimental accessibility of data (with sufficient innovation and diligence on the part of the experimenter) and a reasonably firm theoretical basis. The presence of an intervening fluid in such cases carries with it certain advantages of preparation, manipulation, and interpretation that are not present in an aerosol, for example. The same circumstances can, of course, carry with them certain disadvantages, especially with regard to purification, contamination, specific solvent effects, etc. Fundamentally, the same theoretical considerations apply to both solid dispersions in a liquid medium, areosols, emulsions, and foams. However, in each case, certain extensions and modifications may be required.

Kinetics of Particle Collisions: Fast Coagulation

When one considers the question of the kinetics of coagulation, the discussion must begin with the question of whether there exists some barrier to coagulation between two approaching particles. Since the overall process, in its simplest form, involves the interaction of two particles or units, it can be thought of in terms of chemical kinetic

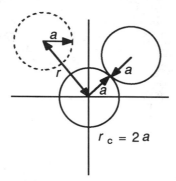

Figure 10.7. Schematic illustration of particle diffusion and flux in colloidal coagulation process.

processes; that is, a bimolecular, second-order process. If no barrier to particle approach, contact, and adherence exists, the particles are said to be *noninteracting* (except on contact, of course). The chemical equivalent would be a reaction with zero activation energy. If the particles are noninteracting and the primary minimum for the interaction energy is sufficiently deep, every collision between particles will be "sticky;" that is, each will result in the formation of a multiple particle (dimer, trimer, etc) floc. In such a case, the rate of coagulation will be controlled entirely by diffusion kinetics, analogous to a diffusion-controlled bimolecular reaction (Figure 10.7).

The first theory relating the rate of coagulation to diffusion was that of Smoluchowski.[3] The complete derivation of the final equation will not be given here. Suffice it to say that the process begins with Fick's first law of diffusion given by

$$J = 4\pi a^2 D(\partial n / \partial a) \tag{10.17}$$

where J is the "flux" or number of collisions per unit time between particles in the system and some "central" or reference particle (Figure 10.8), a is the radius of the central particle, and D is the diffusion coefficient. By way of illustration one can visualize the process in the following way: a particle of radius a is placed with its center at the origin of a rectangular coordinate system, where a will be some characteristic radius of the particle in question (eg, hard sphere, hydrodynamic, etc). Other particles in the system, in the process of random Brownian motion, move through the coordinate system at distances r from the center until one particle center approaches to within a critical distance r_c of the central particle. At that distance, the particles "touch" and coagulation is assumed to occur. For a system of monodispersed particles, $r_c = 2a$; r_c, then, is the *collision diameter* of the particles.

At time $t = 0$, the particle concentration in the system well away from the central particle ($r \rightarrow \infty$) is n_0 (particles per unit volume) while at

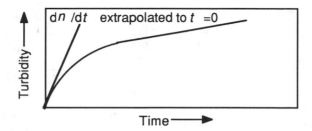

Figure 10.8. Schematic illustration of experimental results for the determination of coagulation rates from turbidity data.

distance r_c, $n = 0$, so that the number of collisions per unit time between the central and a second particle will be

$$J = 4\pi D r_c n_o \qquad (10.18)$$

The diffusion coefficient D is given by

$$D = kT/6\pi\eta a \qquad (10.19)$$

where η is the viscosity of the dispersion medium. If the constraint of holding the central particle stationary is removed, the value of D to be employed in eq. 10.18 becomes $D = D_1 + D_2$ (for particles of different radii), or for identical particles, $2D$. The kinetic equation for the disappearance of primary particles now becomes

$$dn/dt = -8\pi r_c D n^2 \qquad (10.20)$$

where n is the concentration of primary particles and dn/dt is the rate at which they disappear. Based on eq. 10.20, the time for the reduction of the number of particles by one half, the *half-life* , will be

$$t_{1/2} = (8\pi r_c D n_o)^{-1} \qquad (10.21)$$

For water at 25°C

$$t_{1/2} = 3\eta/4kT n_o = 2 \times 10^{11}/n_o \text{ (seconds)} \qquad (10.22)$$

so that for a typical dilute dispersion containing $10^{12\pm1}$ particles cm^{-3}, the half-life is in the range of 0.2 and 1 s. Those represent very short times, shorter than are normally encountered in practice (seconds to minutes, depending on the circumstances) for noninteracting dispersions.

The primary reason for the descrepency is the fact that as two particles approach one another, it is necessary for solvent molecules between the particles to be moved out of the way. This process is accounted for by the viscosity term in eq. 10.22 for large distances of separation, but at smaller distances, of the order of molecular dimensions, the simple viscosity relationships no longer strickly apply, so that the mutual diffusion coefficient D is no longer equal to $D_1 + D_2$. One could say that the "microscopic" viscosity of the solvent increases so that diffusion is slowed and the particles approach at a much reduced velocity. The exact calculation of this *hydrodynamic effect* represents a difficult problem in fluid dynamics. However, a relatively simple formula for two spheres of equal diameter is

$$D_{eff} = D \left[(6s^2 - 20s + 16) / (6s^2 - 11s) \right]$$ (10.23)

where $s = r /a$. Experimentally, it is found that this hydrodynamic effect results in a collision rate of approximately one half that predicted by eq. 10.22. Therefore the measured half-life will be about twice as long.

The simple theory presented so far considers only collisions between single primary particles, not those involving dimer, trimer, etc, flocs. When all particles are taken into consideration, the overall kinetics are still second order overall, but the value of n in the equation is the number of *all* particles present, regardless of their size and composition, and $t_{1/2}$ refers to the time for disappearance of one half of *all* particles, not just the primary ones represented by n_o. Because of the complications inherent in trying to define the necessary values of n, r_c, D, etc, for the multiparticle aggregates, the relevant experiments are normally interpreted for the early stages of the coagulation process only, where the majority of particles are primary units. In practice, the experimental process usually involves measuring the value of n versus time using, for example, light scattering, turbidity, etc, and extrapolating the data back as $t \rightarrow 0$ (Figure 10.8). The regime in which a coagulation process follows the above theory is usually referred to as *fast* or *rapid coagulation*.

Slow Coagulation

If some energy barrier to particle contact and adherence exists --- that is, if some of the collisions are not "sticky" --- the collision process can be seen as analogous to a bimolecular reaction in which there is an activation energy. In that case, a *slow coagulation* process occursthat can be described kinetically by the relationship

$$dn/dt = -8\pi r_c Dn^2 \exp(-V_{max} / kT)$$ (10.24)

where V_{max} is the height of the activation energy barrier opposing coagulation. The rates of fast and slow coagulation (eqs. 10.20 and 10.24) obviously differ by the exponential factor $\exp(-V_{max}/kT)$. Since the coagulation process is slowed down by that factor, dividing eq.10.20 by eq. 10.24 gives what is termed the *stability ratio*, W

$$W = \exp(V_{max} / kT) \tag{10.25}$$

Equation 10.25 would be valid if colloidal diffusion processes were exactly analogous to those for individual molecules. However, the interactions between particles in colloidal systems tend to extend over distances much greater than those involved in the formation of activated complexes (say 10--100 nm vs 0.1--1.0nm). As a result, the effects of those interactions will begin to be felt by the particles well before they approach to the critical distance r_c. Their mutual diffusion rate will therefore be reduced and the collision frequency will drop accordingly. The collision frequency will also be reduced by the hydrostatic effect mentioned above for rapid coagulation.

A more accurate expression for the stability ratio, W, taking into consideration the above retarding effect of the interaction potential on collision frequency, is

$$W = r_c \int_{r=2a}^{r=\infty} (1/r^2) \exp(V(r) / kT)\, dr \tag{10.26}$$

where $V(r)$ is the interaction potential when the two particle centers are separated by a distance r.

For charged particles experiencing electrostatic repulsion, an approximate equation for the stability ratio is

$$W = (1/\kappa r_c) \exp(V_{max}/kT) \tag{10.27}$$

where κ is the thickness of the electrical double layer. A more practically useful form of Eq. 10.34 for purposes of interpretation of data is

$$\ln W = -\ln \kappa r_c + V_{max}/kT \tag{10.28}$$

V_{max} is a roughly linear function of the concentration of electrolyte in the system, c_o, so that a plot of $\ln W$ vs $\ln c_o$ is also roughly linear. That approximate relationship is generally born out by experiment, although the hydrodynamic effect, among others, may cause slight bothersome deviations from the ideal curve shape.

Critical Coagulation Concentration

As a practical application of the relationship in eq. 10.28, one may characterize the stability of a charge stabilized colloidal system by its *critical coagulation concentration* (CCC), the concentration of electrolyte necessary to bring the system into the regime of rapid coagulation. The process involves the extrapolation of the curve of $\ln W$ vs $\ln c_o$ to $\ln W = 0$, which gives $\ln$ (CCC). However, what is the practical use of the ccc and what does it mean in theoretical terms?

Theoretical and practical interest in the effect of added electrolytes on the stability of colloidal dispersions was reported as early as 1856 by Faraday,[4] although it was almost certainly of great interest to others much before that time. Since the earliest days, of theoretical surface and colloid science, the understanding of coagulation phenomena and the formulation of workable theories has been the focus of a major portion of the scientific efforts in the field. Near the turn of the century, the studies by Schultze[5] and Hardy[6] indicated that the primary factor controlling the effect of an electrolyte on a colloid of opposite electrical charge (and for a given concentration) was the valency of the added counterion. It was found that, in general, the valency of ions of the same charge as the colloid was of minor importance.

The studies by Schultze and Hardy led to the formation of the so-called *Schultze - Hardy rule* , which states that the critical coagulation concentration of a colloid is determined primarily by the valency of the counterions. The early relationship indicated that the CCC varied as the inverse sixth power of the valency, although experimental results obtained under carefully controlled conditions of concentration, temperature, etc, give ratios for electrolytes of valency 1, 2, and 3 of 1 : 0.013 : 0.0016. That is, for valency 2, the exponent is - 6.27 and for valency 3, - 5.85. The fact that exponents are found to be fractional almost certainly stems from the fact that the properties of ions in solution vary slightly according to their ionic radius, hydration radius, etc. As a result, their activity in such situations as their effect on the surface tension of a solution, their adsorption at interfaces, the degree of ionization of their salts, their interaction with proteins, and their CCC for a given colloidal system will vary within a given valence. It is found that for monovalent ions, the effectiveness for coagulating negatively charged colloids has the order $Cs^+ > Rb^+ > K^+ > Na^+ > Li^+$, while for divalent cations the order is $Ba^{2+} > Sr^{2+} > Ca^{2+} > Mg^{2+}$. Those and similar series for the effectiveness of ions of a series is commonly referred to as a *lyotropic series* .

Clearly, the stability of an electrostatically stabilized colloid, as measured by its CCC, is a function of the concentration and charge of the counterions in the system. The question is, what type of theory can incorporate all of the observed facts about colloidal stability and be able to serve as a workable predictive model for new systems?

The Deryagin-Landau-Verwey-Overbeek (DLVO) Theory

As indicated above, the overall stability of a colloid will depend on the net form of the interaction energy curve for the system --- the sum of the attractive and repulsive energy terms as a function of the distance of separation of the particles. For the moment, we will consider only two contributing factors, the attractive van der Waals term and the repulsive double-layer term, leaving aside any consideration of entropic or steric stabilization.

For a one-to-one electrolyte of valency z and bulk concentration c_0, the excess charge density, ρ, at a point in the electrical double layer of potential ψ is given by

$$[c^+ - c^-]z = \rho = z c_0 [\exp(-z e \psi / kT) - \exp(+z e \psi / kT)] \quad (10.29)$$

Poisson's equation of electrostatics relates r to the variation of ψ with distance to the charged surface, x, in the form

$$\partial^2 \psi / \partial x^2 = -\rho / \varepsilon \quad (10.30)$$

where ε is the permitivity of the medium. Combination of eqs. 10.29 and 10.30 gives the Poisson-Boltzman equation, which shows that the potential within the EDL will fall off approximately exponentially with distance, for systems with relatively high surface potentials and distances well removed from the surface.

$$\psi = (4kT / z e) \exp(-\kappa x) \quad (10.31)$$

In eq. 10.31, κ is, as previously defined, the thickness of the EDL.

In terms of the interaction of two approaching particles, as the distance of separation decreases, the two electrical double layers begin to overlap. To a first approximation, the two overlapping electrical potentials become additive, resulting in an increased electrical contribution to the total free energy of the system --- a repulsive term, in this case, since the process increases the total energy. The form of the EDL overlap potential is also approximately exponential with distance of separation, H. For two parallel surfaces, the free energy contribution of the electrostatic repulsion term is given by

$$\Delta G_{elec} = (64 c_0 kT / \kappa) \exp(-\kappa H) \quad (10.32)$$

As previously shown, k is dependent on the concentration of ions in the system, and the primary concentration effect in eq. 10.32 comes from the exponential term.

If one sums the attractive van der Waals potential and the repulsive electrostatic potential, the total potential energy expression becomes

$$\Delta G_T = [(64 c_0 kT / \kappa) \exp(-\kappa H)] - [A_H / (12\pi H^2)] \quad (10.33)$$

where A_H is the Hamaker constant for the colloidal material. When eq. 10.33 is solved for various electrolyte concentrations, a series of curves of the general forms shown in Figure 10.9 is obtained. The maxima in the

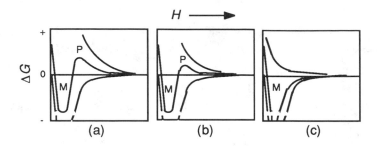

Figure 10.9. Illustration of the effect of electrolyte on interaction energy curves for electrostatically stabilized colloids (see Figure 10.3): (a) at low concentrations, a relatively high energy barrier, P, prevents coagulation; (b) as the concentration is increased, the barrier is lowered, making it easier for the particles to pass into the primary minimum, M; (c) at critically high concentrations, the barrier is removed and the system falls directly into M, leading to complete loss of stability.

curves represent the barriers to coagulation imposed by the electrical double layer. Clearly, as the electrolyte concentration increases, the height of the barrier decreases. As a rule of thumb, one can say that conditions for rapid coagulation (the CCC) will be met when $\Delta G_T = 0$ and $d\Delta G_T/dH = 0$. Mathematically, that is

$$c_o(CCC) \propto 1/(A_H{}^2 z^{\,6})\qquad\qquad(10.34)$$

which agrees with the Schultze-Hardy rule and predicts ratios of 1:0.016: 00014 for ions of valency 1, 2, and 3, and agrees generally with experimental observations noted earlier.

Another consequence of eq. 10.34 is that as the Hamaker constant A_H increases, the CCC's of dispersions with the same electrical characteristics will decrease, that is, the colloidal dispersions will become progressively less stable. As indicated previously, the Hamaker constant in a medium will be a composite of those of the dispersed particle and the medium in vacuum, so that as the characteristics of the medium and dispersed phase become more similar, A_H will decrease, leading to an additional stability for the system, in agreement with observation.

The fact that the DLVO theory predicts (roughly at least) the Schultze-Hardy rule would seem to confirm the validity of the theory. However, several approximations and assumptions are included in the derivation that weaken its claim to complete success. For example, at low surface potentials, the theory predicts that the CCC will be proportional to z^{-2} rather than z^{-6}. A more complete derivation of the theory (ie, taking into consideration such factors as specific adsorption of ions and hydration effects) can remove some of the descripencies so far encountered in the simple theory.

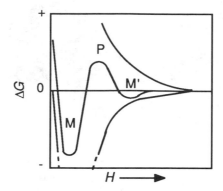

Figure 10.10. Schematic illustration of the secondary minimum, M', in electrostatically stabilized colloids.

An additional important prediction of the DLVO theory is that under certain conditions, a colloid may undergo a form of *reversible flocculation* brought about by the existence of a so-called *secondary minimum* in the potential energy curve. The existence of the secondary minimum has been confirmed experimentally and since it represents a potentially important theoretical and practical aspect of the DLVO theory, it will be discussed briefly below.

The great value of the DLVO theory for the practicing colloid chemist, regardless of the exact nature of the work involved, is that it illustrates dramatically the importance of understanding the electrical properties of a colloid of potential interest, and the importance of understanding the effect of the ionic environment to which that colloid may be exposed.

Reversible Flocculation and the Secondary Minimum

An interesting, and sometimes useful, consequence of the complex interaction between attractive and repulsive terms in the overall equation for colloidal stability is the existence of a so-called *secondary minimum* in the total energy curve, usually at relatively large interparticle distances (Figure 10.10). The primary minimum is, of course, the deep energy well into which the system falls in search of "energetic" stability (the cup in the putting green). Before going over the barrier to enter that well, however, a colloid may encounter a much more shallow energy minimum of the order of a few kT where small, relatively weak aggregates, in this case flocs, may form.

Weak or *secondary minimum flocculation* , unlike flocculation in the primary minimum, is often found to be subject to significant entropy effects, leading to the realization of a number of interesting phenomena. Because of the shallow nature of the secondary minimum, the flocs formed are held together rather weakly and tend to be unstable; that is, they can be broken up by rather small energy inputs such as gentle

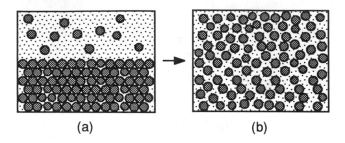

(a)   (b)

Figure 10.11. Illustration of the effect of the secondary minimum in reversible flocculation: (a) a flocculated colloid with two phases---a concentrated flocculated phase and a dilute dispersed phase; (b) after agitation the system returns to a uniformly dispersed, single-phase system.

stirring. In fact, for very small particles ($r < 100$ nm), the minimum may be so shallow that Brownian motion prevents flocculation altogether. For larger particles the secondary minimum may cause observable effects such as an apparent phase separation into a concentrated, highly ordered dispersion, sometimes exhibiting birefrengence, and a second dilute, isotropic phase. In that case, gentle stirring will regenerate the original homogeneous dispersion (Figure 10.11). In other cases a dynamic equilibrium may develop between small flocs and individual colloidal particles. Such an equilibrium may be treated theoretically much like the process of molecular aggregate formation in a vaopr (or a solution), in which the concentration of aggregating molecules is below the saturation point. That is, it is below the concentration necessary for the formation of significant numbers of aggregates of the critical size necessary for condensation or crystallization to occur.

STERIC OR ENTHALPIC STABILIZATION

The mechanism of stabilization discussed above refers only to those systems in which the dispersed particles carry a surface electrical charge, so that the interaction of their respective double layers provides the necessary energy barrier for kinetic stability. Another stabilizing mechanism has been known for centuries but has until recently been much less thoroughly studied and understood. That mechanism involves the presence of a lyophilic colloid, which adsorbs onto the particle surface and provides what is termed *steric* or *entropic* stabilization. (While there exists some controversy as to whether steric and entropic stabilization are the same phenomenon, the two terms will be used interchangeably in the following discussion.) India ink, carbon particles dispersed in an aqueous solution of a natural gum, is a good example of an old, well-known colloid stabilized (at least partially) in that way. Many other instances in which workable (ie, stable) colloids can be prepared with the help of "protective action" provided by added lyophilic colloids (*protective colloids*) are known.

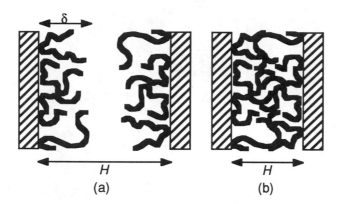

Figure 10.12. The effect of polymer adsorption on colloidal interactions: (a) the distance between particles, H, is greater than twice the adsorbed layer thickness, δ, there is litle or no interaction; (b) when $H < 2\delta$, interpenetration of the adsorbed layers results in repulsive (entropic) interactions.

It has long been recognized that steric "protective agents" need not carry an electrical charge in order to be effective. However, they do have certain requirements, such as (generally) a dual chemical nature with respect to their solubility characteristics and a relatively high molecular weight. Some more details of the necessary characteristics are given below, but in general one can say that a steric stabilizer must have one portion of its molecule that exhibits relatively low solubility in the dispersion medium and/or a high tendency to adsorb onto the particle surface. The net result must be the formation of a relatively thick adsorbed layer that can impose a barrier to close particle approach, which will, of course, improve the stability of the colloid. Since most effective steric stabilizers are macromolecules, the following discussion tends to concentrate on the specific aspects related to their performance. More general information related to the adsorption of polymers at interfaces is provided in Chapter 14.

The Mechanism of Steric Stabilization

Referring to Figure 10.12, if two colloidal particles have an adsorbed layer of a lyophilic polymer, as they approach each other, those layers must begin to interpenetrate. Such interaction can have two effects: an *osmotic effect* due to an increase in the local concentration of the adsorbed species between the two particles, and an *entropic* or *volume restriction* effect because the interacting species begin to lose certain degrees of freedom due to crowding. In both cases, the local system will experience a decrease in entropy, which will, of course, be unfavorable, while the osmotic effect may be accompanied by an unfavorable enthalpic effect due to desolvation of the more closely

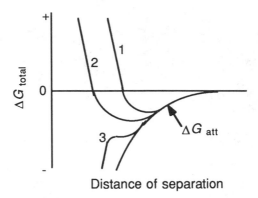

Distance of separation

Figure 10.13. Illustration of the effect of steric stabilization on the total interaction energy: for curves 1 and 2, significant stabilization is attained, although they differ in effectiveness due to differences in adsorbed layer thickness or density; for curve 3, the thickness and/or density of the adsorbed layer is insufficient to prevent coagulation.

packed units. In order to regain the lost entropy, the particles must move, allowing them more freedom of movement, while solvent moves in to"resolvate" the units. The result is an energy barrier retarding the approach of particles and providing an effective mechanisms for stabilization. The process is illustrated schematically in Figure 10.13. In a purely sterically stabilized system (ie, no electrical charges involved) one can evaluate the net interaction energy between particles in a way similar to Figure 10.10. That is, the net energy of interaction will be the sum of the attractive van der Waals forces and the repulsive steric interactions

$$\Delta G^{total} = \Delta G^{steric} - \Delta G^{att} \qquad (10.35)$$

As noted, polymeric protective agents or steric stabilizers must be such as to be strongly anchored to the particle surface at a minimum of one point, or even better, several points. If single-point attachment is involved, the result will be a system with a free-swinging "tail" projecting into the solution providing the protective action (Figure 10.14a). If two or more points are involved the result will be the formation of various loops, and possibly some tails as well (Figure 10.14b). For a given polymer chain length, one can see intuitively that in a system of "tails" the distance the protective layer extends into the solution will be greater than a comparable system of loops. From that one might assume that such a single-point attachment would provide better protection. On the other hand, for loops, once interpenetration begins, there will be twice as many units affected by the volume restriction effect, leading to a stronger entropic effect. One cannot say, therefore, that one configuration is better than another. In most practical systems, both configurations will be involved.

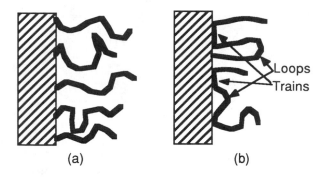

(a)   (b)

Figure 10.14. Illustration of various modes of polymer adsorption: (a) tails only; (b) loops, trains, and tails.

The Effect of Polymer Molecular Weight

In general, for a class of protective lyophilic colloids, a higher molecular weight material can be expected to provide better protection against flocculation, the reason being, of course, that longer chains imply longer loops and tails and the formation of a thicker protective layer around the particle. Like most physical systems, however, there are certain limits that must be adhered to, because "overkill" in the moloecular weight department may lead to what is termed *sensitization* and *bridging flocculation* .

If a very high molecular weight polymer with more than one potential point of attachment to the particle surface is added to a colloidal dispersion there exists the possibility that the various possible points of attachment will encounter two different particles rather than attach to the same particle. That is especially the case where there is a large excess of particles relative to the concentration of polymer. Attachment of the same polymer chain to two particles essentially ties them together and brings them closer, in effect *sensitizing* the particles to flocculation. The process is illustrated in Figure 10.15.

While sensitization and bridging flocculation can be potential hazards in the formulation of sterically stabilized colloids, they also have their positive aspects. For example, the addition of small amounts of a high molecular weight acrylamide polymer to water leads to the flocculation of particulate matter that may be difficult to remove otherwise. Similar applications are found in, for example, the treatment of coal washing effluents and the flocculation of fines from uranium containing calcium phosphate minerals. In most cases of bridging flocculation, the resulting flocs are relatively open and rigid, which means that separation and filtration is relative easy. That fact is also exploited in the use of soil "improvement" polymers, which flocculate the soil particles forming an open structure that allows for the freer movement of moisture and air throughout the soil.

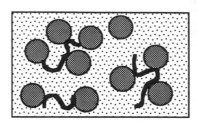

Figure 10.15. Illustration of the mechanism of bridging flocculation: when the ends of widely spaced segments of a high molecular weight polymer bind to two or more particles, the system becomes sensitized and flocculates since the process draws the particles closer together than may otherwise be the case, helping to overcome any inherent stability.

Depletion Flocculation

If one adds a polymer that is not adsorbed or poorly adsorbed on the particles to a colloidal solution, there may occur another phenomenon, termed *depletion flocculation* . In depletion flocculation, as two particles approach, polymer chains that are weakly adsorbed, or simply are located between the particles, become squeezed out of the area of closest approach, leaving "bare" surfaces that are attracted in the normal way. However, there may arise an additional attractive force as a result of the removal of polymer from the intervening region (Figure 10.16).

As polymer is forced out of the area between the approaching particles, the local osmotic balance is displaced; that is, the solution concentration between the particles is less than that in the bulk. Osmosis then forces solvent to flow from between the particles out into the solution. The net effect on the particles is that they are drawn together by the solvent flow (a type of hydrodynamic "suction" effect, if you will), resulting in a loss of stability and flocculation.

The above picture of depletion flocculation is, of course, very schematic and simple minded, but it should serve to illustrate the concepts involved. A more detailed discussion would involve the introduction of complex theories of polymer adsorption and solution phenomena that are beyond the scope of this book.

Solvent Effects in Steric Stabilization

Based on the picture of steric stabilization presented above, it should be clear that the solvent must play a critical role in determining the effectiveness of a given stabilizer--colloid--solvent system. If the nature of the solvent is changed such that it becomes a better solvent for the monomer units acting as anchors for the stabilizer, then the chains will be more weakly adsorbed, providing for the possible onset of depletion flocculation or related phenomena. Conversely, if the solvent is changed from a "good" to a "poor" solvent for the loops and tails, the

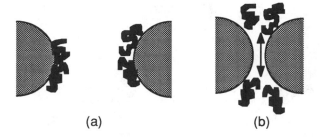

(a)   (b)

Figure 10.16. The origin of depletion flocculation: as two particles approach, weakly adsorbed polymer in the area between the two becomes desorbed, leaving a bare area so that attraction between the two is increased. The effect may be further enhanced by the fact that osmotic forces may cause solvent in the intervening space to flow out, creating a suction effect.

thickness of the protective layer will be reduced as the polymer chains collapse in on the particle surface. The result will be reduced stability against flocculation.

Because of the complex thermodynamics of polymer solutions, most polymers exhibit transitions in their solubility in a given solvent as a function of temperature. In a "good" solvent, the polymer chains will be extended in relatively open, random-coil configurations, giving optimum protective layer thickness. As the temperature is changed, the quality of the solvent may decrease, at some point becoming "poor," and the polymer chains will collapse into a more compact configuration. The point at which the transition from good to poor solvent properties occurs is termed the θ *point* . In terms of colloidal stability, the temperature at which the solvent character changes from "good" to "bad" is referred to as the *critical flocculation temperature* (CFT) of the system.

The θ point can also be attained by the addition of a miscible nonsolvent for the polymer loops and tails. The way in which the interaction potential is affected by the quality of the solvent is illustrated in Figure 10.17. In a poor solvent, the adsorbed layers may even add an additional attractive potential to the curve due to van der Waals attraction between the two.

THE COMPLETE INTERACTION CURVE

The two stabilization mechanisms discussed above, electrostatic and steric, represent the extremes of the range of phenomena that allow us to prepare colloidal systems which are at least kinetically stable. The two mechanisms, however, are not mutually exclusive. In fact, most natural and many technologically important colloids involve a combination of both effects (plus possibly others, yet to be fully determined). A total potential energy curve, then, might contain a number of terms

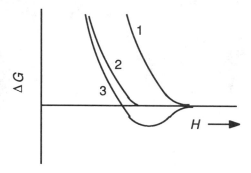

Figure 10.17. Illustration of the role of solvent in steric stabilization: curve 1 represents a good solvent --- effective stabilization is obtained; curve 2 is for a θ solvent; curve 3 represents a poor solvent in which no stabilization occurs.

$$\Delta G^{\text{total}} = \Delta G^{\text{elec}} + \Delta G^{\text{steric}} + \Delta G^{\text{misc}} - \Delta G^{\text{att}} \qquad (10.36)$$

Proteins, for example, are excellent protective colloids. They are polymeric and tend to adsorb on various surfaces, providing good steric stabilization, and may also be charged (depending on the pH), providing electrostatic stabilization as well. Such a dual nature can have certain practical applications. For example, if a colloid (silver iodide, for example) is prepared in the presence of gelatin at a pH above the isoelectric point, a very stable dispersion will result. If the pH is lowered (or raised, depending on the gelatin type used) to the isoelectric point, the gelatin chains tend to collapse onto the particles, reducing the degree of steric stabilization in addition to that lost by neutralization of the charges on the polymer. The system will tend to flocculate (also sometimes called *coacervation*), allowing for the separation of the silver iodide--gelatin complex from most of the water. If the pH is then returned to is initial value, the adsorbed polymer will swell and redisperse the original colloid. The presence of the gelatin prevents the individual particles from making contact so that true flocculation or coagulation cannot occur. Similar processes using both natural and synthetic polymers have found application in several important areas, including photography, microencapsulation, "carbonless" carbon papers, etc.

Obviously, knowing how colloids can be stabilized provides an invaluable tool for the preparation of many useful systems. It also can provide clues to how an unwanted colloid can be destabilized and removed. The above ideas, at times in a slightly different guise, will appear again in the following chapters on emulsions, foams, aerosols, etc.

handwritten notes at top of page:

emulsion

foam → d.p > 90
emulsion → > 50%
colloid → 10 to 85%

1-1000 nm   emulsion > 0.1 μm

CHAPTER 11

EMULSIONS

The preparation, stabilization, and application of emulsions impacts almost every aspect of our lives, from food to pharmaceuticals. Chapter 10 introduced some of the important physical--chemical factors in the preparation and stabilization of dispersions of solid particles in a liquid (sols). Most of those same factors are important in emulsions, although their significance and the approaches used to understand them often differ between the two classes of colloids. Although colloidal in character, emulsions are usually systems whose dispersed phase dimensions fall outside the "normal" defined range for colloids. Commonly encountered emulsions will have average drop sizes of at least several microns, with a rather broad size distribution. In addition, most, but not all, sols have a volume fraction of dispersed material that seldom exceeds 50%. In emulsions, that quantity is seldom less than 10% and sometimes as high as 90%. In foams, to be discussed in the following chapter, the dispersed phase content is often even higher. However, as stated in Chapter 10, since emulsions *look like* colloids and *act like* colloids - size notwithstanding - they *are* colloids.

FUNDAMENTAL CONCEPTS IN EMULSION SCIENCE AND TECHNOLOGY

As is the case in most discussions of interfacial systems and their applications, definitions and nomenclature can play a significant role in the way the material is presented. The definition of *emulsion* to be followed here is that they are heterogeneous mixtures of at least one immiscible liquid dispersed in another in the form of droplets, the diameters of which are, in general, greater than 0.1μm. Such systems possess a minimal stability, generally defined rather arbitrarily by the application or some relevant reference system, which may be enhanced by the inclusion of additives such as surfactants, finely divided solids, polymers, etc. Such a definition excludes foams and sols from classification as emulsions, although it is possible that systems prepared as emulsions may, at some subsequent time, become dispersions of solid particles or foams.

When discussing emulsions, it is always necessary to specify the

221

role of each of the immiscible phases of the system. Since in almost all cases, at least one liquid will be water or an aqueous solution, it is common practice to describe an emulsion as being either oil-in-water (O/W) or water-in-oil (W/O), where the first phase mentioned represents the dispersed phase and the second the continuous phase.

There is, in principle, no reason why one cannot prepare an oil-in-oil emulsion (O/O). However, the generally high miscibility of most organic liquids is an important limitation. More important, however, is the fact that the nature of interfaces dictates that a system tends to attain a situation of minimum energy, in this case minimum interfacial area, so that some additive must be employed to retard that process. Unfortunately, few materials are sufficiently surface active at such oil--oil interfaces to impart the required minimal stability necessary for the preparation and maintenance of such emulsions. Oil-in-oil emulsions of short persistence can, however, constitute an intermediate step in the preparation of nonaqueous emulsion polymers.

There are three major characteristics of an emulsion, factors (at least) that must be considered. These include:

(1) Which of the two liquid phases will be the continuous phase and which the dispersed phase when the emulsion is formed, and what factors can be used to control that result.

(2) What factors control the stability of the system; that is, what factors affect the creaming or sedimentation of the dispersed phase, drop coalescence, flocculation, etc.

(3) What factors control the often complex rheology of emulsified systems and how they can be effectively controlled.

The following discussions will touch upon each of these three questions. While the information presented is far from comprehensive, it will hopefully provide a useful introduction to some of the problems and solutions encountered in the practice of emulsion technology.

EMULSION FORMATION

The preparation of an emulsion requires the formation of a very large amount of interfacial area between two immiscible liquids. If a sample of 10 mL of an oil is emulsified in water to give a droplet diameter of 0.2 μm, the resulting O/W interfacial area will have been increased by a factor of approximately 10^6. The work required to generate one square centimeter of new interface is given by

$$W = \sigma_i \Delta A \qquad (11.1)$$

where σ_i is the interfacial tension between the two liquid phases and ΔA is the change in interfacial area. If the interfacial tension between the oil and water is assumed to be 52 mN m^{-1} (as for a hydrocarbon liquid), the *reversible* work required to carry out the dispersion process will be on the order of 2×10^8 ergs. Since that amount of work remains in the systems as potential energy, the system is thermodynamically unstable and rapidly undergoes whatever transformations possible to reduce that energy, in this case, by reducing the interfacial area. If some material can be added to the system to reduce the value of σ_i to approximately 1 mN m^{-1}, the magnitude of W will be reduced to 3 X 10^6 ergs --- a substantial reduction in *W* - but the system will still be unstable. Only if the interfacial tension (and therefore *W*) is zero can a truly stable system be obtained. Obviously, thermodynamics is the constant enemy of the emulsion maker.

Luckily, although thermodynamics will be the factor controlling the ultimate long-term stability of an emulsion, kinetics can play an important role over the short term, and it is through kinetic pathways that most useful emulsions achieve their needed stability. It is clear, then, that while lowering the interfacial tension between phases is an important factor in the formation and stabilization of emulsions, that may not always represent the most important factor in their preparation and ultimate application.

As anyone familiar with the preparation of practical emulsions knows, the process is still almost as much an "art" as a science. The results obtained for a given oil--water system will depend on the dispersing process used, the characteristics and quantities of additives employed, mixing temperature, order of mixing, and, reportedly, the surface properties of the mixing machinery (ie, which liquid preferrentially wets the container material), etc. Then, of course, there is always the question of the phase of the moon!

Art and magic aside, there are three principal methods of emulsion preparation which are most often employed. A fairly comprehensive coverage of those methods is presented in the work by Becher et al[1] cited in the Bibliography. The three methods most often employed include: (1) physical emulsification by drop rupture; (2) emulsification by phase inversion: and (3) "spontaneous" emulsification. The latter two methods may be described as "chemicall based" processes in that the nature of the final emulsion will be controlled primarily by the chemical makeup of the system (ie, the chemical nature of additives, the ratios of the two phases, temperature, etc), while in the first it will depend more upon the mechanical nature of the process (ie, amount and form of energy input.), as well as the rheological and chemical properties of the components. Other possibilities exist (see Table 11.1), however, most are of limited practical importance.

EMULSIONS AND THE LIQUID-LIQUID INTERFACE

In almost all practical emulsions, some additive (an *emulsifier*) is required to facilitate the formation of drops of the desired size and stability. Normally, one additive, at least, will be a material (defined below) that has the necessary characteristics to facilitate the formation of small droplets and produce the type of emulsion desired (O/W or W/O). The additive, an *emulsifier* and/or *stabilizer*, may perform two primary functions: (1) lower the energy requirements of drop formation (ie, lower

Table 11.1. Some general "mechanical" methods of emulsification.

Method	Energy input[a]	Process[b]	Drop formation[c]
1. Shaking	L	B	T
2. Stirring			
a. Simple	L	B,C	T,V
b. Rotor-stator	M-H	B,C	T,V
c. Vibrator	L	B,C	T,V
d. Scraper	L-M	B,C	V
3. Pipe flow			
a. Laminar	L-M	C	V
b. Turbulent	L-M	C	T
4. Colloid mill	M-H	C	V
5. Ball and roller mill	M	B,C	V
6. Homogenizer	H	B,C	T,V,C
7. Ultrasonic	M-H	B,C	C,T
8. Injection	L	B,C	T,V
9. Electrical	M	B,C	---
10. Condensation	L-M	B,C	---
11. Aerosol to liquid	L - M	B,C	---

[a] L = low, M = medium, H = high.
[b] B = batch, C = continuous.
[c] T = turbulence, V = viscous forces in laminar flow, C = cavitation.

the interfacial tension) and (2) retard the process of drop reversion to separate bulk phases. In order to function properly, it must adsorb at the L--L interface.

In its second function, the additive must form some type of film or barrier (monomolecular --- electrostatic or steric --- or liquid crystalline) at the new L--L interface that will prevent or retard droplet flocculation and coalescence. The process of barrier formation or adsorption must be rapid relative to the rate of drop coalescence or a rather coarse emulsion will result. Also, with the formation of more interface, the adsorption of the emulsifier depletes its bulk concentration, so that attention must be paid to the quantity of the material employed relative to the final result desired, as well as its quality as an emulsifier. As will be seen below, the exact role of an emulsifier in emulsion formation can be quite complex, and is not always completely understood. In any case, its (or, in many cases, their) presence will be vital to successful emulsion formation and stability.

Classification of Emulsifiers and Stabilizers

There are four general classes of materials that can, under the proper circumstances, act as emulsifiers and/or stabilizers for emulsions. The list includes common (non-surface-active according to the definition in Chapter 3) ionic materials, colloidal solids, polymers, and surfactants. Each class varies greatly in its effectiveness in a given role, and in its mode of action.

Adsorbed ions (non-surface-active) will usually do little to affect interfacial tension (except to raise it in some cases) and therefore do little to facilitate emulsification. However, some may, under the proper circumstances, aid in stabilizing the system by imposing a slight electrostatic barrier between approaching drops. Alternatively, they may affect the stability of the system by their action in orienting solvent molecules in the neighborhood of the interface, thereby altering some local physical properties (dielectric constant, viscosity, density, etc, (See Chapter 4) producing a small stabilizing effect.

Small *colloidal materials* (sols), while not affecting interfacial tensions, can stabilize an emulsion by erecting a physical barrier between drops, thereby retarding or preventing drop coalescence. The action of such materials will depend on several factors, the most important of which are the partical size and the specific interfacial interactions between the solid surface and the two liquid phases making up the system.

Polymeric additives may aid in emulsion formation as a result of surface-active properties but are usually more important as stabilizers. Their action may result from steric or electrostatic interactions, from changes in the interfacial viscosity or elasticity, or from changes in the bulk viscosity of the system. In many if not most cases, the function of polymeric stabilizers is a combination of several actions.

Finally, "normal" *monomeric surfactants* are usually added in order to decrease the interfacial tension and impart added stability to the system. The type and quantity of surfactant employed will be determined by the specific properties of the liquid phases, the type of emulsion desired, conditions of use, etc.

Of the possible emulsifiers, most are what are considered true surfactants, in that they are effective at lowering significantly the interfacial tension between the two liquid phases. Other additives such as polymers and colloids function primarily as stabilizers, rather than emulsifiers. Most polymers are not sufficiently effective at lowering interfacial tensions to act in that regard. In addition, because of their molecular size, the adsorption process for polymers is generally very slow relative to the time scale of the emulsification process. The same applies to stabilizing colloids, in which their action requires the wetting of the particles by the two liquid phases to facilitate their location at the interface. The primary function of polymers and sols in emulsions is in the retardation of droplet flocculation and coalescence.

The processes of flocculation and coalescence in the context of emulsion stability will be treated in a bit more detail below. At this point it is useful to point out their role in the determination of the final nature of the emulsion. The process leading to emulsion formation usually begins with the production of preliminary large drops, probably of both liquid phases. The continuous phase-to-be will be determined by many factors, to be outlined below. In any case, droplets of that phase must disappear rapidly during the process through flocculation and coalescence. The ultimate dispersed phase, on the other hand, must maintain (or reduce) its droplet size during and after processing.

The emulsification process is so dynamic and complex that an accurate model and theoretical treatment is almost impossible. With certain limitations its is possible to obtain order-of-magnitude estimates of such steps as droplet formation rate and surfactant transport and adsorption rates. However, the work involved is seldom worth the trouble in practical cases. Flocculation and coagulation rates during preparation are even more difficult to analyze because of the dynamics of the process and the turbidity of the flow involved. Collision rate theory has been found useful in the analysis of emulsion flocculation and coalescence in "quiet" emulsions and to some extent in idealized "turbulent" systems. In reality, however, those events are beyond prediction in the context of emulsion preparation processes. There are simply too many events (simultaneous and sequential) occurring in that time frame and, in essence, chaos rules. Luckily, if conditions are closely controlled, an adequate degree of reproducibility can be obtained so that useful systems can be formulated and produced with reasonable confidence.

ADSORPTION AT LIQUID--LIQUID INTERFACES

The relationship between the adsorption of a molecule at an oil--water interface and the resulting interfacial tension is an important one and warrants a brief review here. The Gibbs equation for a system composed of one phase containing a nonionic solute adsorbing at the interface with a second phase is written

$$\Delta G_i = -1/RT \, (\partial \sigma_i / \partial \ln a)_T \qquad (11.2)$$

where the terms are as previously defined. The equation states that at a liquid--liquid interface, as in the liquid--vapor case, the amount of surfactant adsorbed can be determined from the slope of the σ_i vs ln a curve. In dilute surfactant systems, the concentration, c (mols L^{-1}) can be substituted for activity without serious loss of accuracy. The simple relationship of the Gibbs equation can have significant practical application in the preparation of emulsions, especially in defining the relationship between emulsion droplet size and total surfactant concentration. Using eq. 11.2, the adsorption of surface-active molecules at an interface can be calculated from determinations of the interfacial tension. If ionic species are involved, life is complicated a bit by the presence the counter-ion and a variable surface potential that increases as adsorption proceeds.

It is of interest to try to relate the adsorption characteristics of a surfactant to the stability of an emulsion stabilized solely by an adsorbed monomolecular film. The total number of molecules that can be adsorbed in a given interfacial area will be controlled mainly by the effective "area per molecule" of the adsorbing species. That is, how many of the molecules can fit into the limited space of the interface? For most "normal" surfactant species, the area per molecule is determined primarily by the hydrophilic group and its hydration layer. The relative solubility of the surfactant in the two phases will also affect the result, but that factor is difficult to determine and is most often ignored. A few representative molecular "areas" at the oil water interface are given in Table 11.2, along with various oil phases.

From the data in Table 11.2, it can be seen that the "experimental" area occupied by the sulfate and carboxylate groups are relatively large compared to their "projected" areas calculated from molecular models (0.45 vs 0.25 nm^2 for carboxyl and 0.48 vs 0.28 nm^2 for sulfate). Also, while the effective area per molecule is found to vary by only about 6% for normal hydrocarbon oils with from 7 to 17 carbon atoms, in the presence of benzene, unsaturation, and non-hydrocarbon liquids, the head group "size" is often found to increase by as much as 35%.

Several explanations for those observations have been proposed, although the truth probably lies in a mixture of events. The relatively large sizes of the two head groups considered almost certainly arise due to electrostatic repulsion between the charges, although solvation is undoubtedly involved to some extent. When the ionic strength of the solution is increased, the effective molecular area is seen to decrease due to screening of the charges and possibly due to solvation changes as counterions begin to be more tightly bound at the interface. In the second case, normal hydrocarbon oils are expected to behave essentially the same with respect to their interaction with adsorbed molecules, regardless of their chain length (within limits, of course). However, when benzene, unsaturated compounds, or a non-hydrocarbon liquid is used, there exists the probability of

Table 11.2. Typical molecular areas of common surfactants at aqueous--oil interfaces and saturation adsorption.

Surfactant and oil phase	Area (nm^2)
$CH_3(CH_2)_{11}SO_4Na$ -- n-hexane	0.45
" --n-octane	0.48
" --n-decane	0.49
" --n-heptadecane	0.51
" --benzene	0.65
" --carbon tetrachloride	0.53
" --1-hexene	0.57
Na diisoctylsulfo-succinate -- n-heptane	1.11
Na lauroyl taurate -- "	0.57
$CH_3(CH_2)_{10}SO_3Na$ --benzene	0.57
Na laurate--n-heptane	0.45

significantly more interaction between the oil containing π-electrons or large, polarizable atoms and the hydrophilic head group. Such specific interactions between aromatic molecules and polar species are well established, as in the interfacial tension for benzene--water (35 mN m^{-1}) vs that for octane--water (50.8. mN m^{-1}) at 20°C. For the surfactants with bulky hydrophobic tails or "kinks" in their structures, such as the sulfosuccinates and taurates, the packing density at the interface is locally reduced leading to the observed increase in "size."

Since most nonionic surfactants have large, highly solvated polyoxyethylene head groups, it is more difficult to relate the interfacial area occupied by a given molecule to its structure. The problem is further exacerbated by the fact that such surfactant systems are composed not of a single molecular type, but rather an homologous series of varying composition and molecular weight. In such a case one must rely on some more empirical surfactant characterization system such as those discussed below for obtaining some idea of the activity of a given system.

GENERAL CONSIDERATIONS OF EMULSION FORMATION AND STABILITY

The question of emulsion stability has already been raised in the context of emulsion preparation. However, the preparation process is very dynamic and represents a complicated combination of events that is not easily analyzed. Once prepared, however, and left at rest to "do its own thing," the fates of individual droplets become more readily determined and some semblance of understanding can be dragged from the initial chaos. That is not to say, however, that there exists a good general theory of emulsion stability which one can apply under all, or even most, circumstances.

Even though emulsions as defined have been in use for thousands of years (even longer if natural emulsions are considered) no comprehensive theory of emulsion formation and stabilization has yet been developed that adequately describes, and predicts, the characteristics of many of the complex formulations encountered in practice. Except in very limited and specialized areas, the accurate prediction of such aspects of emulsion technology as droplet size, size distribution, and stability remain more in the realm of art than true science.

Definition of Emulsion Stability

When discussing the stability of an emulsified system (like the colloidal sols of Chapter 10), it is important to have a clear idea of the physical condition of the components and the terminology employed. Four terms commonly encountered in emulsion science and technology related to stability are: "breaking," "coalescence," "creaming," and "flocculation." Although they are sometimes found to be used almost

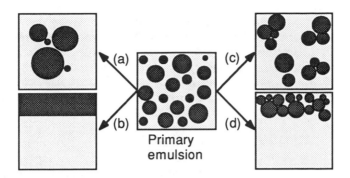

Figure 11.1. Illustration of the "fates" of emulsions: (a) coalescence; (b) breaking; (c) flocculation; (d) creaming.

interchangeably, those terms are in fact quite distinct in meaning so far as the condition of an emulsion is concerned.

Coalescence , for example, refers to the joining of two (or more) drops to form a single drop of greater volume, but smaller interfacial area (Figure 11.1a). Such a process is obviously energetically favorable in all cases in which there exists a positive (even if small) interfacial tension. Although coalescence results in significant microscopic changes in the condition of the dispersed phase (eg, changes in average particle size and distribution), it may not immediately result in a macroscopically apparent alteration of the system.

The *breaking* of an emulsion (Figure 11.1b) refers to a process in which a gross separation of the two phases occurs. The process is a macroscopically apparent consequence of the microscopic process of drop coalescence. In such an event, the identity of individual drops is lost, along with the physical and chemical properties of the emulsion. Such a process obviously represents a true loss in the stability of the emulsion.

Flocculation refers to the mutual attachment of individual emulsion drops to form flocs or loose assemblies of particles in which the identity of each is maintained (Figure 11.1c), a condition that clearly differentiates it from the action of coalescence. Flocculation can be, in many cases, a reversible process, overcome by the input of much less energy than was required in the original emulsification process.

Finally, *creaming* is a process which is related to flocculation in that it occurs without the loss of individual drop identities (Figure 11.1d). Creaming will occur over time with almost all emulsion systems in which there is a difference in the density of the two phases. The rate of creaming will be dependent on the physical characteristics of the system, especially the viscosity of the continuous phase. It does not necessarily represent a change in the dispersed state of the system, however, and can often be reversed with minimal energy input. If the dispersed phase happens to be the more dense of the two phases, the separation process is termed *sedimentation* .

Obviously, both flocculation and creaming represent conditions in which drops "touch" but do not combine to form a single unit of greater volume but smaller total interfacial area. The key to understanding the true stability of emulsions, then, lies on the line separating the processes of flocculation and coalescence.

SOME MECHANISTIC DETAILS
OF STABILIZATION

Even in the infancy of emulsion technology, several thousand years ago, it was recognized that in order to obtain a useful emulsion with any long-term persistence it was necessary to include a third component, which served some "magical" purpose and imparted the required degree of stability. Such additives, as outlined above, include simple inorganic electrolytes, natural resins, and other macromolecular compounds; finely divided, insoluble solid particles, which located themselves at the interface between the two phases; and amphiphilic or surface-active materials, which were soluble in one or both phases and significantly altered the interfacial characteristics of the system. In practice, one commonly finds that a combination of two or more mechanisms is most effective for producing true, long-term emulsion stability.

While some brief comments on the actions of emulsifiers and stabilizers were presented above, it is useful to have a little more detailed concept of just how these materials complete their function in the emulsified system. The actions of the most important systems --- polymers, sols, and surfactants --- will therefore be explained in a bit more detail.

Polymeric Emulsifiers and Stabilizers

In nature as well as in technology, polymeric emulsifiers and stabilizers play a major role in the preparation and stabilization of emulsions. Natural materials such as proteins, starchs, gums, cellulosics, and their modifications, as well as wholly synthetic compounds such as polyvinyl alcohols, polyacrylic acid, and polyvinylpyrrolidone, have several characteristics that make them extremely useful in emulsion technology. By the proper choice of chemical composition, such materials can be made to adsorb strongly at the interface between the continuous and dispersed phases. By their presence, they can reduce interfacial tension and/or form a barrier (electrostatic and/or steric) between drops. In addition, their solvation properties serve to increase the effective adsorbed layer thickness, increase interfacial viscosity, and introduce other factors that tend to favor the stabilization of the system.

The effectiveness of polymeric materials at lowering interfacial tensions is usually quite limited. More important to their function is the fact that polymers can form a substantial mechanical and thermodynamic barrier at the interface which retards the approach and coalescence of individual emulsion droplets. The polymeric nature of the materials means that each molecule can be strongly adsorbed at many sites on the interface. As a result, the chance of desorption is greatly reduced or effectively eliminated, and the interfacial layer attains a degree of strength and rigidity not easily found in systems of monomeric materials. In addition, the presence of polymeric materials in the system can retard processes such as creaming by increasing the viscosity of the continuous phase thereby reducing the rate of droplet encounters that could lead to flocculation or coalescence. The concept of steric stabilization of colloids has already been introduce in Chapter 10 and will not be considered further here. It can be assumed that the mechanisms involved apply equally to solid and liquid dispersed systems, although slight variations on the theme are to be expected.

The subject of the mechanisms and degree of polymer adsorption at interfaces is also discussed in more detail in Chapter 14. For now, suffice it to say that macromolecular additives to emulsion systems constitute a major pathway for attaining workable, long-lived practical emulsions. In fact, their use is essential to many important product types, not the least of which are food colloids, inks, pharmaceuticals, and the photographic industry.

Solid Particles

A second class of effective emulsifying agents and stabilizers commonly encountered is finely divided solid particles. It has been known for some time that particles of true colloidal dimensions (less than 100 nm in diameter, for example) which are partially wetted by both aqueous and organic liquids can form stabilizing films and produce both O/W and W/O emulsions with significant stability. Emulsion stabilization by solid particles relies on the specific location of the particles at the interface to produce a strong, rigid barrier that prevents or inhibits the coalescence of drops. It may also impart a degree of electrostatic repulsion, which enhances the overall stabilizing power of the system. There are three keys to the use of particulate solids as emulsion stabilizers: (1) particle size, (2) the state of stabilizer particle dispersion, and (3) the relative wettability of the particles by each liquid component of the emulsion system.

In practice it is found that he stabilizer particles must be small compared to the emulsion droplet and in a state of incipient flocculation. That is, they should have limited colloidal stability in both liquids, otherwise their tendency to "locate" at the oil--water interface will not be sufficiently strong for them to "complete their mission." For the third condition, the solid must exhibit a significant contact angle at the three-phase (oil--water--solid) contact line, conventionally as measured through the aqueous phase. For maximum efficiency, the stabilizer usually should be preferentially wetted by the continuous phase (but not excessively so). If the solid particles are too strongly wetted by either of the two liquid phases the required stabilizing action will not result. It is usually necessary, therefore, to control such factors closely by controlling the system pH or by the addition of materials that adsorb onto the particles and impart the required surface characteristics (see Chapter 17).

Surfactants

The last major class of emulsifiers and stabilizers is that of the monomeric surfactants which adsorb at interfaces, lower the interfacial tension, and, hopefully, impose a stabilizing barrier between emulsion drops. Surfactants are the most widely studied and perhaps best understood class of emulsifiers and stabilizers. Perhaps because they are more amenable to both experimental and theoretical analysis, they have been used to probe the finer points of emulsified systems. They will therefore be discussed in more detail than polymers and sols.

Because of their effectiveness at lowering interfacial tensions, they are of vital importance to most practical systems, facilitating the formation of small droplets with a minimum of power input. However, because of their relative mobility into and out of the interface, their practical effectiveness as stabilizers acting alone has been questioned. There is no doubt that their presence significantly prolongs the life of most emulsions; however, the assumed role of the monomolecular adsorbed film is slowly being diminished in favor of their more complex activity in the guise of liquid crystals.

Surfactant Structure and Emulsion Performance

It would be nice if the world of emulsion formulation were such that a simple correlation could be obtained between the chemical structure of a surfactant and its performance in practice. Unfortunately, the complicated nature of typical emulsion formulations --- the nature of the oil phase, additives in the liquid phases, specific surfactant interactions, etc --- make correlations between surfactant structure and properties in emulsification processes very empirical.

In the absence of a handy quantitative and absolute method for choosing a surfactant for a given application, it is possible to outline a few rules of thumb that have historically proved useful for narrowing down the possibilities and limiting the amount of experimentation required for the final selection of surfactant(s) for a given application. First and foremost, of course, the surfactant must exhibit sufficient surface activity to ensure significant adsorption at the oil--water interface. That activity must be related to the actual conditions of use and not inferred from its activity in water alone. The presence of materials such as electrolytes and polymers can greatly alter the role of the surfactant in stabilizing an emulsion as well as in controlling the type of emulsion formed.

The surfactant (or surfactants) employed in an emulsion formulation should produce as strong an interfacial film as possible, consistent with their ability to produce the required droplet size under the conditions of emulsification. It is useful, therefore, to choose a surfactant system with maximum lateral interaction among the surfactant molecules

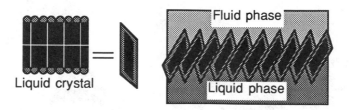

Figure 11.2. Schematic illustration of the probable mechanism by which liquid crystals stabilize liquid--fluid interfaces (ie, emulsions and foams) by the formation of relatively rigid interfacial barriers that retard or prevent interactions leading to coalescence or breaking.

concurrent with efficient and effective lowering of the interfacial tension.

At the molecular level, the choice of surfactant for a given application must take into consideration the type of emulsion desired and the nature of the oil phase. As a general rule, oil-soluble surfactants will preferentially produce W/O emulsions while water-soluble surfactants yield O/W systems. Because of the role of the interfacial layer in emulsion stabilization, it is often found that a mixture of surfactants with widely differing solubility properties will produce emulsions with enhanced stability. Finally, it is usually safe to say that the more polar the oil phase, the more polar will be the surfactant required to provide optimum emulsification and stability. Such rules of thumb, while having great practical utility, are less than satisfying on a scientific level. One would really like to have a neat, quantitative formula for the design of complete emulsion systems. A number of attempts have been made over the years to develop just such a quantitative approach to surfactant selection. A brief discussion of some such approaches is given below.

Liquid Crystals and Emulsion Stability

The mechanical strength of the interfacial film stabilizing an emulsion can have a significant impact on the overall stability of the system. Liquid crystalline phases occur in solutions of surface-active materials as the concentration is increased from that of a dilute solution to a saturated system in which true crystallization occurs (see also Chapter 15).Such phases possess a degree of order that produces substantial changes in the properties of the system relative to those of molecular or micellar solutions, including a higher degree of rigidity, larger structural units, and less fluctuation in composition. In the present context, such phases, if present at the O/W interface, might be expected to impart an added degree of stability to systems in which they are produced (Figure 11.2). In a utilitarian sense, then, surfactant liquid crystals at interfaces may be considered to act in a manner similar to colloidal sols or perhaps the mixed interfacial complexes discussed below.

The presence of liquid-crystalline phases at the oil--water interface

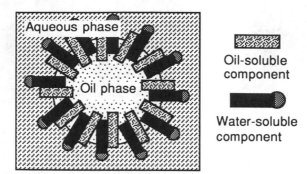

Figure 11.3. Schematic illustration of the formation of interfacial complexes between water- and oil-soluble surfactants, which may produce a greater degree of emulsion stability.

has been shown to produce improvements in the stability of various emulsions, although the exact mechanism of their action is still subject to some question. Even in the absence of complete understanding, the use of liquid crystals at O/W interfaces has been demonstrated in practical applications.

Mixed Surfactant Systems and Interfacial Complexes

It has been found that the presence of two primary surfactant species, one water and the other oil soluble, can greatly enhance the stability of an emulsion system. The effect has been related to the production of very low interfacial tensions and the formation of cooperative surfactant "complexes," which impart greater strength to the O/W interface.

An *interfacial complex* may be defined as an association of two or more surface-active molecules at an interface in a relationship that does not exist in either of the bulk phases (Figure 11.3). Each bulk phase must contain at least one component of the complex, although the presence of both in any one phase is not ruled out. According to Le Chatelier's principle, the formation of an interfacial complex will increase the Gibbs interfacial excess Γ_i for each individual solute involved, and consequently, the interfacial tension of the system will decrease more rapidly with increasing concentration of either component.

The existence of the interfacial complex is distinct from the situation of simple coadsorption of oil-soluble and water-soluble surfactants. In the case of coadsorption, each component is competing for available space in the interfacial region and contributes a weighted effect to the overall energetics of the system. The effect of the complex will be synergistic, with the net effect exceeding that of either component or any simple combination of the two.

A possible beneficial effect of interfacial complex formation, in addition to improved surface energetics, is that such structures may

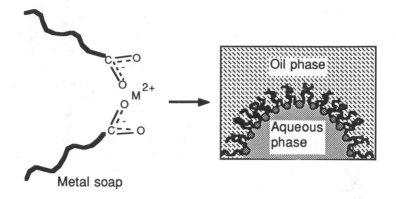

Metal soap

Figure 11.4. Insoluble metal soaps at emulsion interfaces: the limited solubility of metal soaps in water favors the formation of water-in-oil emulsions, in which the adsorbed soaps may form a strong, resilient interfacial film.

possess a greater mechanical strength than a simple mixed interfacial layer. The closer molecular packing density and greater extent of lateral interaction between hydrophobic chains may result in significant decreases in the mobility of molecules at the interface and a decrease in the rate of drop coalescence. Such an effect has often been mentioned in terms of increased interfacial viscosity or elasticity, although the exact role of interfacial rheology in emulsion stabilization is not completely understood.

Emulsion   Type

The idea that surfactant molecules preferentially orient at the oil--water interface not only helps clarify the picture of monomolecular film stabilization, but also sheds light on the problem of explaining the emulsion type obtained as a function of the chemical structure of the adsorbed species. It was recognized early that the nature of the surfactant employed in the preparation of an emulsion could influence the type of emulsion formed. For example, while the alkali metal salts of fatty acid soaps normally produce O/W emulsions under a given set of circumstances, the use of di- and trivalent soaps often results in the formation of W/O systems. The invocation of a monolayer mechanism for the stabilization of emulsion droplets requires the formation of a relatively close-packed surfactant film at the interface. It is clear, then, that the geometry of the adsorbed molecules must play an important role in the effect obtained. For efficiency of packing, it can be seen from Figure 11.4 that the formation of W/O systems with polyvalent soaps seems almost inevitable.

The steric requirements of surfactant molecules have historically been referred to in terms of an "oriented wedge" of surfactant molecules at the interface. The concept lead to the "rule of thumb" that if the hydrophilic head of the surfactant was larger than the tail, the result would be an O/W emulsion. If the relationship were reversed, the emulsion would be of the W/O type --- a very neat and simple relationship which, due to numerous exceptions and lack of theoretical foundation, fell out of favor for some time. More recently, however, consideration of the critical role of the structure of the surfactant has again come into vogue, this time with some theoretical backing. A bit more will be said about molecular geometric considerations below.

A related "rule" concerning surfactant structure and the type of emulsion formed is related to the solubility of the surfactant in the two liquids. The rule states that the liquid in which the surfactant is most soluble will be the continuous phase in the final emulsion. That is, if the surfactant is more soluble in the oil phase, a W/O emulsion will result. A more water-soluble material will produce a O/W system.

That concept was extended from a theoretical standpoint by the postulation that the presence of an absorbed interfacial film requires the existence of two interfacial tensions --- one at the oil--monolayer interface and a second at the water-monolayer interface. Since the two tensions will not, except in very unusual circumstances, be equal, the interfacial layer will spontaneously curve, with the direction of curvature determined by the relative magnitudes of the two tensions. Logically, the film will curve in the direction of the higher interfacial tension so that the phase associated with that interface will become the dispersed phase in the system. Unfortunately, this seemingly useful rule also falls on the sword of too many exceptions. Although such simple views of the role of the adsorbed monolayer in determining the nature of the emulsion can be quite useful, the many exceptions make them less than satisfying from a theoretical point of view.

In addition to the molecular nature of the emulsifier employed, the relative amounts of the two phases in the system might be expected to affect strongly the type of emulsion obtained. If one assumes that an emulsion is composed of rigid, spherical droplets of equal size, simple geometry shows that the maximum volume fraction of dispersed phase which can be obtained is 74.02%. One may speculate, then, that any emulsified system in which that level was exceeded must result in phase inversion to an emulsion of the opposite type. It has been shown, however, that it is possible to prepare emulsions of dispersed-phase volume fractions far exceeding that "theoretical" limit. Looking at the reality of the situation more closely, it is possible to identify several reasons to invalidate such a simple geometric approach.

In the first place, emulsion droplets are not, and will never be, perfectly monodisperse; as a result, it is possible for smaller droplets to locate themselves in the spaces between close packed, larger droplets (Figure 11.5), increasing the total potential packing density of the system. In addition, emulsion droplets are not rigid spheres, but highly deformable so that their shape can be changed from spherical to various elongated or polyhedral shapes to fit the needs of the system. Large excursions from sphericity will be generally unfavorable, of course, since

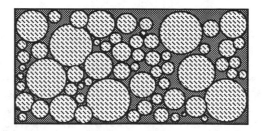

Figure 11.5. Because of the polydispersity of most emulsions, it is possible to obtain a packing density of the dispersed phase greater than that theoretically possible for monodisperse systems. The situation may be further helped by distortion of the drops into polygonal shapes, provided that the interfacial film is sufficiently strong to protect the increased interfacial area.

that will require the formation of additional interfacial area for a given dispersed volume fraction. Such an increase in interfacial area may strain the ability of the adsorbed emulsifier or stabilizer film to the point where droplet coalescence occurs.

Even though the early rules of thumb concerning emulsifier type and emulsion type are rife with exceptions, there remains something inherently satisfying in the ideas they contain, aside from their continued practical utility. However science forever strives to develop the "unified theory of " everthing which allows one to predict exactly what will happen in a given set of circumstances --- and the area of emulsion science and technology is no exception. The first really practical strides to that goal were taken with the introduction of the so-called *hydrophile-lipophile balance* (HLB) number.

The Hydrophile--Lipophile Balance (HLB)

It has been a long-term goal of surface science to devise a quantitative way of correlating the chemical structure of surfactant molecules with their surface activity which would facilitate the choice of material for use in a given formulation. The greatest success along these lines has been achieved in the field of emulsions. The first reasonably successful attempt at that goal was the hydrophile--lipophile balance (HLB) system first developed by Griffin.[1] His work was an attempt to place the choice of the optimum nonionic surfactant for the stabilization of a given emulsion on a quantitative, somewhat theoretical basis. In this system, Griffin proposed to calculate the so-called HLB number of a surfactant from its chemical structure and match that number with the HLB of the oil phase which was to be dispersed. The system employed certain empirical formulas to calculate the HLB number for molecular structures, producing numbers in betwen 0 and 20 on an arbitrary scale.

At the high end of the scale (8--18) lie hydrophilic surfactants which possess high water solubility and generally act as good aqueous solubilizing agents, detergents, and stabilizers for O/W emulsions; at the low end (3--6) are surfactants with low water solubility which act as soublizers of water in oils and good W/O emulsion stabilizers. In the middle are materials that are very surface active, in terms of lowering surface and interfacial tensions, but that generally perform poorly as emulsion stabilizers. The effectiveness of a given surfactant in stabilizing a particular emulsion system would then depend upon the balance between the HLB's of the surfactant and the oil phase involved.

For nonionic surfactants with polyoxyethylene solubilizing groups, the HLB may be calculated from the formula

$$\text{HLB} = (\text{mol \% hydrophilic group}) / 5 \qquad (11.3)$$

In such a calculation, an unsubstituted polyoxyethylene glycol would have an HLB of 20. HLB values for some typical nonionic surfactants are given in Table 11.3. Surfactants based upon polyhydric alcohol fatty acid esters such as glycerol monostearate can be handled by the relationship

$$\text{HLB} = 20(1 - S/A) \qquad (11.4)$$

where S is the saponification number of the ester and A is the acid number of the acid. A typical surfactant of this type, Tween 20 (polyoxyethylene-20-sorbitan monolaurate), with $S = 45.5$ and $A = 276$ would have an HLB of 16.7. For materials which cannot be saponified, an empirical formula of the form

$$\text{HLB} = (E + P)/5 \qquad (11.5)$$

Table 11.3. HLB values for typical nonionic surfactant structures.

Surfactant	Commercial name	HLB
Sorbitan trioleate	SPAN 85	1.8
Sorbitan tristearate	SPAN 65	2.1
Propylene glycol monostearate	"PURE"	3.4
Glycerol monostearate	ATMUL 67	3.8
Sorbitan monooleate	SPAN 80	4.3
Sorbitan monostearate	Span 60	4.7
Diethylene glycol monolaurate	GLAURIN	6.1
Sorbitan monolaurate	SPAN 20	8.6
Glycerol monostearate	ALDO 28	11
Polyoxyethylene(2) cetyl ether	BRIJ 52	5.3
Polyoxyethylene(10) cetyl ether	BRIJ 56	12.9
Polyoxyethylene(20) cetyl ether	BRIJ 58	15.7
Polyoxyethylene(6) tridecyl ether	RENEX 36	11.4
Polyoxyethylene(12) tridecyl ether	RENEX 30	14.5
Polyoxyethylene(15) tridecyl ether	RENEX 31	15.4

can be employed, where E is the weight percent of oxyethylene chains and P is the weight percent of polyhydric alcohol (glycerol, sorbitan, etc) in the molecule.

Although the system proposed by Griffin proved to be very useful from a formulation chemists point of view, its empirical nature did not satisfy the desire of many for a more sound theoretical basis for surfactant characterization. Davies and Rideal[2] suggested that HLB numbers could be calculated based upon group contributions according to the formula

$$HLB = 7 + \Sigma(\text{hydrophilic group numbers}) - \Sigma(\text{hydrophobic group numbers}) \qquad (11.6)$$

Some typical group numbers as listed by Davies and Rideal, as well as other investigators, are listed in Table 11.4.

The use of the HLB system for choosing the best emulsifier system for a given application originally required the performance of a number of experiments in which surfactants or surfactant mixtures with a range of HLB numbers are employed to prepare emulsions of the oil in question, and the stability of the resulting emulsions is evaluated by measuring the amount of creaming that occurrs with time. The use of surfactant mixtures can become complicated by the fact that such mixtures often produce more stable emulsions than a single surfactant with the same nominal HLB number. The HLB of a mixture is usually assumed to be an algebraic mean of the HLB's of the components

$$HLB_{mix} = f_A \times HLB_A + (1 - f_A) \times HLB_B \qquad (11.7)$$

where f_A is the weight fraction of surfactant A in the mixture. While strict adherence to eq. 11.7 is not usually found, the concept remains a useful tool for surfactant formulation purposes, especially in the absence of a better option.

While it goes a long way toward simplifying the choice of surfactants for the preparation of a given O/W emulsion, the HLB system does not always provide a clearcut answer for a given system. It does not, for example, take into consideration the effects of a surfactant on the physical properties of the continuous phase, especially its rheological characteristics. As noted previously, the viscosity of the continuous phase will significantly affect the rate of creaming, as will alterations in the relative densities of the two phases. As a result, it is possible to prepare very stable emulsions with surfactants whose HLB numbers lie well away from the "optimal" which would be predicted by the the strict application of the HLB approach. Regardless of its faults, the HLB system as originally proposed by Griffin and subsequently expanded by others has found extensive practical use. Theoretical explanations of an HLB-like approach based on cohesive energy densities[3] go a long way toward justifying (in a theoretical sense) the empiricism of the original Griffin

Table 11.4. Group numbers for the calculation of HLB's acording to Davies and Rideal.

Group	HLB Number
Hydrophilic	
---SO_4Na	38.7
---COOK	21.1
---COONa	19.1
---N (tertiary amine)	9.4
Ester (sorbitan)	6.8
Ester (free)	2.4
---COOH	2.1
---OH (free)	1.9
---O---	1.3
---OH (sorbitan)	0.5
Hydrophobic	
---CH---	-0.475
---CH_2---	-0.475
---CH_3	-0.475
=CH---	-0.475
---CF_2---	-0.87
---CF_3	-0.87
Miscellaneous	
---(CH_2CH_2O)---	0.33
---$(CH_2CH_2CH_2O)$---	-0.15

method and turning it into a "satisfying" theory.

The Geometrical Approach

As an additional tool for approaching the problem of relating surfactant structure to emulsion formulation (to supplement, but not replace the "classical" approaches) Israelachvilli, Mitchell, and Ninham[4,5] considered the geometrical constraints imposed by the particular molecular characteristics of the surfactant molecule which control the formation of aggregates (eg, micellization), interfacial interactions, etc.

In analyzing the relationships between the aggregation characteristics of a surface-active material --- aggregate size, shape, curvature, etc --- and molecular structure, the authors defined a geometric factor F by the equation

$$F = v/a_o l_c \qquad\qquad (11.8)$$

where v is the molecular volume of the hydrophobic group, a_o is the head-group area, and l_c is the critical length of the hydrophobe. The factor F can be viewed as a type of HLB number, based on _volume fraction_ instead of _weight fraction_ of hydrophobe, and the geometry of the hydrophobic chain.

By use of the geometric considerations (see also, Chapter 15), it can be seen that the value of F determined from molecular geometry should predict the type of emulsion formed by a particular surfactant. For instance, if $F < 1$, the curvature of the oil--water interface should be concave toward the oil phase leading, to an O/W emulsion. For $F > 1$, the reverse would be expected. At $F = 1$, a critical condition is expected where phase inversion would occur, or multiple emulsion formation will be favored.

Phase Inversion Temperature (PIT)

An important class of surfactants for use as emulsifiers and stabilizers is that of the nonionic polyoxyethylene (POE) adducts. This class of materials is solubilized in water through hydrogen bonding with the POE chain. As mentioned in Chapter 4, hydrogen bonding is a temperature sensitive interaction and decreases as the temperature increases. Nonionic materials therefore exhibit an inverse temperature-- solubility relationship leading to the appearance of a _cloud point_ for many examples of the class. The cloud point has already been discussed in general terms, but its existence has interesting ramifications when viewed in the context of emulsions.

Since the cloud point of a surfactant is a structurally related phenomenon, it should also be related to HLB, CMC, etc, as is found to be the case. Clearly, temperature can play an important role in determining the surface activity where hydration (or hydrogen bonding) is the principle mechanism of solubilization. Because of the temperature sensitivity of such materials, their activity as emulsifiers and stabilizers also becomes temperature sensitive. In particular, their ability to form and stabilize O/W and W/O emulsions may change dramatically over a very narrow temperature range. In fact, an emulsion may "invert" to produce the opposite emulsion type as a result of temperature changes. Such a process is termed, _phase inversion_ and the temperature at which it occurs for a given system its _phase inversion temperature_ (PIT).

Shinoda and Arai[6,7] recognized the importance of the temperature effect on surfactant properties and introduced the concept of using the phase inversion temperature as a quantitative approach to the evaluation of surfactants in emulsion systems. As a general procedure, emulsions of oil, aqueous phase, and approximately 5% surfactant were prepared by shaking at various temperatures. The temperature at which the emulsion was found to be inverted from O/W to W/O (or vice versa) was then defined as the PIT of the system. Since the effect of temperature on the solubility of nonionic surfactants is reasonably well understood, the physical principles underlying the PIT phenomenon follow directly.

It is generally found that the same circumstances that affect the solution characteristics of nonionic surfactants --- their CMC, micelle size, cloud point, etc --- will also affect the PIT of emulsions prepared with the same materials. For typical polyoxyethylene nonionics, increasing the length of the POE chain will result in a higher PIT, as will a broadening of the POE chain length distribution. The use of phase inversion temperatures, therefore, represents a very useful tool for the comparative evaluation of emulsion stability. Although the PIT approach to surfactant evaluation is considerably newer than the HLB number, the effects of variables on the relationship between PIT's, surfactant structures, and emulsion stability show an almost linear correlation between the HLB of a surfactant under a given set of conditions and its PIT under the same circumstances. In essence, the higher the HLB of the surfactant system, the higher will be its PIT.

Application of HLB and PIT in Emulsion Formulation

The choice of a particular emulsifier system for an application will depend upon several factors, one of which will be chemically related (optimum HLB, PIT, etc); but others will be driven by the three "e's" - economic, environmental, and esthetic factors. The relative weight of the latter will depend mostly on price and value-added considerations for each individual system. Here we are concerned with the technical aspects of emulsion formation and stabilization, so other factors will be ignored.

In most general applications, the HLB system has been found most useful in guiding the formulator to a choice of surfactant most suited to his needs. Table 11.5 lists the ranges of HLB numbers that have been found to be most useful for various applications.

Table 11.5. HLB ranges and their general areas of application.

Range	Application
3--6	W/O Emulsions
7--9	Wetting
8--18	O/W Emulsions
3--15	Detergency
15--18	Solubilization

Obviously, the ranges in which surfactants of various HLB's can be employed are quite broad. Specific requirements for many systems have been tabulated in the work of Becher[8] and other cited references. While such tabulations can be very useful to the formulations chemist, it must be kept in mind that there is nothing particularly magic about a given HLB number. Many surfactants or mixtures may possess the same HLB, yet subtle differences in their chemical structures or interfacial properties may result in significant differences in performance. Particularly important may be the formation of interfacial complexes, liquid-crystalline phases, etc. Even though the additive nature of surfactant mixtures has not been found to be linear over a wide range of compositions, over the short range of one or two HLB units usually encountered in formulation work linearity can usually be assumed with little risk. It is therefore possible to fine tune a surfactant mixture with a minimum of actual experimental effort.

One approach to the application of surfactant HLB to formulation is to match that of the surfactant to the oil phase being employed. The HLB of the oil can be determined empirically or calculated using the data in Table 11.4. It is usually found that the principle of additivity will hold for mixtures of oils in a way similar to that for surfactants. Therefore, in formulating an emulsion, it is possible to determine the HLB of the oil phase and vary the surfactant or mixture HLB to achieve the optimum performance. The HLB numbers of some commonly used oil phases are given in Table 11.6. It should be pointed out that the values listed in the table are for the formulation of O/W emulsions. When W/O emulsions are required, the appropriate HLB value will usually be smaller.

Table 11.6. HLB numbers for typical oil phases.

Oil Phase	Nominal HLB
Lauric Acid	16
Oleic Acid	17
Cetyl Alcohol	15
Decyl Alcohol	14
Benzene	15
Castor Oil	14
Kerosene	14
Lanolin	12
Beeswax	9
Carnuba Wax	12
Paraffin Wax	10

It will be noted that HLB numbers are most often used in connection with nonionic surfactants. While ionic surfactants have been included in the HLB system, the more complex nature of the solution properties of the ionic materials makes them less suitable for the normal approaches to HLB classification. In cases where an electrical charge is desirable for reasons of stability, it is often found that surfactants that have limited water solubility and whose hydrophobic structure is such as to inhibit efficient packing into micellar structures should be most effective emulsifiers. Surfactants such as the sodium trialkylnaphthalene sulfonates and dialkylsulfosuccinates, which do not readily form large micelles in aqueous solution, have found some use in that context, usually providing advantages in droplet size and stability over simpler materials such as sodium dodecylsulfate.

Clearly, the process of selecting the best surfactant or surfactants for the preparation of an emulsion has been greatly simplified by the development of the more or less empirical approaches exemplified by the HLB and PIT methods. Unfortunately, each method has its significant limitations and cannot eliminate the need for some amount of trial-and-error experimentation. As our fundamental understanding of the complex phenomena occurring at oil--water interfaces, and of the effects of additives and environmental factors on those phenomena, improves it may become possible for a single, comprehensive theory of emulsion formation and stabilization to lead to a single, quantitative scheme for the selection of the proper surfactant system.

Some Other Factors Affecting Stability

A discussion of emulsion stability must include not only possible mechanisms of stabilization, but also some comments concerning the time frame of the stability requirements, because there exist external and internal factors unrelated to interfacial and colloidal phenomena that work unrelentingly to destroy the most "stable" system. The rates of colloidal degradation of emulsions vary immensely, so that it is not possible to define a single number that can be used as a measure of acceptable or unacceptable persistence. In any emulsion, especially one that is unstabilized or only very poorly stabilized, the breaking process will involve the coalescence of droplets brought together by the action of Brownian motion, convection currents, and other random disturbances. Their stability may be measured on the order of seconds or minutes.

Emulsions that contain more effective stabilizing additives such as one of those described above may be stable for hours, days, months, or even years. In such systems the action of random or induced motion and droplet collision will continue, but the rheological properties of the continuous phase will slow down such processes and/or interfacial layers will posses sufficient strength and rigidity so that coalescence will occur on a relatively long time scale.

In addition to the mechanical actions and interfacial energy considerations which will act to reduce the degree of dispersion of an emulsion, there are other considerations which act to limit the stability of emulsions. One such factor is the phenomenon, commonly termed "Ostwald ripening," in which large drops are found to grow at the expense of smaller ones. Such growth, whether in a crystal or an emulsion, results from differences in the chemical potential (and therefore solubility) of molecules in small particles relative to those in larger ones. Such differences arise from the fact that the pressure (or chemical potential), Δp, of a material inside a drop is inversely proportional to the drop radius, r, as given by that old friend, the Laplace equation. In terms of the Kelvin equation and solubility, the effect of radius on the process is given as

$$\ln (S_1/S_2) = \sigma_i\, V/RT\, (1/r_1 - 1/r_2) \qquad (11.9)$$

where S_1 and S_2 are the solubilities of the particles of principle radii r_1 and r_2 and V is the molar volume of the phase inside the drops or crystals. The effect of the Kelvin relationship is often readily apparent in foam systems where the solubility of gases in the liquid phase can be substantial. In emulsion systems, on the other hand, the solubility of the dispersed phase may be so low that diffusion from small to large droplets will be exceedingly slow. Even in such circumstances, the process will occur, but at such a rate that it will not be apparent for long periods. In the present context, it is often possible to reduce greatly the rate of droplet growth due to Ostwald ripening by employing emulsifiers and stabilizers which form a barrier to the passage of dispersed phase molecules into the continuous phase. This process can be especially important in multiple emulsion systems discussed later.

Other external factors affecting the stability of emulsions include the actions of bacteria and other microorganisms and freezing, especially in O/W emulsions. During the process of freezing, the formation of ice crystals in the continuous phase forces the emulsion droplets together under significant pressures, often resulting in the rupture of the interfacial film and drop coalescence. It is obvious, then, that stability to such action will require an interfacial film of considerable strength. Even though the protection of an emulsion from breaking due to freezing action is of considerable economic importance, there has been relatively little fundamental research published in the area.

Bacterial action can be of importance in areas such as food, pharmaceutical, and cosmetic emulsions, or other systems which contain components subject to biological degradation such as proteins, natural gums, etc. Such systems are obviously of great economic importance, so that a great deal of research has been devoted to the problem. In cases where biological stability is important, some advantages can be gained by the proper choice of surfactant in the stabilizing formulation, since many such materials show significant microbiocidal activity. Alternatively, antimicrobial additives, antioxidants, etc, can be added as extra protection to prolong the life of the emulsion.

In addition to direct degradation, bacterial action affecting emulsion stability may be of a "second hand" nature. The breaking or curdling of milk emulsions, for example, involves changes in the pH of the system. As the bacteria in the system propogate, they produce acidic waste products. The lowering of the pH of the system by those products decreases the degree of ionization of the milk protein (casein) stabilizing the O/W emulsion so that at a certain pH, the casein is no longer able to function as an efficient emulsion stabilizer and the emulsion breaks. It is sometimes observed that milk or cream that appears perfectly good in the bottle breaks when poured into a cup of coffee. In that case the coffee, which is acidic, simply finishes the job begun by the bacteria.

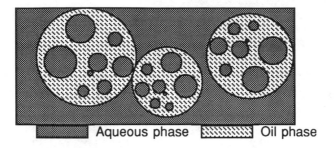

Figure 11.6. Schematic illustration of a water-in-oil-in-water multiple emulsion.

MULTIPLE EMULSIONS

While a great deal of information has been published over the years on the theoretical and practical aspects of emulsion formation and stabilization, until recently little has been said about more complex systems generally referred to as multiple emulsions. Multiple emulsions, as the name implies, are composed of droplets of one liquid dispersed in larger droplets of a second liquid, which is then dispersed in a final continuous phase. Typically, the internal droplet phase will be miscible with or identical to the final continuous phase. Such systems may consist of a W/O/W dispersion as indicated in Figure 11.6, where the internal and external phases are aqueous, or O/W/O multiple emulsions which have the reverse composition. Although known for almost a century, such systems have only recently become of practical interest for possible use in controlled drug delivery, emergency drug overdose treatment, wastewater treatment, and separations technology. Other useful applications will no doubt become evident as our understanding of the physical chemistry of such systems improves.

Because they involve a great variety of phases and interfaces, multiple emulsions must be inherently unstable, even more so than conventional emulsions. Their surfactant requirements are such that two stabilizing systems must be employed --- one for each oil--water interface. Each surfactant or mixture must be optimized for the type of emulsion being prepared but must not interfere with the companion system designed for the opposite interface. Long-term stability, therefore, requires careful consideration of the characteristics of the various phases and surfactant solubilities.

Nomenclature for Multiple Emulsions

For systems as potentially complex as multiple emulsions, it is very important that a clear and consistent system of nomenclature be employed. For a W/O/W system, for example, in which the final continuous phase is aqueous, the primary emulsion will be a W/O emulsion, which is then emulsified into the final aqueous phase. The surfactant or emulsifier system used to prepare the primary emulsion is referred to as the *primary surfactant* . To avoid further ambiguity as to components or their locations in the system, subscripts may be used. For example, in a W/O/W system the aqueous phase of the primary emulsion would be denoted as W_1 and the primary emulsion as W_1/O. After the primary emulsion is further dispersed in the second aqueous phase W_2, the complete system may be denoted $W_1/O/W_2$. In the case of an O/W/O multiple emulsion in which the oil phases are different, the notation becomes $O_1/W/O_2$. Additional refinements to fit even more complex systems, including the "order" of multiple emulsions, have been suggested.

Preparation and Stability of Multiple Emulsions

In principle, multiple emulsions can be prepared by any of the many methods for the preparation of conventional emulsion systems, including sonication, agitation, and phase inversion. Great care must be exercised in the preparation of the final system, however, because vigorous treatments normally employed for the preparation of primary emulsions will often break that system if used in secondary emulsion formation, resulting in loss of the identity of the primary phase.

Multiple emulsions reportedly have been prepared conveniently by the phase inversion technique mentioned earlier; however, such systems will generally have a limited persistence. It generally requires a very judicious choice of surfactants or surfactant combinations to produce a system that has useful characteristics of formation and stability. A general procedure for the preparation of a W/O/W multiple emulsion may involve the formation of a primary emulsion of water-in-oil using a surfactant suitable for the stabilization of such W/O systems. Generally, that involves the use of an oil soluble surfactant with a low HLB (2--8). The primary emulsion will then be emulsified in a second aqueous solution containing a second surfactant system appropriate for the stabilization of the secondary O/W emulsion (HLB 6--16). As noted above, because of the possible instability of the primary emulsion, great care must be taken in the choice of the secondary dispersion method. Excessive mechanical agitation such as in high-speed mixers and sonication could result in coalescence of the primary emulsion and the production of essentially "empty" oil droplets. The evaluation of the yield of filled secondary emulsion drops, therefore, is very important in assessing the value of different preparation methods and surfactant combinations.

The nature of the droplets in a multiple emulsion will greatly depend upon the size and stability of the primary emulsion. Florence and Whitehill[9] have suggested the existence of three main classes of droplets in W/O/W emulsions, based upon the nature of the oil-phase droplets. Type A systems (Figure 11.7a) are characterized as having one large internal drop essentially encapsulated by the oil phase. Type B (Figure 11.7b) contain several small, well-separated internal drops, and

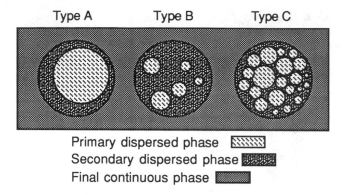

Figure 11.7. Classification of multiple emulsions according to the nature of drop dispersion in the primary emulsion.

type C (Figure 11.7c) contain many small internal drops in close proximity. It is understood that any given system will in all probability contain all three classes of drops, but one will be found to dominate depending primarily on the surfactant system employed.

Primary Emulsion Breakdown

There are several possible pathways for the breakdown of multiple emulsions. A few are shown schematically in Figure 11.8. Although all possible mechanisms for droplet coalescence cannot be illustrated conveniently in a single figure, a consideration of just a few possibilities can help to clarify the reasons for instability in a given system. Even though there may be a number of factors involved, one of the primary driving forces will be, as always, a reduction in the free energy of the system through a decrease in the total interfacial area. As has been noted previously, a major role of surfactants at any interface is to reduce the interfacial energy through adsorption. In a typical multiple-emulsion system, the primary mechanism for short-term instability will usually be droplet coalescence in the primary emulsion. It becomes important, then, to select as the primary emulsifier a surfactant or combination of surfactants that provides maximum stability for that system, whether W/O or O/W.

A second important pathway for the loss of "filled" emulsion droplets is the loss of internal drops by the rupture of the oil layer separating the small drops from the continuous phase. Such an expulsion mechanism would be expected to account for the loss of larger internal droplets. Unless the two phases are totally immiscible (in fact a rare situation), there will always exist the possibility that osmotic pressure differences between the internal and continuous portions of the system will cause material transfer to the bulk phase. The high pressures in the smaller droplets would be expected to provide a driving force for the loss of material from smaller drops in favor of larger neighbors (Ostwald

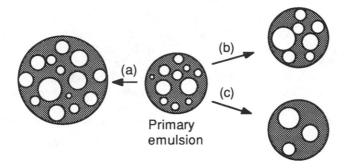

Figure 11.8. Important pathways for the breakdown of the primary emulsion: (a) coalescence of secondary emulsion drops; (b) coalescence of primary emulsion drops; (c) loss of primary emulsion dispersed phase to external phases.

ripening), as well as to the continuous phase. Finally, the presence of an oil-soluble surfactant always raises the possibility of reverse micelle formation and the subsequent solubilization of internal aqueous phase in the oil. Such a solubilization process also represents a convenient mechanism for the transport of material. In the context of a critical application such as controlled drug delivery, in which the mechanism of delivery is diffusion controlled, such breakdown mechanisms would be very detrimental to the action of the system since they could result in a rapid release of active solute with possibly dangerous effects.

The above mechanisms of emulsion breakdown, as well as others, must be addressed in order to understand and control a particular multiple emulsion system of interest. In all cases, the final stability of the system will depend on the nature of the oil phase of interest, the characteristics of the primary and secondary emulsifier systems, and the relationship between the internal and continuous phases.

The Surfactants and Phase Components

Choice of surfactants for the preparation of multiple emulsions can, in principle, be made from any of the four classes of surfactants discussed in Chapter 3. The choice will be determined by the characteristics of the final emulsion type desired, the natures of the various phases, additives, solubilities, etc. In many applications (eg, foods, drugs, cosmetics) the choice may be further influenced by such questions as toxicity, interaction with other addenda, and biological degradation. For that reason, nonionics have received a great deal of attention for such applications. In a given system, different types of surfactant may produce different types of multiple emulsions (A, B, or C types), so that such questions must also be considered.

Clearly, multiple emulsions represent a fertile field of research in both applied and academic surface science. Although there are an ever-increasing number of publications appearing on the subject, the area remains somewhat empirical in that each system is highly specific. As yet there are few general rules that can guide the interested formulator in the selection of the optimum surfactants for his application. A great deal remains to be done to gain a better understanding of the colloidal stability of such complex systems, and the effects of the various components in each phase on the overall process of preparation and stabilization. A sound understanding of the role of surfactant and other addenda in simple emulsions and an intuitive feel for the effect of the multiple interfaces present can serve as a good starting guide.

CHAPTER 12

FOAMS

It has been noted previously that emulsions and foams are related by the fact that each represents a physical state in which one fluid phase is finely dispersed in a second phase, the state of dispersion and the long-term stability (persistence) normally being dependent on the composition of the system. In emulsions, each phase is a liquid so that such factors as mutual solubility and the solubility of additives in each phase must be considered. In foams, the dispersed phase is a gas so that problems of solubility are less critical, although as will be seen, the transfer of the dispersed gas from one bubble to another or out into the adjacent atmosphere is important.

Because of the forces involved in their formation and stabilization, foams will have a definite structure. On a macroscopic scale, if the volume of liquid in the foam is large relative to the volume of gas, individual bubbles will be spherical and not interact to any significant extent. Such systems are sometimes referred to as *gas emulsions* . In what some prefer to label "true" foams, the bubbles are so closely packed that they are no longer spherical but take on the shape of polyhedrons separated by thin bilayer or lamellar liquid films.

In 1953 Manegold[1] proposed the classification of foams into two morphological classes: (1) the *kugelschaum* or spherical foams, which consist of widely separated spherical bubbles; and (2) *polyederschaum* or polyhedral foams, which consist of bubbles that are nearly polyhedral in shape, having narrow lamellar films of very low curvature separating the dispersed phase. The two types of foam are illustrated schematically in Figure 12.1. The morphological classification of foams is useful because each type of foam undergoes distinct changes with time, leading to collapse or a final "persistent" configuration.

The second morphological class of foams is the *polyederschaum* (the "true" foams). Wherever three bubbles meet, they will form three equal angles of 120°. The three interfacial tensions involved in the bubbles must be equal so that three identical forces will be involved where they meet. Mechanical quilibrium therefore requires that the three

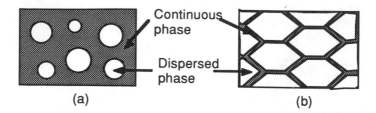

Figure 12.1. The two basic classes of foams: (a) spherical foams, in which the continuous phase greatly exceeds the dispersed phase and the bubbles maintain an essentially isolated, spherical structure; (b) polyhedral foams, in which the dispersed phase constitutes a major portion of the system and crowding distorts the bubbles from spherical to polyhedral in shape.

angles between them be equal. In an ideal foam, with all bubbles of the same size, the foam would assume the shape of pentagonal dodecahedrons. In almost all foams, however, there are a variety of different volumes present and their shapes are far from the ideal.

THE IMPORTANCE OF FOAMS

The presence of foam in an industrial product or process may or may not be desirable. Foams have wide technical importance, as such, in the fields of fire fighting, polymeric foamed insulation, foam rubbers and foamed structural materials such as concrete, whipped cream, shaving cream, and many areas of the baking industry. They also have certain esthetic utility in many detergent and personal care products, although their presence may not add much to the overall effectiveness of the process. Foams also serve useful purposes in industrial processes such as mineral separation (froth flotation), in the secondary recovery of petroleum by fluid displacement, and for environmental reasons in some electroplating operations. In the latter case, the presence of a foam blanket over the electroplating solution helps prevent solution splattering and the loss of volatile materials, therefore reducing the costs of maintaining an acceptable working environment. Unwanted foams, on the other hand, may be a significant problem in many technical processes, including sewage treatment, coatings applications, surfactant manufacture, extraction processes, and crude oil processing. Some of the above uses of foams will be discussed in more detail under specific subject headings.

By understanding the basic laws governing foam formation and the physical and chemical characteristics of materials that produce and sustain foams, or prevent and destroy them, the investigator or operator is well equipped to maximize (or minimize) the desired foaming effect. In the following sections, some of the basic physical principles of foam formation and stabilization will be covered along with some practical approaches to problems of foam characterization and control. Many of the formulas and concepts have been introduced in preceding chapters but for the sake of clarity are presented again in the present context.

FOAM FORMATION

Like other colloidal systems, foams may be formed either by dispersion or condensation processes. In the former process, the incipient gas phase may initially be present as a condensed phase, where small volumes of the future gas are introduced into the continuous liquid phase and converted into gas by some mechanism such as heating, pressure reduction, etc. Alternatively, the gas may be introduced directly as fine bubbles or incorporated mechanically. In the case of condensation, the gas phase is introduced at the molecular level and allowed to "condense" within the liquid to form bubbles.

The formation of the "head" on a glass of beer is a classic example of foam formation by condensation. In such a system, when the can, bottle, or tap is opened, carbon dioxide produced by fermentation in the container and solubilized under pressure is liberated. The solution becomes supersaturated and the excess gas forms a dispersed phase which rises to the top and forms the head. Many industrial processes for the formation of solid foams employ a similar process in which a "blowing agent" is added to the polymerizing system creating the foam.

The simplest way to form a nearly ideal foam is to introduce the gas into the liquid through a capillary tube. In that way individual bubbles of equal (almost) size will break off from the capillary tip under the action of surface tension. The process, however, must be slow in order to insure that interfacial equilibrium is achieved for each bubble, otherwise monodispersity will not be achieved. A much more rapid, but less controllable procedure is to bubble gas into the system through a porous plug. In that process a highly polydisperse foam will result since many small bubbles will have the opportunity to coalesce while still attached to the plug. Even less consistent results will be obtained for foams produced by agitation.

In all of the methods for the formation of foams, the initial bubbles will be separated by relatively thick layers of the continuous phase to produce spherical foams. However, in most cases, gravity will transform them into the polyhedral structure, with the foam at the top of the container and a reservoir of liquid accumulated at the bottom. As the bubbles rise, the external hydrodynamic pressure will decrease and the bubble volume increase, reducing the internal pressure of each bubble, although that internal pressure will still be greater than that externally. There will therefore be a mechanical driving force impelling the bubble to release the excess pressure by rupture. The fact that in many cases a reasonable long-lived foam is established indicates that some mechanism acting within the narrow lamellar films separating the bubbles is sufficient to withstand that mechanical pessure.

BASIC PROPERTIES OF FOAMS

The continuous phase of foams may consist of a entirely liquid components or a mixture of various liquids and solutes. The long-term stability of a given foam will depend on that composition, but in the special additives, even a completely liquid lamellar film will have some degree of rigidity. When a "stable" polyhedral foam structure is formed, it represents a transient minimum in the surface energy of the system --- a metastable configuration that could, in theory, remain for a significant period of time. It would require a "push" from some external source to cause it to increase its surface area and "break." Under normal circumstances, such "pushes" are common enough in the form of dust particles, air movement, convection due to temperature differencials, etc. Therefore the weaker foams will rapidly fall out of that metastable state and revert to a phase-separated state. If the liquid phase is "fortified" by the addition of various components that can increase that lamellar rigidity, greatly enhanced stability will result. Details of such action will be given below.

A primary characteristic of foams is that they have very low densities. An aqueous foam with bubble diameters of about 1 cm and lamellar thicknesses of 10^{-3} cm has a density of approximately 0.003 g cm^{-3}. That low density makes foams very useful in a number of applications, including fire fighting foams and various separation techniques. In the former case, the light foam is easily transported, can be produced rapidly to cover a large area, and is composed primarily of water, which is relatively inexpensive and easy to obtain and has the added advantage that it serves as an efficient mechanism for the removal of heat from the system. The foam blanket, of course, performs the role of any foam in relation to the fire --- it prevents contact between the air and the combustible material below. For fire fighting, the use of spherical rather than polyhedral foams may have some advantages since the higher water content would aid in the removal of heat from the system.

Related to the low density of foams is the characteristic that they have a large surface area for a given weight of foam. For the example above, the foam will have a surface area of about 2000 cm^2 g^{-1}. Under properly controlled conditions, specific finely powdered mineral ores will attach to foam bubbles and be carried to the surface of a solution where they can be skimmed off and the desired mineral significantly enriched. Unwanted materials such a rock and dirt will be left to sink to the bottom of the container. The procedure, known generally as *flotation* , is a valuable tool in many mineral purification processes. Similar events have been postulated to occur in many detergency processes. Techniques based on the flotation principle have been proposed for the purification or removal of surface-active soluble materials from solution, although they seem to have found little large-scale use.

FOAM STABILITY OR PERSISTENCE

Foams are inherently unstable systems. Because they are encountered in so many technological areas, foams have been the subject of a significant amount of discussion in the literature. A number of reviews have been published over the years that cover most aspects of foam formation and stabilization (see Bibliography). While the theoretical aspects of stabilization are reasonably well worked out, a great deal remains to be understood concerning the practical details of foam formation, persistence, and prevention.

Like almost all systems containing two or more immiscible phases, foams involve thermodynamic conditions in which the primary driving force is to reduce the total interfacial area between the phases --- that is, they are thermodynamically unstable. In spite of their ultimate tendency to collapse, foams can be prepared that have a lifetime (persistence) of minutes, days, or even months. There are three fundamental physical mechanisms for the collapse of a foam: (1) the diffusion of the gas phase from one bubble (small, high internal pressure) to another (larger, lower internal pressure) or into the bulk gas phase surrounding the foam; (2) bubble coalescence due to rupture of the lamellar film between the gas phase (slower than mechanism 1 and occurring even in stabilized systems); and (3) rapid hydrodynamic drainage of liquid between bubbles leading to rapid collapse (in the absence of any stabilizing mechanisms discussed below). Each mechanism can be important for a given foam system and will be addressed in turn. In most nonrigid systems, however, all three mechanisms are operative to some extent during some phase of the collapsing process so that an analysis of the system and its stability may not be a simple undertaking.

The first mechanism occurs due to the difference in gas pressure inside the bubble as a result of differences in curvature of the lamellar films. As a simple example, consider the system of two contacting bubbles shown in Figure 12.2. The Laplace equation states that the pressure difference, Δp, on either side of a curved interface will be given by,

$$\Delta p = \sigma \left(1/r_1 + 1/r_2 \right) \qquad (12.1)$$

where r_1 and r_2 are the major radii of curvature of the system and σ the interfacial tension. For the two-bubble system let R_1 be the radius of the larger and R_2 that of the smaller bubble. The common interfaces or partition will have a radius R_0. From eq. 12.1, the pressure in the larger bubble will be atmospheric pressure, $p_a + 4\sigma/R_1$. The factor of 4 arises here because the bubble wall involves two interfaces, each of which contributes to the total. The small bubble will likewise have a total pressure of $p_a + 4\sigma/R_2$.

At equilibrium, the common interface (the *septum*) must be

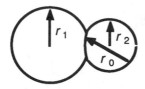

Figure 12.2. Schematic illustration of the joining, but not coalescence, of two bubbles of different size. At mechanical equilibrium, the septum between the two bubbles will be concave toward the smaller bubble, with a radius of curvature, $R_0 = R_1 R_2 / (R_1 - R_2)$.

concave toward the smaller bubble with a radius of curvature given by

$$4\sigma/R_0 = 4\sigma/R_2 - 4\sigma/R_1 \qquad (12.2)$$

or $R_0 = R_1 R_2/(R_1 - R_2)$. If the septum were completely impermeable to the gas, such a situation would be mechanically stable. In reality, such impermeability does not exist in foams and gas will diffuse spontaneously from the region of high to that of low pressure. As a result, the smaller bubble will shrink while the larger will grow. If the difference in the two radii R_1 and R_2 is initially small, diffusion will be slow. With time, however, the difference will increase, as will the rate, meaning that in a foam the average bubble size and polydispersity will increase with aging.

The second major reason for film drainage and ultimate rupture is capillary flow. When several bubbles are in contact, especially when the foam has reached a more or less stable polyhedral structure, the liquid region of multiple bubble contact will have a much greater curvature (ie, smaller *radius* of curvature) than the lamellar films, which may be almost planar. Those regions of high curvature toward the gas phase, referred to as Plateau borders (Figure 12.3), act as a capillary "pump" to evacuate the region between bubbles. Since the gas pressure within the bubble must be the same throughout, the liquid pressure within the Plateau borders will be lower than in the more parallel areas. That pressure difference will drive the liquid into the Plateau borders, thinning the film and advancing the process of ultimate film rupture. Liquid will also be drained from the lamellae due to gravitational forces; as a result the lamellae will become thinner and thinner until a critical thickness may be reached at which time the system can no longer sustain the pressure and collapse occurs.

Of the three mechanisms, hydrodynamic drainage due to gravity is usually the most rapid and, if the foam is particularly unstable, leads to total collapse before other mechanisms can become important. In those cases, once the loss of liquid from the lamellar layer produces a critical thickness of 5 -- 15 nm the liquid film can no longer support the pressure of the gas in the bubble and film rupture occurs. As a model for gravity drainage, a film may be treated as a vertical slit of thickness δ and width

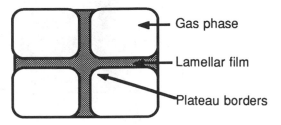

Gas phase

Lamellar film

Plateau borders

Figure 12.3. Illustration of the critical regions of high curvature, the Plateau border regions, at which the reduced Laplace pressure acts to suck fluid from the lamellar films between bubbles, leading to rupture.

w between two parallel walls. Assuming a constant thickness of the lamellar film between bubbles, the rate of film drainage will be

$$dV/dt = g\,\rho w\,\delta^3/12\eta \qquad (12.3)$$

where g is the gravity constant, ρ is the density difference between the liquid and the gas around the foam, and η the viscosity of the liquid. In reality, of course, the slit or film will not be of constant size --- δ will decrease as drainage occurs --- so that eq. 12.3 is not quantitatively valid. Qualitatively, however, the relationship provides a clue as to how one may influence the drainage process. Also, the lamellar film in a foam is not usually a rigid structure, so that the mechanical characteristics of the wall may affect the rate of drainage, especially in thinner films.

The overall question of foam stability and bubble coalescence requires the consideration of both the static and dynamic aspects of bubble interactions. In the initial stages of film drainage, where relatively thick lamellar films exist between gas bubbles, gravity can make a significant contribution to the drainage of liquid from between foam bubbles. Once the films have thinned to a thickness of a few hundred nanometers, however, gravity effects become negligible and interfacial interactions begin to predominate. When the two sides of the lamellar film are in sufficiently close proximity, interactions can occur involving the interfacial forces discussed in previous chapters. Such forces (per unit area), acting normal to (across) the lamellar film, are collectively exhibited as the so-called *disjoining pressure* of the system $\pi(\delta)$. The net interaction energy between bubbles as a function of distance of separation will have a form similar to that in Figure 12.4, where the minima at distances of separation h' and h'' correspond to metastable states in which $\pi = 0$ and the films have some degree of equilibrium stability. In the plane parallel regions of the lamellar film, the Laplace (or capillary) pressure given by eq. 12.1 will be zero. In the Plateau border regions, however, that will not be the case and mechanical equilibrium requires that

$$\Delta p = -\pi(\delta) \qquad (12.4)$$

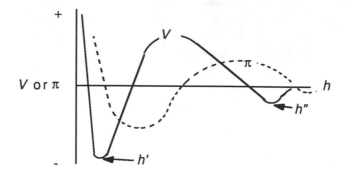

Figure 12.4. Illustration of the relationship between foam bubble interaction energy (V) and disjoining pressure (π) as a function of distance of separation of the two interfaces involved (h).

That is, the internal pressure of the bubbles is just balanced by the interfacial forces acting across the lamellar film. The most important interfacial interactions contributing to $\pi(\delta)$ are electrostatic repulsion between charged interfaces and steric interactions between adsorbed species. Those topics have already been discussed in the context of colloidal stability (Chapter 10) and will not be treated further here.

Internal thermodynamic and hydrodynamic factors aside for the moment, it should be remembered that foams are sensitive to a number of environmental stresses, that act to bring about bubble coalescence and foam collapse. Those include vibration, the presence of solid particles, organic contaminants, and temperature differentials. It will therefore be important to take such factors into consideration when carrying out foaming studies or formulating foam systems.

Stabilization Mechanisms

Practical mechanisms for extending the persistence of foams can include one or several of the following conditions: (1) a high viscosity in the liquid phase, which retards hydrodynamic drainage as well as providing a cushion effect to absorb shocks resulting from random or induced motion; (2) a high surface viscosity, which also retards liquid loss from between interfaces and dampens film deformation prior to bubble collapse; (3) surface effects such as the Gibbs and Marangoni effects which act to "heal" areas of film thinning due to liquid loss; (4) electrostatic and steric repulsion between adjacent interfaces due the adsorption of ionic and nonionic surfactants, polymers, etc, which can oppose drainage through the effects of the disjoining pressure; and (5) retardation of gas diffusion from smaller to larger bubbles.

The addition of surfactants and/or polymers to a foaming system can alter any or all of the above system characteristics and therefore enhance the stability of the foam. They may also have the affect of lowering the surface tension of the system, thereby reducing the work required for the initial formation of the foam, as well as producing smaller, more uniform bubbles.

THE PRACTICAL CONTROL OF FOAMABILITY AND PERSISTENCE

For a liquid to produce a foam of any degree of utility, it must: (1) be able to expand its surface area so as to form a membrane around gas bubbles; (2) possess the correct rheological and surface properties to retard the thinning of the lamellae leading to bubble coalescence; (3) and/or be able to retard the diffusion of trapped gas from small to large bubbles or to the surrounding atmosphere. Foaming does not occur in pure liquids because such a system lacks mechanisms for the completion of any of those three tasks. When surface-active molecules or polymers are present, however, rheological effects and adsorption at the gas--liquid interface serve to retard the loss of liquid from the lamellae and, in some instances, to produce a more mechanically stable system.

Theories related to such film formation and persistence, especially film elasticity, derive from a number of experimental observations about the surface tension of liquids. First, as is well known from the Gibbs adsorption equation, the surface tension of a liquid will decrease as the concentration of the surface-active material in solution increases (assuming positive adsorption) up to the point of surface saturation. Second, the instantaneous (dynamic) surface tension at a newly formed surface is always higher than the equilibrium value; that is, there is a finite time requirement during which the surface-active molecules in the solution (or bulk molecules, if no solute is present) must diffuse to the interface in order to lower the surface tension. The time lag in reaching the equilibrium surface tension due to diffusion is generally known as the *Marangoni effect* . The two surface tension effects due to adosrption and diffusion are usually complementary, and are often being discussed as the combined *Gibbs-Marangoni* effect.

The fundamental impact of surfactant concentration and diffusion rate in lamellar films can be viewed roughly as follows (Figure 12.5): as the lamellar film between adjacent bubbles is stretched as a result of gravity, agitation, drainage, or other mechanical action, new surface will be formed having a lower transient surfactant concentration, and a local surface tension increase will occur. A surface tension gradient along the film will be produced, causing liquid to flow from regions of low σ toward the new stretched surface, thereby opposing film thinning. Additional stabilizing action is thought to result from the fact that the diffusion of new surfactant molecules to the surface must also involve the transport of associated solvent into the surface area, again countering the thinning effect of liquid drainage. The mechanism can be characterized as producing a "healing" effect at the site of thinning.

Even though the Gibbs and Marangoni effects are complementary,

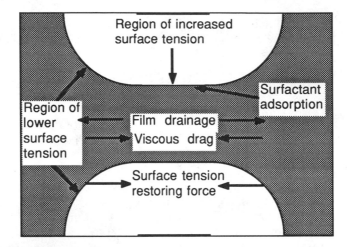

Figure 12.5. Schematic illustration of the "healing" action of the Gibbs-Marangoni effects in foam stabilization, as related to surfactant adsorption and action.

they are generally important in different surfactant concentration regimes. The Marangoni effect is usually of importance in fairly dilute surfactant solutions and over a relatively narrow concentration range. In the absence of external agitation, the amount of surfactant adsorbed at a new interface can be estimated by

$$n = 2(D/\pi)^{1/2}ct^{1/2}(N/1000) \qquad (12.5)$$

where n is the number of molecules per square centimeter, D is the bulk diffusion constant (cm^2 s^{-1}), c is the bulk concentration of the surfactant (mols L^{-1}), t is the time in seconds, and N is Avogadro's number. Using eq. 12.5, it is possible to estimate the time required for the adsorption of a given amount of surfactant at a new interface compared to the rate of generation of that interface. If the surfactant solution is too dilute, the surface tension of the solution will not differ sufficiently from that of the pure solvent for the restoring force to counteract the effects of casual thermal and mechanical agitation. As a result, the foam produced will be very transient. In line with the Marangoni theory, there should be an optimum surfactant concentration for producing the maximum amount of foam in a given system, under defined circumstances. Such effects have been verified experimentally in terms of the Ross-Miles test (see Table 12.1). It can be seen from the data that the optimum concentration is usually within a factor of 2 of the critical micelle concentration.

In the case of the Gibbs effect, it is proposed that the rise in surface tension occurring as the film is stretched results from a depletion of the surfactant concentration in the bulk phase just below the newly formed interface. Obviously, in systems such as foams, where the available bulk

Table 12.1. Typical surfactant concentrations required to attain maximum foam height (MFH) (Ross-Miles method, 60°C).[3]

Surfactant	CMC (mM)	Concentration for MFH (mM)
$C_{12}H_{25}SO_3^-Na^+$	11	13
$C_{12}H_{25}SO_4^-Na^+$	9	5
$C_{14}H_{29}SO_3^-K^+$	3	3
$C_{14}H_{29}SO_4^-Na^+$	2.3	3
$C_{16}H_{33}SO_3^-K^+$	0.9	0.8
$C_{16}H_{33}SO_4^-Na^+$	0.7	0.8
p -$C_8H_{17}C_6H_4SO_3^-Na^+$	16	13
p -$C_{10}H_{21}C_6H_4SO_3^-Na^+$	3	4.5
p -$C_{12}H_{25}C_6H_4SO_3^-Na^+$	1.2	4
o -$C_{12}H_{25}C_6H_4SO_3^-Na^+$	3	4
$(C_8H_{17})_2CHSO_4^-Na^+$	2.3	4

phase in the narrow lamellae may be small compared to the amount of interface being formed, the effect will be enhanced. As with the Marangoni effect, if the surfactant concentration in the bulk phase is too low, a surface tension gradient of sufficient size to produce the necessary "healing" action will not be produced. Conversely, if the concentration is too large, well above the CMC, the available surfactant will be such that no gradient is formed.

Quantitatively, the Gibbs effect can be described in terms of a coefficient of surface elasticity E which Gibbs defined as the ratio of the surface stress to the strain per unit area[2]

$$E = 2A (\delta\sigma/\delta A) \tag{12.6}$$

Since the elasticity is the resistance of the film to deformation, the larger the value of $E,$ the greater will be the ability of the film to sustain shocks (its resiliency, so to speak) without rupture. As mentioned earlier, when a film of a pure liquid is stretched, no significant change in surface tension occurs and the elasticity as defined by eq. 12.6 is be zero. This is the theoretical basis for the observation that pure liquids do not foam. The relationship between surface elasticity and surface transport is important since, if a film has a significant value of $E > 0,$ stretching the film will produce an increase in the local surface tension and induce flow of subsurface liquid into the stretched area, again acting to restore the original thickness of the lamellae.

Two surfactant-related processes, then, must be considered in conjunction with these foam-stabilizing mechanisms. One is the rate of surface diffusion of surfactant molecules from regions of low to high surface tension. The second is the rate of adsorption of surfactant from the underlying bulk phase into the surface. In each case, a too rapid arrival of surfactant molecules at the new surface will destroy the surface tension gradient and prevent the restoring action of the Gibbs-Marangoni "healing" process. Conversely, a very low bulk concentration will result in equally ineffective action.

Polymers and Foam Stabilization

Very stable foams can be prepared if surface-active polymers are included in the formulation. When polymers (and proteins in particular) are adsorbed at the liquid--air interface, they assume configurations significantly different from their equilibrium situation in the bulk solution; proteins become partially denatured. The relatively dense, somewhat structured adsorbed polymer layer imparts a significant degree of rigidity or mechanical strength to the lamellar walls, producing an increase in the stability of the final foam. The presence of polymer also aids stability in the initial stages after foam formation since the liquid viscosity increase that results from its presence slows the process of film drainage. Polymers generally are not effective in the context of the Gibbs-Marangoni effect since their diffusion rates are much slower that that of low molecular weight surfactants.

The presence of polymers in foaming systems can cause particularly difficult problems where foam stability is not desirable. Such is particularly the case in waste-treatment facilities, where the presence of proteins can cause extreme problems. If polyvalent ions such as Al^{3+} are present, the problem is exacerbated still more. Proteins chelate strongly with such ions and form a surface film so rigid that it may approach the strength of a solid foam. Obviously, such a situation will be detrimental to the overall treatment process. In other situations, such as breads and other baked products, the formation of rigid foam walls can be particularly advantageous.

FOAM FORMATION AND SURFACTANT STRUCTURE

The relationship between the foaming power of a surfactant and its chemical structure can be quite complex. The correlation is further complicated by the fact that there is not necessarily a direct relationship between the ability of a given structure to produce foam and its ability to stabilize that foam. One usually finds that the amount of foam produced by a surfactant under a given set of circumstances will increase with its bulk concentration up to a maximum, which occurs somewhere near the CMC. It would appear, then, that surfactant CMC could be used as a guide in predicting the initial foaming ability of a material, but not necessarily the persistence of the resulting foam.

Any structural modification that leads to a lowering of the CMC of a class of surfactants, such as increasing the chain length of an alkyl sulfate, can be expected to increase its efficiency as a foaming agent. Conversely, branching of the hydrophobic chain or moving the

Table 12.2. Foaming characteristics of typical anionic and nonionic surfactants in distilled water (Ross-Miles, 60°C).[3]

Surfactant	Concentrtation (wt %)	Foam height (mm) Initial	After (min)
$C_{12}H_{25}SO_3^-Na^+$	0.25	---	205 (1)
$C_{12}H_{25}SO_4^-Na^+$	0.25	220	175 (5)
$C_{14}H_{29}SO_3^-Na^+$	0.11	---	214 (1)
$C_{14}H_{29}SO_4^-Na^+$	0.25	231	184 (5)
$C_{16}H_{33}SO_3^-K^+$	0.033	---	233 (1)
$C_{16}H_{33}SO_3^-Na^+$	0.25	245	240 (5)
$C_{18}H_{37}SO_4^-Na^+$	0.25	227	227 (5)
o-$C_8H_{17}C_6H_4SO_3^-Na^+$	0.15	148	---
p-$C_8H_{17}C_6H_4SO_3^-Na^+$	0.15	134	---
o-$C_{12}H_{25}C_6H_4SO_3^-Na^+$	0.25	208	---
p-$C_{12}H_{25}C_6H_4SO_3^-Na^+$	0.15	201	---
t-$C_9H_{19}C_6H_4O(CH_2CH_2O)_8H$	0.10	55	45 (5)
t-$C_9H_{19}C_6H_4O(CH_2CH_2O)_9H$	0.10	80	60 (5)
t-$C_9H_{19}C_6H_4O(CH_2CH_2O)_9H$	0.10	110	80 (5)
t-$C_9H_{19}C_6H_4O(CH_2CH_2O)_{13}H$	0.10	130	110 (5)
t-$C_9H_{19}C_6H_4O(CH_2CH_2O)_{20}H$	0.10	120	110 (5)

hydrophilic group to an internal position, all of which increase the CMC, will result in a lower foaming efficiency. Typical foaming characteristics for several anionic and nonionic surfactants are given in Table 12.2. In the table, foaming efficiency and persistence were determined according to the Ross-Miles procedure.

The ability of a surfactant to perform as a foaming agent is dependent primarily on its effectiveness at reducing the surface tension of the solution, its diffusion characteristics, its properties with regard to disjoining pressures in thin films, and the elastic properties it imparts to interfaces. The amount of foam that can be produced in a solution under given conditions (ie, for a set amount of work input) will be related to the product of the surface tension and the new surface area generated during the foaming process (eq. 12.1). Obviously, the lower the surface tension of the solution, the greater will be the surface area that can be expected to be developed by the input of a given amount of work. Maintenance of the foam, however, may be as important as original formation.

It is often observed that the amount of foam produced by the members of an homologous series of surfactants will go through a maximum as the chain length of the hydrophobic group increases. This is probably due to the conflicting effects of the structural changes. In one case, a longer chain hydrophobe will result in a more rapid lowering of surface tension and a lower CMC. However, if the chain length grows too long, low solubility and slow diffusion and adsorption may become problems.

It has been found in many instances that surfactants with branched hydrophobic groups will lower the surface tension of a solution more rapidly than a straight-chain material of equal carbon number. However, since the branching of the chain increases the CMC and reduces the amount of lateral chain interaction, the cohesive strength of the adsorbed layer, the film elasticity, will be reduced, yielding a system with higher initial foam height but reduced foam stability. Similarly, if the hydrophilic group is moved from a terminal to an internal position along the chain, higher foam heights, but lower persistence, can be expected. In all such cases, comparison of foaming abilities must be compared at levels above their CMC's.

Ionic surfactants can contribute to foam formation and stabilization as a result of the presence at the interface of the electrical double layer that can interact with the opposing interface in the form of the disjoining pressure. Additional stabilizing effect may be gained from the fact that the ionic group requires a significant degree of solvation, with the associated solvent molecules adding to the steric (or entropic) contribution to the disjoining pressure π. Not surprisingly, it is found that the effectiveness of such surfactants as foaming agents can be related to the nature of the counterion associated with the adsorbed surfactant molecules. The effectiveness of dodecylsulfate surfactants as foam stabilizers, for example, decreases in the order $NH_4^+ > (CH_3)_4N^+ > (C_2H_5)_4N^+ > (C_4H_9)_4N^+$. Such a series may reflect a change in the solvation state of the surfactant from the essentially totally dissociated ammonium counterion, producing a maximum disjoining pressure and requiring significant solvation, to the more tightly ion-paired tetrabutylammonium counterion with greatly reduced π and different solvation requirements.

Nonionic surfactants generally produce less initial foam and less stable foams than ionics in aqueous solution. Because such materials must by nature have rather large surface areas per molecule, it becomes difficult for the adsorbed molecules to interact laterally to a significant degree, resulting in a lower interfacial elasticity. In addition, the bulky, highly solvated, nonionic groups will generally result in lower diffusion rates and less efficient "healing" via the Gibbs-Marangoni effect. Polyoxyethylene nonionic surfactants in particular exhibit a strong sensitivity of foaming ability to the length of the POE chain. At short chain lengths, the material may not have sufficient water solubility to lower the surface tension and produce foam. A chain that is too long, on the other hand, will greatly expand the surface area required to accommodate the adsorbed molecules and will also reduce the interfacial elasticity. This characteristic of POE nonionic surfactants has made it possible to design highly surface-active, yet low-foaming surfactant formulations. Even more dramatic effects can be obtained by the use of "double-ended" surfactants in which both ends of the POE chain are substituted. In many cases only a single methyl group on the end of a surfactant chain will significantly reduce foaming where such a result is desired.

If the solubility of a surfactant is highly temperature dependent, as is the case for many POE nonionics, it will be found that foaming ability will increase in the same direction as its solubility. Nonionic POE surfactants, for example, exhibit a decrease in foam production as the temperature is increased and the cloud point is approached. Long-chain carboxylate salts, on the other hand, which may have limited solubility in water and poor foaming properties at room temperature, will be more soluble and will foam more as the temperature increases.

Liquid Crystals and Foam Stability

As we have seen, the stability of foams depends on a wide variety of factors involving several aspects of surface science. The potential importance of liquid crystal formation to emulsion stability was pointed out in the previous chapter. Not surprisingly, an equally important role for such structures has been identified in foaming applications. Although the phenomenon of LC stabilization of aqueous foams has been recognized for some time, their role in nonaqueous foaming systems has been less well documented. Friberg, et al [4,5] have shown that the presence of a liquid-crystalline phase can also serve as a sufficient condition for the production of stable foams in organic systems.

The role of the liquid crystal in stabilizing a foam can be related to its effect on several mechanisms involved in foam loss, including hydrodynamic drainage, the mechanical strength of the liquid film, and the diffusion rate of entrapped gas. The effect of the liquid crystal phase on film drainage can be considered to be twofold. In the first place, the more ordered, multilayer nature of the phase imparts a much higher viscosity to the film than a normal surfactant monolayer, thereby preventing or slowing the process of liquid drainage. In addition, it has been found that liquid-crystalline phases tend to accumulate in the Plateau border areas. Their presence there results in an increase in the size of the areas, a larger radius of curvature (Figure 12.6) and thus a smaller Laplace pressure forcing film drainage. The second stabilizing function for the liquid crystal can be related to the Gibbs-Marangoni effects, in that the presence of a large quantity of surfactant at the Plateau borders allows them to act as a reservoir for surfactant molecules needed to maintain the high surface pressures useful for ensuring foam stability.

The production of a liquid crystal phase not only can add to the stability of the foam from a surface chemical standpoint, but it can also significantly enhance the mechanical strength of the system. When thinning reaches the point at which bubble rupture can become important, the mechanical strength and rigidity of such structures can help the system withstand the thermal and mechanical agitation that might otherwise result in film rupture and foam collapse.

Finally, because the liquid crystal structure is more highly ordered and, potentially, more dense than a normal fluid, the diffusion rate of gas

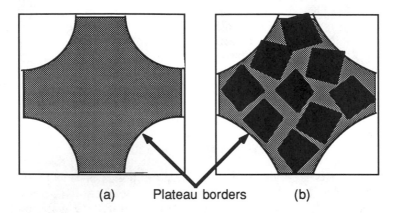

(a)   Plateau borders   (b)

Figure 12.6. The accumulation of liquid crystals in the Plateau border region (a) increases the radius of curvature in the region (b), thereby reducing the negative Laplace pressure and enhancing the stability of the foam.

molecules between bubbles may be expected to be slowed significantly.

The Effects of Additives on Surfactant Foaming Properties

As we have seen, the foaming properties of a surfactant can be related to its solution properties through the CMC. It is not surprising, then, that additives in a formulation can affect foaming properties in much the same way that they affect other surfactant solution properties. The presence of additives can affect the stability of a foam by influencing any of the mechanisms already discussed for foam stabilization. It may, for example, increase the viscosity of the liquid phase or the interfacial layer, or it may alter the interfacial interactions related to Gibbs-Marangoni effects or electrostatic repulsions. By the proper choice of additive, a high-foaming surfactant can be transformed into one exhibiting little or no foam formation. Conversely, a low-foaming material may produce large amounts of foam in the presence of small amounts of another surface-active material, which itself has few if any useful surfactant properties. It is theoretically possible, then, to custom build a formulation to achieve the most desirable combination of foaming action to suit the individual needs of the system. The addition of small amounts of such additives has become the primary way of adjusting the foaming characteristics of a formulation in many, if not most, practical surfactant applications.

Additives that alter the foaming properties of a surfactant through changes in its micellization characteristics can be divided into three main

Table 12.3. The effect of the structure of organic additives on the CMC's and foaming characteristics of sodium 2-*n*-dodecylbenzene sulfonate solutions.[3]

Additive	CMC (g L^{-1})	ΔCMC (%)	Foam Vol. (ml, 2 min.)
None	0.59	---	18
Laurylglycerol ether	0.29	- 51	32
Laurylethanolamide	0.31	- 48	50
n-Decyl glycerol ether	0.33	- 44	34
Laurylsulfolanylamide	0.35	- 41	40
n-Octylglycerol ether	0.36	- 39	32
n-Decyl alcohol	0.41	- 31	26
Caprylamide	0.50	- 15	17
Tetradecanol	0.60	0	12

classes: (1) inorganic electrolytes, which are most effective with ionic surfactants; (2) polar organic additives, which can affect all types of surfactants; and (3) macromolecular materials. The latter materials can affect the foaming properties of a system in many ways, some unrelated to the surface properties of the surfactant itself. Electrolyte additives can act to increase foamability by reducing the CMC of ionic foaming agent. On the other hand, an excessive amount of electrolyte may, and probably will, greatly reduce foam persistence by reducing the electrostatically induced disjoining pressure.

From a practical point of view, the most important class of additives is the polar organic materials. They have received a great deal of attention both academically and industrially because of the relative ease of application and control of the additive. Some of the earliest uses of polar organic materials as profoaming additives was in the area of foam stabilizers for heavy-duty laundry formulations. As a general rule it was found that additives that lowered the CMC of a surfactant could stabilize foams in the presence of materials that were normally detrimental to foam formation and persistence. The ability of additives to increase foamability and foam stability by lowering the CMC of the primary surfactant could be related to the extent of such lowering. Straight-chain hydrocarbon additives whose chain length is approximately the same as that of the surfactant are generally the most effective at lowering the CMC and increasing initial foam height. Bulky chains on the additives produce much smaller effects on foaming properties. The effectiveness of polar additives of various types as foam stabilizers is found to be in the approximate order: primary alcohols < glyceryl ethers < sulfonyl ethers < amides < *N*-substituted amides. This is essentially the same order found for the effects of such materials on the CMC's of surfactants. The effects on CMC and foam stability of the addition of polar additives to sodium dodecylbenzene sulfonate are given in Table 12.3.

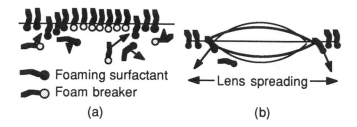

Figure 12.7. Mechanisms of action of foam inhibitors and breakers: (a) foam-breaking materials that *replace* stabilizing surfactant in the interface; (b) lens-forming liquids that *displace* adsorbed surfactant.

Not only does foam stabilization by additives seem to go hand-in-hand with the effect of the additive on the CMC of the surfactant, but there is also a correlation with the relative amount of additive that is located in the interfacial film. The greater the mole fraction of additive adsorbed at the interface, the more stable is the resulting foam. Many of the most stable foaming systems were found to have surface layers composed of as much as 60--90 mol % additive.

FOAM INHIBITION

Although the presence of certain additives can enhance the foaming ability and persistence of a surfactant system, the similar materials can also significantly reduce foam formation or persistence. Such materials may be termed f*oam inhibitors* , which act to prevent the formation of foam, or *foam breakers* , which increase the rate of foam collapse. Foam breakers may include inorganic ions such as calcium, which counteract the effects of electrostatic stabilization or reduce the solubility of many ionic surfactants, organic or silicone materials that act by spreading on the interface and displacing the stabilizing surfactant species, or materials that directly interfere with micelle formation.

A foam breaker that acts by spreading may do so as a monolayer or as a lens (Figure 12.7). In either case, it is assumed that the spreading foam breaker sweeps away the stabilizing layer, leading to rapid bubble collapse from the outside of the foam. The rate of spreading of the defoamer will, of course, depend on the nature of the adsorbed layer present initially. If the foaming agent can be desorbed rapidly, the defoamer will spread rapidly, resulting in fast foam collapse. If the foamer does not desorb rapidly, on the other hand, spreading will be retarded or even halted. Foam collapse will then be a much slower process, relying on the thinning of the lamellae by other drainage mechanisms.

In some cases it is found that the action of defoaming agents may depend on the concentration of the surfactant present. If the surfactant concentration is below the CMC, the defoamer will usually be most effective if it spreads as a lens on the surface rather than as a monolayer film. Above the CMC, however, where the defoamer may be solubilized, the micelles may act as a reservoir for extended defoaming action by adsorption as a surface monolayer. If the solubilization limit is exceeded, initial defoaming effects may be due to the lens-spreading mechanism with residual action deriving from solubilized material.

So far the discussion of foams and defoaming has centered on aqueous systems. The action of defoamers in organic systems is essentially the same as that in aqueous phases. Unfortunately, the choices of possible candidates is much more restricted, limited principally to silicones and fluorocarbon materials. In potentially critical systems such as lubricating oils, some of the few materials having the required characteristics of limited solubility and adequate surface tension lowering properties, act as foaming agents below their solubility limit but inhibit foam formation when that limit is exceeded. In other words, they act as foaming agents by adsorption and surface tension lowering until their solubility is exceeded, at which time they became antifoamers acting via the lens mechanism.

Chemical Structures of Antifoaming Agents

Materials that are effective as defoaming agents can be classified into eight general chemical classifications, with the best choice of material depending on such factors as cost, the nature of the liquid phase, the nature of the foaming agent present, and the nature of the environment to which it may be subjected. One of the most common classes of antifoaming agents consists of the polar organic materials composed of highly branched aliphatic alcohols. As noted earlier, linear alcohols in conjunction with surfactants can result in increased foam production and stability due to mixed monolayer formation and enhanced film strength. The branched materials, on the other hand, reduce the lateral cohesive strength of the interfacial film, which increases the rate of bubble collapse. The higher alcohols also have limited water solubility and are strongly adsorbed at the air--water interface, displacing surfactant molecules in the process.

Fatty acids and esters with limited water solubility are also often used as foam inhibitors. Their mode of action is similar to that of the analogous alcohols. In addition, their generally low toxicity often makes them attractive for use in food applications. Organic compounds with multiple polar groups are, in general, found to be effective foam inhibitors. The presence of several polar groups generally acts to increase the surface area per molecule of the adsorbed antifoamer and results in a loss of stabilization.

Metallic soaps of carboxylic acids, especially the water-insoluble polyvalent salts such as calcium, magnesium, and aluminum, can be effective as defoamers in both aqueous and nonaqueous systems. In water, they are usually employed as solutions in an organic solvent, or as a fine dispersion in the aqueous phase. Water-insoluble organics containing one or more amide groups are found to be effective antifoaming agents in a number of applications, especially for use in boiler systems. It is generally found that greater effectiveness is obtained with materials containing at least 36 carbon atoms (eg, distearoylethylenediamine, $C_{38}H_{78}N_2O_2$) compared to simple fatty acid amides.

Alkyl phosphate esters are found to possess good antifoaming characteristics in many systems due to their low water solubility and large spreading coefficients. They also find wide application in nonaqueous systems such as inks and adhesives. Organic silicone compounds are also usually found to be outstanding antifoaming agents in both aqueous and organic systems. Because of their inherently low surface energy and limited solubility in many organics, the silicone materials are one of the two types of materials that are available to modify the surface properties of most organic liquids.

The final class of materials that have found some application as antifoaming agents are the fluorinated alcohols and acids. Due to their very low surface energies they are active in liquids where hydrocarbon materials have no effect. They are, in general, expensive, but their activity at very low levels and in very harsh environments may overcome thier initial cost barrier.

CHAPTER 13

AEROSOLS

The previous chapters have introduced several classes of colloids and some of the important surface aspects of their formation, stabilization, and destruction. Emulsions, foams, and dispersions are the most commonly treated and intensely studied examples of colloidal systems. Those constitute by far the majority of practical and ideal systems one encounters. There exists one other class of colloids, however, which although less important in an absolute sense, can be of great practical importance in certain situations. That class is generally referred to as the *aerosols* .

Aerosols constitute systems in which there exists a condensed phase of one material (solid or liquid) that is colloidally dispersed in a gaseous phase. There are two subclasses of aerosols depending on whether the dispersed phase is a liquid or a solid. There cannot be, of course, a dispersion of one gas in another. Where the dispersed phase is a liquid, the system is commonly referred to as a *mist* or a *fog* . For solid aerosols, one may commonly refer to a *dust* or *smoke* . Each class of aerosol has its own characteristics of formation and stabilization and will be discussed briefly below.

LIQUID AEROSOLS: MISTS AND FOGS

Mists and fogs are colloidal dispersions of a liquid in a gas. They may therefore be thought of as being roughly the inverse of a foam system. The interactions controlling their stability, however, are not in general the same as those involved in foam stabilization, because mists and fogs do not normally possess the thin lamellar films encountered in most foams, whether of the spherical or the polyhedral type.

Liquid aerosols may be formed by one of two processes, depending on whether the dispersed system begins as a liquid or undergoes a phase change from vapor to liquid during the formation process. In the first case, since the dispersed material does not change phases, the aerosol is formed by some process that changes the dispersity or unit size of the liquid. To this class belong the spray mists such as those formed at the bottom of a waterfall or by ocean waves, mists produced by vigorous agitation, and those formed by some direct *spraying* or *atomization* process. The term "atomization" is a somewhat unfortunate choice because it has nothing to do with the nature of the process being described; however, the term entered the aerosol field many years ago and is still encountered. A more apt term is *nebulization* . Liquid aerosols can also be formed directly by the application of high electrical potentials to the liquid.

The second class of mists or fogs is that produced by some process in which the incipient liquid phase is introduced as a vapor and forms droplets as a result of some equilibrium condensation process or the liquid is produced as a result of some chemical reaction. The former mechanism includes, of course, cloud and fog formations, while the latter corresponds to some "chemical" fogs and mists.

SPRAYING AND RELATED MECHANISMS OF MIST AND FOG FORMATION

Liquid aerosol formation by spraying is a very important industrial process, even though the fundamental details of the process are still not very well understood. Major applications include: paint application; fuel injection in diesel, gasoline, and jet engines; spray drying of milk, eggs, etc; the production of metal and plastic powders; medicinal nose and throat sprays; the application of pesticides to crops; and many more. In all of those applications, it is vitally important that the characteristics of the aerosol produced be optimized to produce the desired end result. Unfortunately, because of the nature of the process, it is usually necessary to arrive at the optimum spraying system by trial-and-error techniques, based upon previous experience in the field.

Aerosol spays are usually formed by one of five basic processes. These include:
  1. Directing a jet of liquid against a solid surface thereby breaking the liquid up into fine droplets
  2. Ejecting a jet of liquid from an orifice into a stream of air or gas
  3. Ejecting a stream of liquid from a small orifice under high pressure
  4. Dropping liquid onto a solid rotating surface from which small droplets are ejected by centrifugal force
  5. The application of high electrical potentials to liquid drops

Simple Theories of Drop Formation

Notwithstanding the practical importance of spraying processes, the chemistry and mechanics of the processes are still not very well understood. Numerous attempts have been and are being made to quantify and understand the processes involved in order to get a better practical handle on the matter. Most of those treatments are quite complex and beyond the scope of this book. However, it may be instructive to work through a relatively simple approach in order to see how surface tension forces come into play.

Spray production by methods involving ejection of a liquid through an orifice, impacting a stream of liquid against a flat surface, and ejection from a spinning disk by centrifugal force perhaps represent the simplest situations because they involve knowledge of only one material velocity --- that of the liquid. Spray production by the action of an incident air stream on a jet of liquid involves, of course, the velocity of both the liquid and the air.

If a liquid is forced through an orifice under a high pressure, the velocity of the liquid in the channel of the orifice becomes so high that turbulent flow is encountered. That is, the liquid will not flow smoothly in lines parallel to the walls of the orifice but will flow in complex patterns with eddies, swirls, and vortices. When the liquid leaves the orifice in this turbulent or, to use a more fashionable term, *chaotic* state, the angular forces in the vortices will act against the surface tension of the liquid to strip off units of liquid to form droplets (Figure 13.1).

Figure 13.1. Schematic illustration of drop formation in a liquid jet leaving an orifice under pressure.

For a simple classical analysis of the situation, assume that as the liquid leaves the orifice it not only has a linear velocity due to the pressure forcing it through the system, but also some angular velocity ω resulting from its chaotic flow pattern. Liquid will therefore rotate within the jet with a period of $2\pi/\omega$. The rotation creates a local centrifugal force. For a column of exiting liquid of radius r and height dz that force is given by

$$F_\omega = 2/3\,\pi\,\rho r^3 \omega^2 dz \tag{13.1}$$

where ρ is the density of the liquid. The pressure disrupting the jet will be given by

$$P = 2/3\,\pi\,\rho r^3 \omega^2 dz\ /2\pi r\,dz\ = 1/3\,\rho r^2 \omega^2 \tag{13.2}$$

The surface tension forces keeping the jet together will be σ/r. The second radius of curvature for the jet being infinitely large. The critical radius at which a continuous jet of liquid becomes unstable and breaks up to form droplets will be

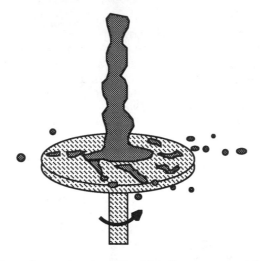

Figure 13.2. Drop formation by a liquid falling on a rapidly rotating disk.

$$r_c = (3\sigma/r\,\omega^2)^{1/3} \qquad\qquad (13.3)$$

It is difficult, of course, to determine the value of ω in a flowing system, so that experimental verification of such an analysis is not a trivial matter. However, if one assumes that ω is proportional to the injection pressure, the product of the pressure and r^3 should be constant. In practice, the agreement is not quite there. If one were to use the *excess pressure* --- that is, the pressure in excess of that at which chaotic flow begins --- the agreement might logically be expected to improve.

When a liquid is thrown from a rotating plate by centrifugal force, one can visualize the liquid being formed into filaments that break down as stretching begins to overcome cohesive surface tension forces and drops begins to form (Figure 13.2).

Since theories for predicting the drop size of a spray produced based on the characteristics of the liquid and the apparatus are complex and sometimes unsatisfactory, it is usually necessary to measure sizes for each given situation. In general, however, the following rules hold for most fluid ejection systems.

1. Increasing the surface tension of the liquid will decrease the average drop size.
2. The higher the viscosity of the liquid, the greater will be the average radius.
3. The higher the injection pressure, the smaller will be the average drop radius.

If an electrical potential is applied to drops of a liquid as they form at the tip of a capillary tube, in a gas or in the presence of a second immiscible liquid phase, the drop may disintegrate to form an aerosol (or an emulsion in the latter case). The breakup of the drop may be attributed to the action of an electrical pressure building up in the drop as a result of the applied voltage which acts against the surface tension forces holding the drop together. If a drop of radius r and surface tensions σ is given a charge of q by the applied voltage, the energy of the drop, E, is given by

$$E = 3\sigma V/r + 2\pi r^2 Q^2/3V \qquad (13.4)$$

where V is the total volume of the liquid in the drop and Q ($= nq$, the number of aerosol drops formed multiplied by the charge on each drop) is the total charge on the volume of liquid, V. The value of the aerosol drop radius formed can be evaluated by setting $dE/dr = 0$, so that

$$r^3 = 9 \, V^2 \sigma / 4\pi Q^2 \qquad (13.5)$$

From the above "superficial" treatment of mist drop formation by typical ejection processes, it can be seen that theoretical analysis can be used in predicting an approximate result based on a given set of circumstances. However, much more complex analyses are necessary to obtain more than a ball park figure, and even then the results are unlikely to be worth the trouble. In liquid aerosol formation, as in many such areas, experience is often the best guide.

Liquid Aerosol Formation by Condensation

In order for a vapor to condense under conditions far from its critical point, certain conditions must be fulfilled. If the vapor contains no foreign substances that may act as *nucleation sites* for condensation, the formation of aerosol drops will be controlled by the degree of saturation of the vapor, analogous to the situation for homogeneous crystal formation.

It will be remembered that the formation of a new phase by homogeneous nucleation involves first the formation of small clusters of molecules, which then may disperse or grow in size by accretion until some critical size is reached, at which point the cluster becomes recognizable as a liquid drop. The drop may then continue to grow by accretion or by coalescence with other drops to produce the visible aerosol drops. Normally, extensive drop formation is not observed unless the vapor pressure of the incipient liquid is considerably higher than its saturation value; that is, unless the vapor is *supersaturated*.

The barrier to the condensation of the liquid drop is related to the high surface energy possessed by a small drop relative to its total free energy. Thermodynamically, a simple argument can be given to illustrate the process. If one considers the condensation process as being

$$n \, \mathbf{A} \, (gas, \, P) \leftrightarrow \mathbf{A}_n \, (liquid \, drop)$$

where n denotes the number of molecules of $\mathbf{A}$ involved in the process,

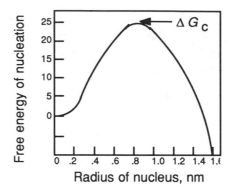

Figure 13.3. Illustration of the relationship between the free energy of nucleation and the radius of the nucleating drop.

then in the absence of surface tension effects, the free energy change of the process will be given by

$$\Delta G = -nkT \ln P / P_o \qquad (13.6)$$

where P is the pressure or activity of **A** in the vapor phase and P_o is that in the liquid phase. The ratio P / P_o is often referred to as the *degree* of supersaturation of the system. A drop of radius r, being a liquid, must have a surface energy equal to $4\pi r^2 \sigma$, so that the actual free energy change on drop formation will be

$$\Delta G = -nkT \ln P / P_o + 4\pi r^2 \sigma \qquad (13.7)$$

   Both elements to the right in eq. 13.7 can be written in terms of the drop radius, r. If ρ is the density of the liquid and M its molecular weight, the equation becomes

$$\Delta G = -(4/3)\pi r^3 (\rho/M) RT \ln P / P_o + 4\pi r^2 \sigma \qquad (13.8)$$

In eq. 13.8, the two terms are of opposite sign and have a different dependence on r. A plot of ΔG vs r exhibits a maximum as illustrated in Figure 13.3 for a hypothetical material with a density of one, molar volume of twenty, and pressure or activity ratio of 4 at a given temperature. The radius at which the plot is a maximum may be defined as the critical radius, r_c, which can be determined from eq. 13.8 by setting $(\Delta G)/dr = 0$. That transformation gives the old faithful Kelvin equation, which on rearrangement leads to

$$r_c = 2\sigma V_m / RT \ln P / P_o \qquad (13.9)$$

where V_m is the molar volume of the liquid.

For water at 25°C and supersaturation (P/P_o) of 6, eq. 13.9 predicts a critical radius of 0.58 nm, corresponding to a cluster size of about 28 water molecules. It is difficult to say whether a drop of that small size actually has the same properties as the bulk liquid. It is probable, in fact, that the relatively high surface-to-volume ratio in such an assembly will result in an actual surface tension greater than the "true" bulk value of 72 mN m^{-1}. If larger values for σ is used, the value of r_c decreases. The same occurs as the degree of supersaturation increases. The uncritical quantitative use of eq. 13.9 can be misleading in that it predicts critical cluster sizes for homogeneous nucleation that are unlikely to occur with much frequency if left to the chance of random fluctuation processes. Qualitatively, however, the equation is useful in explaining the difficulty of formation of liquid aerosols by direct condensation in highly purified systems.

If one combines eqs. 13.8 and 13.9 it is possible to obtain a value for the free energy of formation of a cluster of the critical radius for drop formation, ΔG_{max}.

$$\Delta G_{max} = 4\pi r_c^2 \sigma/3 = 16\pi\sigma^3 M^2 / [3\rho^2 (RT \ln P/P_o)^2]$$

$$= 16\pi\sigma^3 V_m^{2/3} (RT \ln P/P_o)^2 \qquad (13.10)$$

Conceptually, one can think of the nucleation process in the following terms. If the pressure or activity of the vapor, P, is small relative to P_o, then ΔG for a given cluster of molecules will increase with each added molecule. That is, the tendency will be for clusters smaller than r_c to return to the vapor phase. Statistically, one might expect to encounter clusters of all sizes due to random fluctuation processes; however, all but the smallest would be very uncommon. There would therefore be little likelihood of obtaining the critical radius necessary for drop formation to occur. However, as the degree of supersaturation increases, the value of r_c decreases and random fluctuations begin to result in more clusters of radius r_c. Once that point is reached, the clusters begin to grow spontaneously to form drops. When a specific supersaturation pressure is exceeded, there will develop a steady parade of clusters of the required critical dimensions, resulting in the formation of a visible mist or fog.

SOLID AEROSOLS: DUST AND SMOKE

Aerosols composed of solid particles suspended in a gas are commonly referred to as dust or smoke, the exact terminology depending on, according to preference, the size and sedimentation rate of the particles, or the method of aerosol formation. In some areas, aerosols formed through some dispersion process are termed *dust* while those arising from a condensation process are *smoke* . Alternatively, some prefer to label as dust aerosols of sufficient particle size to have relatively rapid (eg, noticible over a short time span) sedimentation rates in air, while smokes would be of smaller, lighter particles. Regardless of the terminology employed, it is clear that solid aerosols constitute a very important, and usually undesirable, component of many modern processes.

The majority of industrial dusts and smokes are produced by processes of dispersion in which small particles are stripped off from larger solid masses. Smoke from burning wood or coal, dust in sugar refineries and grain mills, mining dusts, and some exhausts from internal combustion engines all fall into this general category of aerosol formation. Condensation aerosols of solid particles are less common but are typified by that formed by the reaction of ammonia with hydrogen chloride in dry air or by burning magnesium.

Because of the practical importance of solid aerosols produced by dispersion processes, they have received a great deal of attention from the point of view of how they can be reduced and suppressed. In that work, the emphasis is usually placed on process changes that will produce fewer, larger (and therefore more rapidly sedimenting) particles, or, as in the case of coal-burning furnaces, retarding the upward movement of carbon aerosol particles so that they can be more completely burned before leaving the stack. More will be said about the suppression and destruction of smokes and dusts below.

Solid aerosols produced by condensation processes, while not as important from a practical standpoint, are often of more theoretical interest because it is easier to control their nucleation and growth rate, particle size and size distribution, and rate of disappearance. They can therefore be used more readily to study the various theories of aerosol formation and destruction. The condensation methods normally employed in such studies can be divided into two classes --- *chemical* and *physical* condensation processes. A typical physical method may involve the heating of a material of relatively low volatility (eg, stearic acid) sufficiently to produce a relatively high degree of supersaturation and passing the vapor into a stream of cold gas, rapidly condensing the vapor to the solid state (but probably going through a liquid aerosol phase). Many of the elements of tobacco smoke are also of such a "physical" origin. Chemical methods would be illustrated by the NH_3--HCl reaction mentioned before, or by the photochemical oxidation of iron pentacarbonyl in air to produce a smoke of ferric oxide.

While dusts and smokes are generally a nuisance, they do have their uses. Particularly important are pesticide dusts (and sprays) which may be applied to wide areas by airplane or surface dispersal techniques. In addition, some industrial catalysts, whose activity depends on their having a large surface to volume ratio, are effectively employed in the form of an aerosol dust or smoke. Finely dispersed solid materials are often found to exhibit combustion properties quite distinct from those observed for a large solid block of the same material. Dust made from a low grade of coal, which would not normally be suitable for certain fuel applications, may perform appreciably like a more expensive liquid fuel. On the other hand, normally easily handled materials such as sugar or flour, not normally considered significant fire hazards, when encountered as a fine dust (as in a silo) may constitute a grave explosion hazard that must always be controlled.

Aerosols, whether solid or liquid, constitute an important but often overlooked class of colloidal systems. Because of their complex nature and the practical difficulties of obtaining "model" systems, their theoretical treatment has perhaps lagged behind that of related systems. However, their importance in many applied areas (both for good and ill) remains undiminished. Just as experience is often the best teacher in the art and science of aerosol production, so it is when the question is aerosol suppression or destruction.

THE DESTRUCTION OF AEROSOLS

Aerosols, like their cousins foams, emulsions, and dispersions, may be viewed as either advantageous or detrimental, depending on the situation. The previous discussion introduced some of the fundamental aspects of aerosol formation, where desired. Of equal or perhaps greater practical importance is the question of the suppression of aerosol formation, or the destruction of unavoidable aerosols. Perhaps the best approach to that problem is through an understanding of some of the general principles involved in their stabilization and destruction. In that context the mechanisms of destruction involved will be essentially the same as those for other colloidal systems --- flocculation and coalescence.

The theory for the coagulation of noninteracting colloidal particles has already been introduced in Chapter 10 and will not be repeated here. However, its application to aerosol systems requires some elaboration since the collision frequency for aerosol particles is not adequately reflected by the continuum theory employed in the case of fully condensed systems.

The methods of destroying aerosols, whether liquid or solid, are numerous. One of the oldest and most important from a practical standpoint is the use of a spray (usually water) to "wash" the aerosol from the gas phase. For most dusts and smokes, every collision between an aerosol particle and a water droplet will result in coagulation of the two particles. Such fusion will obviously increase the size of the average particle and sedimentation due to gravity will become more efficient.

The rate of fall of an aerosol particle in a gas can be estimated by

$$U = 2gr^2 (\rho - \rho_1) / 9\eta \qquad (13.11)$$

in which U is the rate of fall in still gas or vapor, ρ is the density of the particle (liquid or solid), ρ_1 that of the gas phase, g the acceleration due to gravity, and η the viscosity of the gas. Equation 13.11 holds if the size of the aerosol is much greater than the mean free path of a molecule of the gas. If the drops are very small, the relationship becomes

$$U = [2gr^2(\rho - \rho_1)\,/9\eta]\,[1 + K(l\,/r)] \qquad (13.12)$$

where K is a constant (usually ≈ 0.9) and l is the mean free path. For a drop of water of radius 10^{-4} cm falling in air, U will equal about 0.015 cm s^{-1}, meaning that it will take about 2 h for the drop to fall 1 m. If the aerosol in question coalesces during its descent, the value of U will obviously increase with the square of the radius of the new particle. For example, if eight drops of equal volume coalesce to form a single drop, U will increase by a factor of 4.

The above discussion, of course, applies only in still air or other gas. If there exists a net movement of the gas phase, the velocity of that movement must be incorporated into the equation. Also, eqs. 13.11 and 13.12 assume that the viscosity of the aerosol (a liquid in this case) is much greater than that of the gas phase. If such is not the case, eq. 13 can be applied.

$$U = [2gr^2(\rho - \rho_1)\,/9\eta]\,[(3\eta_1 + 3\eta)\,/\,(3\eta_1 + 2\eta)] \qquad (13.13)$$

where η_1 is the viscosity of the liquid drop.

A second mechanism for the destruction of a mist or fog is by changing the degree of "saturation" of the surrounding gas. If the gas contains less than the required amount of vapor of the liquid phase, the drop will evaporate. The rate of evaporation of a drop of liquid of radius r and mass m is given by

$$- dm\,/dt = 4\pi Dcr\,[1 + k\,(2\rho_1 Ur\,/\eta)^{1/2}] \qquad (13.14)$$

In eq. 13.14, k is a constant depending on D, η, and ρ_1, U is the rate of fall as given above, D is the diffusion coefficient (cm^2 s^{-1}), c is the concentration of the saturated vapor next to the drop (g cm^{-3}), and the other terms are as previously defined. For water at 20°C, the value of k is 0.229. A hypothetical drop of radius 10^{-4} cm that falls at 0.015 cm s^{-1}, in dry air, will evaporate well before it could fall one meter. In a practical sense, then, a mist or fog can be dissipated by increasing the temperature of the surrounding gas, thereby lowering the degree of vapor saturation and increasing the rate of evaporation.

A practical and relatively efficient means for the removal of aerosol particles is via *electrostatic precipitation* . Many solid aerosols (and some liquids as well) are formed with or easily attain a net electrical charge on their surface. If such does not occur spontaneously, it can usually be induced by some mechanism such as ionization by electrons or gas ions.

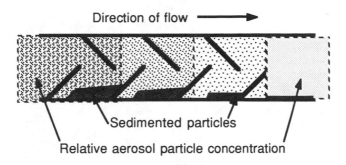

Figure 13.4. Schematic illustration of the use of baffles for trapping aerosols. As the particles strike the baffles, they lose momentum and fall out of the flow to be collected and removed.

The rate of ionization of the particle and its final surface charge will be proportional to the square of the radius of the drop. The velocity of a charged particle in a given electric field will be inversely proportional to its radius, so that in the electrostatic precipitation process, the final velocity of the particle in the field will be proportional to its radius. Like sedimentation, electrostatic precipitation will remove the larger particles first.

Additional practical methods of aerosol removal include filtration and centrifugatiion. Centrifugation, of course, is basically the same as sedimentation except that the force of gravity is replaced by artificial forces of greater strength. Equations 13.11 and 13.12 continue to apply in that case.

Filtration methods, while usually relying on separation due to size, can also involve some degree of a centrifugation effect. For example, if a stream of aerosol is passed through the tortuous pathway of, say, a charcoal filter, as the stream goes through various twists and turns at a steady speed, centrifugal forces will be imposed on the particles, forcing them out along the "radius" of the curve and into the filter walls, where precipitation will occur. The centrifugal force can be estimated assuming the filter pathway to be a spirally wound tube. The magnitude of the force is computed based on the velocity of the aerosol and the radius of the hypothetical coil.

A similar effect is seen when a stream of aerosol is forced to pass through a system of baffles. In such a system, as the aerosol is forced to change direction due to the presence of the baffle plates, it must move through a curved path, setting up a net centrifugal force perpendicular to the stream, normally toward the baffle walls. As a result the baffles will collect aerosol particles and the gas stream leaving the system will contain much less of the aerosol material (Figure 13.4). Other methods of aerosol destruction have been developed over the years, and the problem remains one of great practical importance.

CHAPTER 14

POLYMERS At INTERFACES

As pointed out in Chapter 1, a colloid is characterized by a certain size range of particles (about 1 nm - 1 μm.) While that range is not rigid, since most commonly encountered emulsions are larger on average, it serves as a good reference point with which to classify various heterogeneous systems. Beginning with the size criterion only, we then defined two types of colloids which fall into that size range - the *lyophobic* ("solvent hating") colloids discussed in Chapters 10 and 11, and *lyophilic* ("solvent loving") colloids. The lyophobic colloids are normally formed by the reduction or comminution of coarse particles to achieve the desired particle size or by the controlled growth (by crystallization, condensation, etc) from solutions of small molecules or ions.

The lyophilic colloids are composed of either large molecules (in relation to the size of the solvent molecules) or reversible associated or aggregated structures (*association colloids*) formed spontaneously in solutions of certain types of molecules. The large molecules (usually with molecular weights ranging from about 5000 to several millions) are, of course, macromolecules or polymers and include a broad range of materials such as naturally occurring proteins, carbohydrates, gums, and other biocolloids, modified biopolymers such as gelatin and rayon, and the completely synthetic materials such as polyethylene, nylon, polycarbonates, etc. The best known association colloids are aqueous solutions of surfactants, although certain dyes and drugs may also form association structures. That very important class of lyophilic colloids will be discussed in Chapter 15.

This chapter will introduce in a very brief way the basic concepts of polymers as lyophilic colloids. The treatment of the solution concepts themselves will be severely limited, with most space dedicated more specifically to the effects of polymers on surface and interfacial properties and lyophobic colloidal stability.

THE SOLUBILITY OF MACROMOLECULES

Small molecules of different structures will mix to form homogeneous solutions if the mixing process results in a decrease in the Gibbs free energy of the system. The total free energy change will, of course, have contributions from both the enthalpy of mixing, ΔH_{mix}, and the entropy of mixing, ΔS_{mix}. For such systems, ΔS_{mix} will always contribute to the decrease in ΔG_{mix}, so that a positive heat of mixing will not necessarily prevent the formation of a homogeneous solution. For a solution of macromolecules, however, the concentration of the solute must be relatively small (in terms of moles per unit volume) so that the entropy of mixing will always be small. Solubility or miscibility in such systems, then, will be determined almost exclusively by ΔH_{mix}. A negative ΔH_{mix} will result in a high degree of miscibility, while an even slightly positive value will result in almost complete insolubility. If ΔH_{mix} is near zero, the line between solubility and insolubility for a macromolecular system can be crossed from both sides, for example, by a small change in temperature or in the composition of a mixed solvent.

Due to the size of macromolecules, rates of diffusion and conformation changes accompanying dissolution (or precipitation) may be very slow, and significant amounts of time may be required for the system to reach equilibrium. The fine line between miscibility and immiscibility, and the relatively long equilibrium times involved may be very important to the comportment of macromolecules at various surfaces and interfaces of great theoretical and practical importance.

Statistics of Polymer Chain Conformations in Solution

The conformation of polymer chains (in effect, their size) in solution is an important characteristic of a given system in solution and at interfaces. The detailed analysis of polymer conformations is a very complex process requiring powerful computer facilities. A simplified treatment based upon random flight (or random walk) statistics allows for the approximation of chain dimensions adequate for most practical situations.

The bonds connecting atoms in a polymer chain have a specific length and are separated by an angle of about 110°. Assuming free rotation about each bond, the polymer in solution can assume a large number of conformations. The mechanical and thermodynamic properties of a polymer or polymer solution will be influenced by the average or most common conformation, usually stated in terms of the *average end-to-end distance* of the polymer chain ($<r^2>^{1/2}$) or the root-mean-square distance of a chain element from the center of gravity

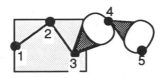

Figure 14.1. Schematic illustration of the method for calculating polymer chain dimensions in which atoms 1-5 are the first five atoms in the chain.

of the coil ($<s^2>^{1/2}$).

If one draws a plane through the first three atoms of a polymer chain as in Figure 14.1, the next bond (bond 3--4) may align itself in any direction on a cone having a specific angle related to the bond angle, as shown. Bond 4--5 may then do the same based upon the possibilities for 3--4, and so on down the chain. Obviously, such a process rapidly produces a very large number of possible orientations so that an atom just a few steps down the chain may assume a position completely independent of that of the first two atoms. By assuming the statistical independence of the direction of interconnected chain elements it is possible to derive an expression for the end-to-end distance for a chain with N elements (eg, monomer units) of length A. Without going through the complete derivation process, the result is

$$<r^2>^{1/2} = (NA^2)^{1/2} \qquad (14.1)$$

Equation 14.1 indicates that the root-mean-square end-to-end length is proportional to the square root of the stretched chain length, which means it will also be proportional to the molecular weight.

The root-mean-square distance of a chain element from the center of gravity of the coil will then be given by

$$<s^2>^{1/2} = (NA^2 / 6)^{1/2} \qquad (14.2)$$

Other statistical data about a polymer chain can be determined using the same model but will not be discussed further here.

Some criticisms of the simple random walk model include: (1) it is valid only for polymers of very high molecular weight; (2) it does not give sufficient weight to conformations in which the chain is stretched almost to its full length; and (3) it does not treat the problem of the interpenetration of various chain elements. The latter conflict can be avoided by a modification of the process to give a "self-avoiding walk," which predicts a slightly expanded coil with $<r^2>^{1/2}$ no longer being directly proportional to $N^{1/2}$. However, in many important instances, the consequences of such statistics are not greatly affected by such corrections.

The dimensions of a polymer chain in solution are important to the rheological properties of the system. More specific to the question of colloidal stability, however, such dimensions play a vital role in the ability of an *adsorbed* polymer to stabilize (or destabilize) a lyophobic colloid as discussed in Chapter 10.

ADSORPTION OF POLYMERS AT INTERFACES

Macromolecular species have played an indispensible role in the stabilization of colloidal systems since the first prelife protein complexes came into existence. Man has consciously (although usually without knowing why) been making use of their properties in that context for several thousand years. Today macromolecules play a vital role in many important industrial processes and products including: as dispersants, stabilizers, and flocculants; as surface coatings for protection, lubrication, and adhesion; for the modification of rheological properties; and, of course, their obvious importance to biological processes.

In order to understand the role of polymers in their various surface and colloidal applications, it is necessary to understand when, where, why, and how they adsorb at surfaces and interfaces. While the interactions that control polymer adsorption at the monomer level are the same as those for any monomolecular species, the size of the polymer molecule introduces many complications of analysis that must be treated in a statistical manner, which means that we never really know what the situation is but must make educated guesses based on the best available evidence.

In contrast to monomolecular species, it is highly unlikely that all of the monomeric segments of a polymer chain will be simultaneously in contact with a surface or interface. For an isolated polymer chain, the statistics based on allowed bond lengths, angles, etc, will dictate that there will be some equilibrium configuration which will describe the average situation. At an interface, that configuration will result from a balance of the net energy change on adsorption (whether positive or negative), the decrease in the entropy of the chain which must accompany adsorption, and the gain in entropy due to freeing solvent molecules. The latter effect is especially important because it explains why some polymers will adsorb at surfaces even when the adsorption process is endothermic overall. As a result of the statistics of chain configuration on adsorption, the adsorbed chain configuration will likely include loops, tails, and trains of monomer units (Figure 14.2).

For a high molecular weight polymer, the equilibrium configuration will produce an adsorbed "layer" of typically 3--30 nm. In general, one can assume that adsorption will be monomolecular, since the thickness of the first polymer layer will make attraction for a second layer very small. The exceptions would be for polymers of low molecular weight, or for systems in which the polymer is close to the point of becoming immiscible.

Because of the large size of polymer chains, it takes a relatively long time for polymer adsorption to reach equilibrium. In a related sense, one may assume that the adsorption of a high molecular weight polymer

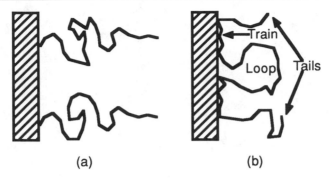

(a)   (b)

Figure 14.2. Schematic illustration of the modes of polymer adsorption at solid surfaces: (a) simple terminal adsorption or grafting; (b) random adsorption to produce loops, trains, and tails.

will be irreversible. Although each polymer segment may be adsorbed reversibly, one must assume that many segments of a given chain will be adsorbed at any given moment, and the probability of all adsorbed segments of that chain being desorbed at the same time become almost vanishingly small.

That is not the case for low molecular weight fractions in which there are only a few points of attachment. That idea explains the commonly observed phenomenon that when a polymer of broad molecular weight distribution is added to a colloidal system, the low molecular weight fraction adsorbs rapidly (ie, it is more mobile), but slowly becomes displaced by high molecular weight chains which, once attached, will not be desorbed to any significant extent. It also helps explain the fact that high molecular weight polymers generally provide better stabilization for a given system than a low molecular weight polymer of the same composition. When particles with adsorbed polymer of low molecular weight approach, hydrodynamic forces may force the desorption of the protecting molecules (Depletion Flocculation, Chapter 10), decreasing its protective "power," while high molecular weight molecules will not be so easily displaced (see Chapter 10 for a discussion of this phenomenon).

Just like their monomeric counterparts, adsorbed polymers exhibit characteristic adsorption isotherms. For macromolecular systems, those isotherms tend to be of the high-affinity type illustrated in Figure 14.3. In general, one finds that up to point A in the figure, all of the polymer in the system is adsorbed, while beyond that point more molecules can be accommodated only by changing the configuration of the adsorbed chains so as to reduce the surface area occupied by each one. That process is normally aided by the presence of favorable lateral interactions among neighboring chains.

At point B, the accommodation of additional chains has reached its maximum. The relative crowding of the chains at point B means that the molecules will not be able to attain their true statistical equilibrium

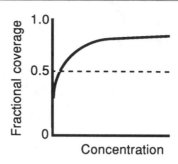

Figure 14.3. The shape of a "typical" high-affinity isotherm for the adsorption of high molecular weight polymers.

conformations; instead, a pseudo-equilibrium will be attained. In the region between A and B, it is usually found that the thickness of the adsorbed polymer layer increases, since more units of each chain are forced further away from the adsorbent surface (Figure 14.4).

POLYMER -- SURFACTANT INTERACTIONS

Surfactants constitute some of the functionally most important ingredients in cosmetic and toiletry products, foods, coatings, pharmaceuticals, and many other systems of wide economic and technological importance. In many, if not most, of those applications, polymeric materials, either natural or synthetic, are present in the final product formulations or are present in the targets for their use. Other surfactant applications, especially in the medical and biological fields, also potentially involve the interaction of polymers (including proteins, nucleosides, etc) with surfactant system.

Interactions between surfactants and natural and synthetic polymers have been studied for many years with varying degrees of understanding and experimental control. Although the basic mechanisms of surfactant--polymer interaction are reasonably well known, there still exists substantial disagreement as to the details of some of the interactions at the molecular level. Observations on changes in the interfacial, rheological, spectroscopic, and other physicochemical properties of surfactant--polymer systems indicate that such interactions, regardless of the exact molecular explanation, can significantly alter the macroscopic characteristics of the system, and ultimately its application.

It is generally recognized that surfactant--polymer interactions may occur between individual surfactant molecules and the polymer chain (ie, simple adsorption), or in the form of polymer-aggregate complexes. In the latter case, there may be complex formation between the polymer chain and micelles or premicellar aggregates. Other associations may result in the formation of so-called "hemimicelles" along the polymer chain. The term "hemimicelle" is relatively new to the field of surfactant science but is now encountered in several contexts, although the exact definition of the

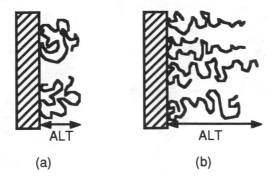

Figure 14.4. Illustration of changes in polymer chain configuration and adsorption layer thickness (ALT) with increased adsorption: (a) low levels of adsorption, smaller ALT; (B) higher levels of adsorption, larger ALT.

term is somewhat elusive. For now, it can be defined simply as a surfactant aggregate formed in the presence of a polymer chain or solid surface having many of the characteristics of a micelle, but being intimately associated with the locus of formation; hemimicelles do not exist as such in solution, although there is ample evidence of the formation of premicellar aggregates in some systems. The formation of such structures in surfactant--polymer systems is often illustrated as resembling a string of pearls or water droplets on a spider's web (Figure 14.5).

The forces controlling surfactant interactions with polymers are identical to those involved in other solution or interfacial properties, namely van der Waals or dispersion forces, the hydrophobic effect, dipolar and acid--base interactions, and electrostatic interactions. The relative importance of each type of interaction will vary with the natures of the polymer and surfactant so that the exact characters of the complexes formed may be almost as varied as the types of material available for study.

Experimental methods for investigating polymer--surfactant interactions vary widely, but they generally fall into two categories: those that measure macroscopic properties of a system (viscosity, conductivity, dye solubilization, etc) and those that detect changes in the molecular environment of the interacting species (nuclear magnetic resonance, optical rotary dispersion, circular dichroism, etc). A comparison of the experimental results of various studies can be complicated by variations in the sensitivity of experimental techniques and the physical manifestations of the interactions occurring, as well as differences in the purity and characterization of the experimental components. The results of each experimental approach, while being useful in understanding the "symptoms" of surfactant--polymer interactions, do not always provide an unequivocal distinction among the possible mechanisms at the molecular level. Newer techniques such as small-angle neutron scattering, which come close to "photographing" the relative relationships

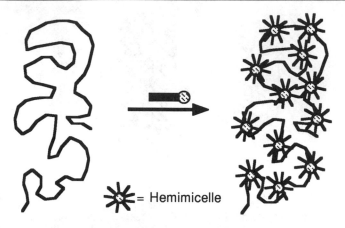

Figure 14.5. A surfactant--polymer "string of pearls" complex.

among polymer and surfactant units, promise to clarify many questions now in dispute.

Mechanisms of Polymer--Surfactant Complex Formation

The most generally accepted model for surfactant--polymer interaction is based on a stepwise sequence of binding between surfactant monomers (S) and the polymer chain (P), with each step being governed by the law of mass action, and with unique rate constants, k, controlling each step.

$$P + S \ldots PS \qquad\qquad (k_1)$$
$$PS + S \ldots PS_2 \qquad\qquad (k_2)$$
$$PS_2 + S \ldots PS_3 \qquad\qquad (k_3)$$
$$PS_{n-1} + S \ldots PS_n \qquad\qquad (k_n)$$

The values of the various interaction constants and their dependence on experimental conditions (eg, temperature, solvent, ionic strength, pH) serve as a basis for formulating feasible descriptions of the molecular processes involved in the interactions. The combination of macroscopic and molecular information can provide valuable insight into the overall process. In the above model, it is assumed that the stepwise binding process occurs initially through surfactant monomeric units. That is, there is no significant direct association of micelles or other aggregates with the polymer chain. The formation of such aggregate--polymer complexes is not excluded, however, since they may form on the chain as the total concentration of bound surfactant increases. Alternatively, if polymer is added to a solution already containing micelles, it may be that a form of adsorption of polymer onto or into the micellar structure may occur.

Surfactant--polymer interactions, as do all surfactant-related phenomena, involve a complex balance of factors encouraging and retarding association and can be understood only if those factors can be reasonably estimated. The dominating forces can be broken down into the categories of either Coulombic attractions and repulsions, dipolar interactions (including hydrogen bonding or acid--base), dispersion forces, and the hydrophobic effect. Combinations are, of course, possible and even likely, adding to the fun of interpreting the experimental results. While the electrostatic processes are fairly straightforward, involving the interaction of charged species on the polymer with those in the surfactant molecule, the remaining interactions are less easily quantified and can be quite complex. Polymers in particular add their own new twists since they may possess in solution secondary and tertiary structures that may be altered during the surfactant binding process in order to accommodate the bound surfactant molecules, thereby adding new energy terms to the total energy balance. The nature of the surfactant--polymer complex may significantly alter the overall energetics of the system so that major changes in polymer chain conformation will result. Any and all of those changes may result in major alterations in the macroscopic and microscopic properties of the system (eg, the denaturation of proteins by detergents).

Forces opposing the association of molecules include thermal energy, entropic considerations, and repulsive interactions among electrical charges of the same sign. It is clear that the strength and character of surfactant--polymer interactions depend upon the properties of both components and the medium in which the interactions occur. However, even in systems where identical mechanisms are active for different surfactant and/or polymer types, the macroscopic symptoms of those interactions may be manifested in such a way that entirely different conclusions could easily be drawn.

Just as in the case of surfactants, four general types of polymer can be defined related to the electronic nature of the species: anionics, cationics, nonionics, and amphoterics. Not surprisingly, each polymer type will exhibit characteristic interactions with each surfactant class, with variations occurring within each group. It is little wonder, then, that surfactant--polymer interactions can become the subject of some very interesting discussions. With the understanding that a great deal remains to be learned about the subject as a whole, the following brief comments will introduce a few of the observed facts about this field of study.

Nonionic Polymers

The largest volume of published work in the field of surfactant--polymer interactions has involved surfactants and nonionic polymers such as polyvinylpyrrolidone (PVP), polyvinyl alcohol (PVA), and polyethylene oxide (POE) in water. In general, the results indicate that the more hydrophobic the polymer, the greater is the interaction of anionic surfactants with it. For a given anionic surfactant interacting with

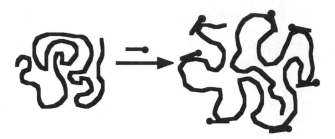

Native chain conformation   Expanded coil

Figure 14.6. Illustration of the possible effect of hydrophobic surfactant--polymer interactions: if the polymer is uncharged or of the same charge as the surfactant, coil expansion is to be expected; if of opposite charge, collapse will occur.

typical polymers, it has been found that adsorption progresses in the order polyvinyl alcohol (PVA) < polyethylene glycol (PEG) < methyl cellulose (MC) < polyvinyl acetate (PVAc) < polypropylene glycol (PPG) ≤ polyvinylpyrrolidone (PVP). In such systems, the primary driving force for surfactant--polymer interaction will be van der Waals forces and the hydrophobic effect. Dipolar and acid--base interactions may be present, depending upon the exact nature of the system. Ionic interactions will be minimal or nonexistent.

   If the primary mechanism of ionic surfactant--nonionic polymer interaction is hydrophobic, the adsorption of surfactant molecules will produce changes in the polymer chain conformation, expanding the coil due to repulsions between the ionic surfactant head groups (Figure 14.6). The properties of the complex solution (eg, viscosity) will be altered as a result of such changes. If a neutral salt is then added to such a system, repulsion between neighboring groups will be screened and the expanded coil will contract or collapse, again affecting various macroscopic properties of the solution. Such expansion and collapse of surfactant--polymer complexes as a function of the extent of surfactant adsorption may be seen as being analogous to the solution behavior of polyelectrolytes as a function of the degree of dissociation and electrolyte content.

   The bulk of the work on cationic surfactant--nonionic polymer interactions has involved the use of long-chain alkylammonium surfactants in aqueous solution. It has been found that the interactions between such species become stronger as the chain length of the surfactant increase. The drive to substitute surfactant--polymer for surfactant--water and polymer--water interactions, with the resulting increase in system entropy due to the released water molecules, becomes a dominating factor. The nature of the cationic head group seems to have some effect on polymer--surfactant interactions. For example, the viscosity of aqueous solutions of dodecylpyridinium thiocyanate--PVAc changes very little with variations in the surfactant concentration, whereas solutions of dodecylammonium thiocyanate--PVAc show considerable viscosity increases with increasing surfactant concentration. Such a result might be interpreted as reflecting a reduced extent of surfactant interaction with the polymer chain due to the greater hydrophilicity of the pyridinium ring relative to that of the simple ammonium group. The relative binding strengths between nonionic polymers and cationic or anionic surfactants are difficult to compare. The general trend is that anionics will exhibit stronger interactions with a given nonionic polymer than analogous cationic surfactants, all other things (eg, chain length of the tail) being equal.

The interactions between nonionic surfactants and nonionic polymers have been much less intensively studied that those for ionic surfactants. The limited number of reports available indicate that there exists little evidence to indicate extensive surfactant--polymer association in such systems. Considering the size of the hydrophilic groups of most nonionic surfactants, their low CMC's, and the absence of significant possibilities for head group--polymer interactions, the apparent absence of substantial interactions is not conceptually hard to accept. An assertion that binding does not occur under any circumstance, however, would be foolish, given the complexities of polymer and surfactant science in general. In food colloids especially, it has been shown qualitatively that many nonionic surfactants (eg, monoglycerides, sorbitan esters, etc) form rather strong complexes with starches and proteins, although the inherent complexity of such systems makes quantification of such effects difficult or impossible.

Ionic Polymers and Proteins

In practice, it is commonly found that surfactants will interact more strongly with charged polymeric species than with the nonionic examples discussed above. Practically all natural polymers, including proteins, cellulosics, gums, and resins, carry some degree of electrical charge. Many of the most widely used synthetic polymers do as well. When one compares the possibilities for interactions between ionic polymers (polyelectrolytes) and surfactants with those for nonionic polymers, it is readily obvious that the presence of discrete electrical charges along the polymer backbone introduces the possibility (and probability) of significant electrostatic interaction, in addition to the nonionic factors mentioned previously. The polymers may be positively or negatively charged, or they may be amphoteric. In any case, they are commonly referred to as *polyelectrolytes* because of the multiple charges carried by each polymer molecule. The presence of charges on a polymer complicates the understanding of the solution properties of the polyelectrolytes. The potential for surfactant--polyelectrolyte interactions does so even more.

Polyelectrolytes, whether natural or synthetic, are of particular interest to surfactant users because of their application as viscosity enhancers (thickening agents), dispersing aids, stabilizers, gelling agents, membranes, binders, etc. They are also encountered, of course, as fibers and textiles. Common synthetic polyelectrolytes include polyacrylic and methacrylic acids and their salts, cellulosic derivatives such as carboxymethyl cellulose, polypeptides such as poly-L-lysine, sulfonated polystyrenes and related strong-acid containing polymers, and polymeric polyammonium salts and quaternized polyamines. Natural polyelectrolytes would include cellulose, various proteins, gum arabic, lignins, etc. In most cases, the charge on the polymer is fixed as either positive or negative, so that possible interactions with surfactants of a given charge type can be reasonably well defined. While such factors as pH, electrolyte content, and the nature of the polymer counterion will affect the extent of interaction in given systems, the sense of the interaction (eg, anion--anion, anion--cation) will not change except where protonation or deprotonation of weak acids and bases occurs. Other polymers, proteins in particular, may be amphoteric in nature, the net character of the charge being determined by pH.

Not surprisingly, interactions between surfactants and polymers of similar charge are usually found to be minimal, with electrostatic repulsion serving to inhibit the effectiveness of any noncoulombic attractions. This is especially true for polymers having relatively high charge densities along the chain. When opposite charges are present, however, the expected high degree of interaction is usually found to occur. In aqueous solution, the result of surfactant binding by electrostatic attraction is normally a reduction in the viscosity of the system, a loss of polymer solubility, at least to the point of charge reversal (see Figure 14.7), and a reduction in the effective concentration of surfactant, as reflected by surface tension increases over what would be measured for that surfactant concentration in the absence of polymer.

Many naturally occurring random-coil polyelectrolytes of a single charge type, including some carbohydrates, pectins, and keratins, are anionic and exhibit the same general surfactant interactions as their synthetic cousins. Proteins, on the other hand, are amphoteric polyelectrolytes, which possess a net charge character (anionic or cationic) that depends on the pH of the aqueous solution. Unlike most synthetic polyelectrolytes, natural polyelectrolytes such as proteins and starch often have well-defined secondary and tertiary structures in solution that can affect, and be affected by, surfactant binding. When secondary and tertiary structures are present, complications arise due to alterations in those structures during surfactant adsorption. The denaturation of proteins by surfactants is, of course, just such a process of the disruption of higher orders of structure in the dissolved polymer molecule.

The question of exactly how a surfactant interacts with a protein molecule has been the subject of a great deal of discussion. In the case of interactions between bovine serum albumin (BSA) and sodium dodecylsulfate (SDS) the initial binding involves the electrostatic

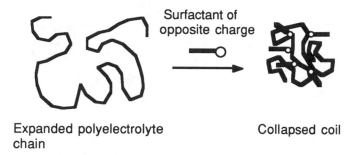

Expanded polyelectrolyte   Collapsed coil
chain

Figure 14.7. Illustration of the effect of addition of ionic surfactant to a solution of polyelectrolyte of opposite charge.

association of oppositely charged species, especially at bound surfactant levels (surfactant molecules per polymer chain) of less than 10. As such binding occurs, the electronic character of the protein changes, possibly resulting in changes in its secondary and tertiary structure. Such changes may then lead to the exposure of previously inaccessible charge sites for further electrostatic binding or of hydrophobic portions of the molecule previously protected from water contact by the higher level protein structure. Ultimately, as charge neutralization occurs, precipitation of the protein will result.

As the charges on a polymer are neutralized by surfactant adsorption, association between the hydrophobic tail of the surfactant and similar areas on the polymer becomes more favorable, again changing the net electrical character of the polymer complex. Reversal of the native charge of the protein may be the result at sufficiently high surfactant--polymer ratios (Figure 14.8). Macroscopically, the above events may lead to dramatic changes in the viscosity of the system due to, first, collapse of the polymer coil, followed by a rapid expansion after charge reversal has taken place. In addition, a minimum in the solubility of the polymer may be encountered as evidenced by precipitation followed by repeptization.

When the bound surfactant level is high, exceeding approximately 20 surfactant molecules per high molecular weight polymer chain, evidence supports the view that both the head group and the hydrophobic portion of the surfactant molecule become involved in the binding process. In fact, there is some evidence that the bound surfactant molecules may be associated into micelle-like structures, forming a "string of pearls" along the polymer chain. Alternatively, micelles may act as sites for polymer adsorption, much as is found for more "permanent" colloidal systems (Figure 14.5). If such structures are present, they have the potential for altering the rheological properties of the system to a much greater extent than single molecular binding by bridging several protein molecules to produce a large polymer aggregate. Behavior suggesting such complex formation has been found for deionized bone gelatin in the presence of several anionic surfactants, and the

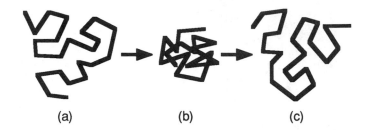

(a)   (b)   (c)

Figure 14.8. Charge reversal in polyelectrolyte--surfactant systems: (a) native polymer coil; (b) collapsed coil after addition of surfactant; (c) reexpansion of coil after adsorption of an excess of surfactant over the amount needed to completely neutralize the charge on the polymer.

mechanism has been suggested to explain the effect of some surfactants on the plasticity of bread doughs (via the formation of cross links between protein chains in the gluten fraction of wheat flour). It is generally found that the extent of interaction as reflected by increases in the viscosity (or plasticity) of a system is highly dependent upon the length of the hydrocarbon tail of the surfactant. For a series of sodium alkyl sulfates, the effect increases rapidly in the order $C_8 < C_{10} < C_{12} < C_{14} < C_{16}$, etc.

The interactions between cationic and nonionic surfactants and proteins has received less attention than the anionic case. Some alkylbenzene-polyoxyethylene surfactants appear to undergo limited binding with proteins, although there is little evidence for sufficient interaction to induce the conformational changes found in the case of anionic materials. The limited number of results published on protein--cationic surfactant systems indicates that little cooperative association occurs in those systems, even though the native protein charge may be of the opposite sign.

Although a great deal is known about the interactions between polymers and surfactants, there is a distinct lack of good experimental data in the form of adsorption isotherms. While it is clear that the surfactant binding processes are controlled by the same basic forces as the other solution and surface properties of surfactants, the location of binding sites on the polymer molecule, the relative importance of the surfactant tail and head group, and the exact role of the polymer structure remain to be more accurately defined. In any case, anyone proposing to use a surfactant in a formulation containing polymers, or in an application where surfactant--polymer interactions will occur, must always consider the effect of each on the performance of the other.

Polymers, Surfactants, and Solubilization

As will be discussed in more detail in Chapter 16, a useful characteristic of many micellar systems is their ability to solubilize water-insoluble materials such as hydrocarbons, dyes, flavors, or fragrances. Surfactant--polymer complexes have been shown to solubilize materials at surfactant concentrations well below the CMC of the surfactant in the absence of polymer. The effectiveness of such complexes differs quantitatively from that of conventional micelles.

In many instances it is found that complexes of surfactant with polymer solubilize various materials at lower total surfactant concentrations and have a greater solubilizing capacity (eg, solubilized molecules per molecule of surfactant) than a surfactant solution alone. Unfortunately, our present state of knowledge in this area is not sufficient to allow quantitative predictions about the potential solubilizing properties of surfactant--polymer complexes based solely on chemical composition, although it is known that the effectiveness of a given combination depends upon the nature of the polymeric component and the polymer--surfactant ratio.

Emulsion Polymerization

Surfactant--polymer systems have additional technological significance since surfactants are normally used in emulsion polymerization processes, often involving the solubilization of monomer (as well as low molecular weight oligomers) in micelles prior to particle formation and growth. Surfactants have also been shown to increase the solubility of some polymers in aqueous solution. The combined actions of the surfactant as a locus for latex particle formation (the micelle), solubilizer, and particle stabilizer (by adsorption,) might lead one to expect quite complex relationships between the nature of the surfactant and that of the resulting latex, which is sometimes the case. Within a class, it is usually found that surfactants with high CMC's produce latexes with larger particle sizes and broader size distributions, although no conclusive trend has been found for nonionic POE surfactants as a function of oxyethylene content.

The ability of surfactants to associate with (or adsorb onto) polymer chains may also affect the ultimate properties and stability of the resulting polymer, especially when the macromolecule exhibits some affinity for or reactivity with water. The best documented case of such a relationship involves polyvinyl acetate latexes, which have been found to differ greatly in stability depending on the surfactant used in their preparation. It is known, for example, that polyvinyl acetate can be dissolved in concentrated aqueous solutions of SDS, while cationic and nonionic surfactants have little or no solubilizing effect. In that case, solubilization presumably does not occur in the micelle, but extensive adsorption of surfactant onto the polymer chain is required. The fact that surfactants such as SDS can promote the solubilization of polyvinyl acetate has been used to suggest reasons for the observed increase in the rate of hydrolysis of polymers prepared with that surfactant relative to materials prepared with other, less strongly interacting surfactants.

The assumption is that the solubilizing surfactant (SDS) can adsorb onto and solubilize the surface polymer units causing swelling and greater exposure to water and catalyst for hydrolysis. There may also be a parallel loss of surfactant available for particle stabilization in the conventional colloidal sense. The nonsolubilizing surfactant, on the other hand, would remain essentially fixed at the surface and available to perform its function as a colloid stabilizer.

In cases where there is little affinity of the polymer for water, as for styrenes or alkyl acrylates and methacrylates, little effect of surfactant on water solubility would be expected. The action of the surfactant on such latex systems is then be limited to its action as a monomer solubilizer during preparation and an adsorbed stabilizer afterwards.

The complex relationships that can exist between polymers and surfactants raises a great many questions concerning the interpretation of data obtained from such systems. They also open the door to possible new and novel applications of such combinations, however, and will no doubt provide many interesting hours of experimentation and thought for graduate students and industrial researchers in the future.

CHAPTER 15

ASSOCIATION COLLOIDS: MICELLES, VESICLES, AND MEMBRANES

Previous chapters have discussed the formation of colloidal particles by various mechanisms including commutation, nucleation and growth, emulsification, etc. There exists another very important class of colloids that differ significantly from those discussed previously. Their formation, for example, does not result from the input of energy such as in commutation or emulsification; it is a spontaneous association process resulting from the energetics of interaction between the individual units and the solvent medium, as is crystallization. However, the size, shape, and basic nature of the associated structure is controlled by a complex series of factors distinctly different from those involved in crystallization. The size, in particluar, will be much more limited than that of a normal crystal. This class of colloids is generally referred to as *association* or *self-assembled colloids* .

The general class of association colloids can be further divided into several subgroups, which include micelles, vesicles, micro-emulsions, and bilayer membranes. Each subgroup of association colloids plays an important role in many aspects of colloid and surface science, both as theoretical probes that help us to understand the basic principles of molecular interactions, and in many practical applications of those principles, including biological systems, medicine, detergency, crude oil recovery, foods, pharmaceuticals, and cosmetics. Before undertaking a discussion of the various types of association colloids, it is important to understand the energetic and structural factors that lead to their formation.

Association colloids form as a result of the unique character of the class of materials already described in Chapter 3 --- the surface-active agents or surfactants. Because of their chemical composition, surfactants have a "love--hate" relationship with most solvents, which results in a constant tug of war between forces tending toward a comfortable accommodation with a given solvent environment, and a driving desire to escape to a more energetically favorable situation. Surfactants, in other

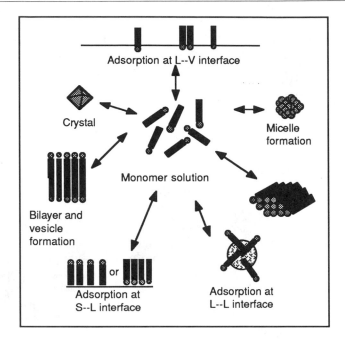

Adsorption at L--V interface

Crystal

Micelle
formation

Monomer solution

Bilayer and
vesicle
formation

Adsorption at
S--L interface

or

Adsorption at
L--L interface

Figure 15.1. Schematic illustration of the "fates" of surfactants in solution and at interfaces.

words, seem to feel that the grass is always greener on the other side of the fence, and as a result, they spend much of their time sitting on the "fence" between phases.

The adsorption of surfactants at interfaces has been, and will be, discussed in specific contexts. However, surfactants also have a life of their own within a given liquid environment --- the formation of various associated structures such as micelles, vesicles, etc. The exact behavior of a given surfactant in solution will depend on a number of internal (molecular) and external factors which will be discussed in turn below. At this point, however, it will be useful to take a general look at the possibilities open to a surface-active molecule.

SURFACTANT SOLUBILITY, KRAFFT TEMPERATURE, AND CLOUD POINT

The nature of surfactant molecules, which have both lyophilic and lyophobic groups, is responsible for their tendency to reduce the free energy of a system by adsorption at various interfaces. However, when all available interfaces are saturated, the overall energy reduction may continue through other mechanisms as illustrated in Figure 15.1.

The physical manifestation of one such mechanism is the crystallization or precipitation of the surfactant from solution --- that is, bulk-phase separation. An alternative is the formation of molecular aggregates or micelles that remain in "solution" as thermodynamically stable, dispersed species with properties distinct from those of the monomeric solution. Before turning our attention to the primary subject of micelles, it is useful to understand the relationship between the ability of a surfactant to form micelles and its solubility.

A primary driving force for the industrial development of synthetic surfactants in this century was the problem of the precipitation of fatty acid soaps in the presence of multivalent cations such as calcium and magnesium (hard-water soap films). While most common surfactants have a substantial solubility in water, that characteristic can change significantly with changes in the length of the hydrophobic tail, the nature of the head group, the valency of the counterion, and the solution environment. For many ionic materials, for instance, it is found that water solubility increases as the temperature increases. It is often observed that the solubility of the material will undergo a sharp, discontinuous increase at some characteristic temperature, commonly referred to as the Krafft temperature, T_k. Below that temperature, the solubility of the surfactant is determined by the crystal lattice energy and heat of hydration of the system. The concentration of the monomeric species in solution will be limited to some equilibrium value determined by those properties. Above T_k, the solubility of the surfactant monomer increases to some point at which *micelle* formation begins and the associated species becomes the thermodynamically favored form. The concentration of monomeric surfactant will again be limited, as discussed below.

The micelle may be viewed simplistically as structurally resembling the solid crystal or a crystalline hydrate, so that the energy change in going from the crystal to the micelle will be less than the change in going to the monomeric species in solution. Thermodynamically, then, the formation of micelles favors an increase in solubility. The concentration of surfactant monomer may increase or decrease slightly at higher concentrations (at a fixed temperature), but micelles will be the predominant form of surfactant present above a critical surfactant concentration --- the *critical micelle concentration* , or CMC. The total solubility of the surfactant, then, will depend not only on the solubility of the monomeric material, but also on the "solubility" of the micelles. A schematic representation of the temperature--solubility relationship for ionic surfactants is shown in Figure 15.2.

The Krafft temperature of ionic surfactants varies as a function of both the nature of the hydrophobic group and the ionic character of the head group. Nonionic surfactants, because of their different mechanism of solubilization, do not exhibit a Krafft temperature. They do, however, have a characteristic temperature--solubility relationship in water in that they usually become less soluble as the temperature increases. In some cases, phase separation is found to occur, producing a cloudy suspension of surfactant aggregates. The temperature at which that occurs is referred to as the *cloud point* .

Many nonionics, especially polyoxyethylenated (POE) materials

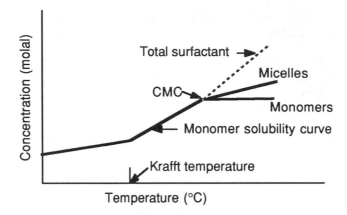

Figure 15.2. The temperature--solubility relationship for typical ionic surfactants illustrating the important characteristics such as the Krafft temperture, the monomer solubility curve, and the "limitung" monomer concentration at the critical micelle concentration.

with a weight fraction of POE less than about 0.8, exhibit sharp, characteristic cloud points in water. As the solution temperature is increased, the clear, homogeneous micellar solution becomes turbid and a two phase system results. The more dense lower phase consists of a surfactant-rich micellar phase, while the upper layer is a dilute solution of monomeric surfactant containing few if any micelles. The turbidity of the concentrated surfactant phase stems from the presence of very large micelles which scatter the visible light passing through the solution.

The cloud point of a surfactant depends on its chemical structure, with longer POE chains tending to increase the cloud point for a given hydrophobic group. For a given *average* POE chain length, the cloud point is changed by: broadening the distribution of POE chain lengths, branching in the hydrophobic chain, nonterminal substitution of the POE chain along the hydrophobe, substitution of the terminal ---OH by a methoxyl group, as well as other structural changes.

The cloud point of a given surfactant can also be altered by the addition of various classes of materials. For example, the addition of neutral electrolyte usually lowers the cloud point, with the effect of a given salt depending on the hydrated radii of both ions. The addition of nonpolar organic materials that can be solubilized in the interior of the micelle (see Chapter 16) normally raises the cloud point, while polar materials have the opposite effect.

The existence of the cloud point phenomenon in nonionic surfactant systems carries with it a number of potential consequences --- both esthetic and functional --- that must always be kept in mind. The appearance of cloudiness, while not necessarily altering the surface activity of a system, may detract from the subjective acceptability of a product. Functionally, the transition from small to large micellar aggregates may significantly alter the solubilizing capacity of a system, as well as altering the availability of free surfactant needed to complete a necessary function, etc.

SURFACTANT LIQUID CRYSTALS

Most discussions of surfactants in solution concern themselves with relatively low concentrations so that the system contains what may be called "simple" surfactant species such as monomers and their basic aggregates or micelles. Before entering into a discussion of micelles, however, it is important to know that although they have been the subject of exhaustive studies and theoretical considerations, they are only one of the several states in which surfactants can exist in solution. A complete understanding of surfactants requires a knowledge of the complete spectrum of possible states of the surfactant, including liquid-crystalline phases, which can be important in the stabilization of emulsions and foams, as well in other areas.

As illustrated in Figure 15.1, the range of possible states for surfactants in the presence of solvents is quite wide. The possibilities range from the highly ordered crystalline phase to the dilute monomeric solution which, although not completely structureless, has order only at the level of molecular dimensions. Between the extremes lie a variety of phases whose natures depend intimately on the chemical structure of the surfactant, the total bulk-phase composition, and the environment of the system (temperature, pH, cosolutes, etc). Knowledge of those structures, and the reasons for and consequences of their formation, influences both our academic understanding of surfactants and their technological application.

When surfactants are crystallized from water and other solvents that can become strongly associated with the polar head group, it is common for the crystalline form to retain a small amount of solvent in the crystal phase. In the case of water, the material would be a *hydrate* . The presence of solvent molecules associated with the head group allows for the existence of several unique compositions and morphological structures that, although truly crystalline, are different from the structure of the dry crystal.

As water or other solvent is added to a crystalline surfactant, the structure of the system will undergo a transition from the highly ordered crystalline state to one of greater disorder usually referred to as a *liquid crystalline* or *mesophase* . Such phases are characterized by having some physical properties of both crystalline and fluid structures. These phases will have at least one dimension that is highly ordered and, as a result, will exhibit relatively sharp x-ray diffraction patterns and optical birefringence. In other dimensions, the phases will behave in a manner more similar to nonstructured fluids.

Two general classes of liquid crystalline structures or mesophases are encountered whether one is considering surfactants or other types of

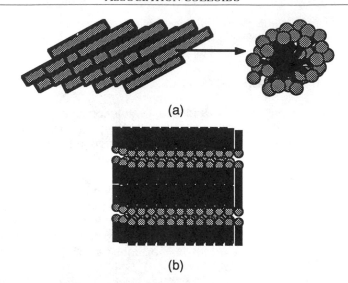

(a)

(b)

Figure 15.3. Schematic illustration of the two most important classes of surfactant liquid crystals: (a) hexagonal packing of rodlike micelles; (b) lamellar packing of monomeric units.

material. These are the *thermotropic* liquid crystals, in which the structure and properties are determined by the temperature of the system, and *lyotropic* liquid crystals, in which the structure is determined by specific interactions between the surfactant molecules and the solvent. With the exception of the natural fatty acid soaps, experimental data suggests that all surfactant liquid crystals are lyotropic.

Although liquid crystal theory often predicts the existence of as many as 18 distinct structures for a given molecular composition and structure, Nature appears to have been kind in that only three of those possibilities have been identified in simple, two-component surfactant--water systems. The three liqui-crystalline phases usually associated with surfactants include the *lamellar*, *hexagonal*, and *cubic*. Of the three, the cubic phase is the most difficult to define and detect. It may have a wide variety of structural variations that involve components of the other mesophases. The remaining two types are more easily characterized and, as a result, better understood (Figure 15.3).

The lamellar liquid crystal can be viewed as a mobile or "plasticized" derivative of the typical surfactant crystalline phase. The hydrophobic chains in these structures possess a significant degree of randomness and mobility, unlike in the crystalline phase in which the chains are usually locked into the all-trans configuration (for terminally substituted *n*-alkyl hydrophobes). The level of disorder of the lamellar phase may vary smoothly or change abruptly, depending on the specific system. It is therefore possible for a surfactant to pass through several distinct lamellar phases. Because the basic unit is bilayered, lamellar phases are normally uniaxial. The lamellar phase resembles the bilayer and multilayer membranes to be discussed later, although they are formed as a result of changes in solvent concentration rather than the specific molecular structural features of the surfactant.

The hexagonal liquid crystal is a high-viscosity fluid phase composed of a close-packed array of cylindrical assemblies of theoretically unlimited size in the axial direction. The structures may be "normal" (in water) in that the hydrophilic head groups are located on the outer surface of the cylinder, or "inverted," with the hydrophile located internally.

Surfactant liquid crystals are normally lyotropic. The characteristics of the system, then, are highly dependent on the nature and amount of solvent present. In a phase diagram of a specific surfactant, the liquid-crystalline phases may span a broad region of compositions and may, in fact, constitute by far the major fraction of all possible compositions. With the continued addition of water or other solvent, the system will eventually pass through the regions of the mesophases into the more familiar isotropic solution phase. The solution is the most highly random state for mixtures of condensed matter and, as a result, tends to have fewer easily detected structural features. Surfactant solutions, however, are far from devoid of structure; it is only the scale of the structure that changes as dilution occurs.

MICELLES

The most intensely studied and debated type of association colloid is also perhaps the simplest in terms of the structure of the aggregate --- the *micelle* . The number of publications related to micelles, micelle structures, and the thermodynamics of micelle formation is enormous. Extensive interest in the self-association phenomenon of surface-active species is evident in such wide-ranging chemical and technological areas as organic and physical chemistry, biochemistry, polymer chemistry, pharmaceuticals, petroleum recovery, minerals processing, cosmetics, and food science. Even with the vast amounts of experimental and theoretical work devoted to the understanding of the aggregation of surface-active molecules, no complete theory or model has emerged that can unambiguously satisfy all of the evidence and all of the interpretations of that evidence for the various association structures.

The solution behavior of surface-active molecules reflects the unique "split personality" of such species. The pushing and pulling that the molecules undergo in aqueous solution (or nonaqueous solution, for that matter) result from a complex mixture of effects including: (1) the interactions (both attractive and repulsive) of the hydrocarbon portion of the molecule with water; (2) the attractive interaction between hydrocarbon tails on separate molecules; (3) the solvation of the hydrophilic head group by water; (4) the interactions between solvated

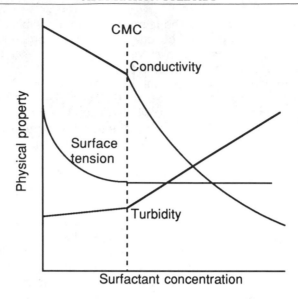

Figure 15.4. Some important manifestations of micelle formation: an abrupt change in solution conductivity, a discontinuity in the surface tension--concentration curve, and a sudden increase in solution turbidity.

head groups (generally repulsive), and between the head groups and co-ions, in the case of ionic materials; and (5) geometric and packing constraints deriving from the particular molecular structure involved.

It is generally accepted that most surface-active molecules in aqueous solution can aggregate to form micellar structures with an average of from 30 to 200 monomers in such a way that the hydrophobic portions of the molecules are associated and mutually protected from extensive contact with the bulk of the water phase. Not so universally accepted are some of the ideas concerning micellar shape, the nature of the micellar interior, surface "roughness," the sites of adsorption into (or onto) micelles, and the size distribution of micelles in a given system. Although ever more sophisticated experimental techniques continue to provide new insights into the nature of micelles, we still have a lot to learn.

Manifestations of Micelle Formation

Early in the study of the solution properties of surface-active agents, it became obvious that the bulk solution properties of such materials were unusual and could change dramatically over very small concentration ranges. The measurement of bulk solution properties such as surface tension, electrical conductivity, or light scattering as a function of surfactant concentration will produce curves that normally exhibit relatively sharp discontinuities at comparatively low concentration

(Figure 15.4). The sudden change in a measured property is interpreted as indicating a significant change in the nature of the solute species affecting the measured quantity. In the case of the measurement of equivalent conductivity (top curve), the break may be associated with an increase in the mass per unit charge of the conducting species. For light scattering (bottom curve), the change in solution turbidity indicates the appearance of a scattering species of significantly greater size than the monomeric solute. These and many other types of measurement serve as evidence for the formation of aggregates or micelles in solutions of surfactants at relatively well-defined concentrations.

The results of studies of surfactant solution properties were classically interpreted in terms of a spherical association of surfactant molecules --- the micelle. The structure was assumed to be an aggregate of from 50--100 molecules with a radius approximately equal to the length of the hydrocarbon chain of the surfactant. The interior of the micelle was described as being essentially hydrocarbon in nature, while the surface consisted of a layer or shell of the head groups and associated counterions, solvent molecules, etc.

Modern studies using techniques unavailable just a few years ago have produced more detailed information about the microscopic nature of the association structures. Micelles are not static species, however. They are very dynamic in that there is a constant, rapid interchange of molecules between the aggregates and the solution phase. It is therefore unreasonable to assume that surfactant molecules pack into a micelle in such an orderly manner as to produce a smooth, perfectly uniform surface structure. If one could photograph a micelle with ultra-high-speed film, freezing the motion of the molecules, the picture would certainly show an irregular molecular cluster more closely resembling a cocklebur than a golf ball.

Although the classical picture of a micelle is that of a sphere, most evidence suggests that spherical micelles are not the rule and may in fact be the exception. Due to geometric packing requirements (to be discussed below) ellipsoidal, disk-shaped, and rodlike structures may be the more commonly encountered micellar shapes (Figure 15.5). However, from the standpoint of providing a concept of micelles and micelle formation for the nonspecialist, the spherical model remains a useful and meaningful tool.

Classical Thermodynamics of Micelle Formation

In the literature on micelle formation two primary models have gained general acceptance as useful (although not necessarily accurate) models for understanding the energetics of the process. The two approaches are the *mass-action* model, in which the micelles and monomeric species are considered to be in a kind of chemical equilibrium, and the *phase separation* model, in which the micelles are considered to constitute a new phase formed in the system at and above the critical micelle concentration. In each case, classical thermodynamic

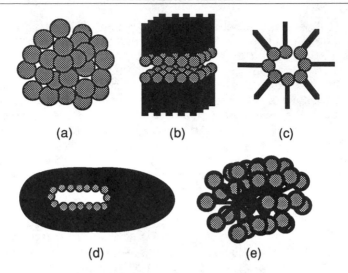

Figure 15.5. Illustration of five important micelle shapes, as interpreted from experimental data: (a) spherical; (b) lamellar; (c) inverted; (d) disk; (e) cylindrical.

approaches are used to describe the overall process of micellization.

In the mass-action model, it is assumed that an equilibrium exists between the monomeric surfactant and the micelles. For the case of nonionic (or un-ionized) surfactants, the monomer--micelle equilibrium can be written

$$nS \leftrightarrow S_n \tag{15.1}$$

with a corresponding equilibrium constant, K_m, given by

$$K_m = [S_n] / [S]^n \tag{15.2}$$

where brackets indicate molar concentrations and n is the number of monomers in the micelle, the *aggregation number*. Theoretically, one must use activities rather than concentrations in eq. 15.2; however, the substitution of concentrations for activities is generally justified by the fact that the critical micelle concentration occurs at such low concentrations that activity coefficients can be assumed to be unity.

It is usually observed that the critical micelle concentration or CMC for a surfactant is relatively sharp and characteristic of a given surfactant. Although the detailed theory of micelle formation can become quite complex, the sharpness of the CMC can be explained conceptually in terms of the law of mass action. If C_t denotes the total concentration of surfactant in solution, C_s the *fraction* of surfactant present as free molecules ($[S]/C_t$), and C_m that in the aggregated state, eq. 15.2 may be written

$$K_m = C_m / [C_s]^n \qquad (15.3)$$

In the process of micelle formation, there will be some value of C, C_{eq}, at which the number of surfactant molecules in the micellar form will be equal to that in the form of free surfactant molecules. At that concentration, $C_m = C_s = 1/2 C_{eq}$. Using eq. 15.3, one can then write that

$$K_m = (1/2 \, C_{eq})^{-(n-1)} \qquad (15.4)$$

At any value of C_t, the relationship between C_s and C_m can be found by substitution of eq. 15.4 into 15.3

$$C_m / (C_s)^n = (1/2 \, C_{eq})^{-(n-1)} \qquad (15.5)$$

where $C_t = C_s + C_m$. Rearrangement of eq. 15.5 gives

$$C_m / C_{eq} = 1/2 \, (2C_s / C_{eq})^n \qquad (15.6)$$

Using eq. 15.6 as a starting point, one can now estimate how the various concentrations vary in the neighborhood when $C_t = C_{eq}$ for a given aggregation number, n. Aggregation numbers for many surfactants lie in the range of 50--100; in Table 15.1 are given the percentages of molecules in the associated state for n = 50, 75, and 100, calculated according to eq. 15.6. The results indicate that, while the CMC for a given system may not represent a truly sharp change in conditions, once the formation of micelles begins, any increase in surfactant concentration will be directed almost completely to the formation of more micelles. It is also obvious that the larger the aggregation number for a given system, the sharper will be the transition from monomolecular solution to predominantly micelles.

The alternative approach to modeling micelle formation is to think in terms of a phase separation model in which, at the CMC, the concentration of the free surfactant molecules becomes constant (like a solubility limit or K_{sp}), and all additional molecules go into the formation of micelles. Analysis of the two approaches produces the same general result in terms of the energetics of micelle formation (with some slight differences in detail), so that the choice of model is really a matter of preference and circumstances. There is evidence that the activity of free surfactant molecules does increase above the CMC, which tends to support the mass-action model; however, for most purposes, that detail is of little consequence.

Table 15.1. Percentage of total surfactant molecules in micellar form near $C_s = C_{eq}$ calculated according to eq. 15.6.

	% C_t in micellar form		
Cs / Ceq	$n = 50$	$n = 75$	$n = 100$
0.45	0.57	0.04	0.003
0.47	4.6	1.01	0.22
0.49	27	18	12
0.495	38	32	27
0.50	50	50	50
0.505	62	68	73
0.51	73	81	88
0.52	87	95	98
0.53	95	99	99.7
0.54	98	99.7	99.95
0.55	99.1	99.9	99.99

Free Energy of Micellization

From eq. 15.2, the standard free energy for micelle formation per mole of micelles is given by

$$\Delta G_m° = -RT \ln K_m = -RT \ln S_n + nRT \ln S \qquad (15.7)$$

while the standard free energy change per mole of free surfactant is

$$\Delta G_m°/n = -(RT/n) \ln S_n + RT \ln S \qquad (15.8)$$

As shown above, at (or near) the CMC, $S \approx S_n$, so that the first term on the right side of eq. 15.8 can be neglected, and an approximate expression for the free energy of micellization per mole of surfactant will be

$$\Delta G_m° = RT \ln CMC \qquad (15.9)$$

The situation is complicated somewhat in the case of ionized surfactants because the presence of the counterion and its degree of association with the monomer and micelle must be taken into consideration. For an ionic surfactant the mass-action equation is

$$n S^x + (n - m)C^y \leftrightarrow S_n{}^\alpha \qquad (15.10)$$

The degree of dissociation of the surfactant molecules in the micelle, α, the micellar charge, is given by $\alpha = m / n$. The ionic equivalent to eq. 15.2 is then

$$K_m = [S_n] / [S^x]^n [C^y]^{n-m} \qquad (15.11)$$

where m is the concentration of free counterions, C (eg, those not bound to the micelle). The standard free energy of micelle formation will be

$$\Delta G_m{}^o = RT / n \{n \ln [S^x] + (n-m) \ln [C^y] - \ln [S_n]\} \qquad (15.12)$$

At the CMC $[S^{-(+)}] = [C^{+(-)}]$ = CMC for a fully ionized surfactant and eq. 15.12 can be approximated as

$$\Delta G_m{}^o = RT(1+ m / n) \ln \text{CMC} \qquad (15.13)$$

When the ionic micelle is in a solution of high electrolyte content, the situation described by eq. 15.12 reverts to the simple nonionic case given by eq. 15.9.

In general, but not always, micelle formation is found to be an exothermic process, favored by a decrease in temperature. The ehthalpy of micellization, ΔH_m, given by

$$- \Delta H_m = RT^2 [d \ln \text{CMC}/ dT \ ] \qquad (15.14)$$

may therefore be either positive or negative, depending on the system and conditions. The process, however, always has a substantial positive entropic contribution to overcome any positive enthalpy term, so that micelle formation is primarily an entropy-driven process.

More recent approaches have employed more complicated treatments with more rigorous statements of the physical phenomena involved. However, they yield little information of value so far as understanding a given practical system is concerned. A different, and perhaps more useful, approach emphasizes the importance of molecular geometry in defining the characteristics of an aggregating system. Such a geometrical approach would seem to be especially useful for applications in which the chemical structure of the surfactant is of central importance.

Molecular Geometry and the Formation of Association Colloids

The theoretical developments based upon the effects of geometry on molecular aggregation have shown that physical characteristics of surfactants such as CMC, aggregate size and shape, and polydispersity can be quantitatively described without relying upon a detailed knowledge of the energetics of the various molecular interactions. It is also useful in that it applies equally well to micelles, vesicles, and bilayer membranes, the latter lying outside the normal models of association processes. For that reason, the geometrical approach warrants a somewhat closer look.

The classical picture of micelles formed by simple surfactant systems in aqueous solution is that of a sphere with a core of essentially liquid like hydrocarbon surrounded by a shell containing the hydrophilic head groups along with associated counterions, water of hydration, etc. Regardless of any controversy surrounding the model, it is usually assumed that there are no water molecules included in the micellar core, since the driving force for micelle formation is a reduction of water--hydrocarbon contacts. Water will, however, be closely associated with the micellar surface; as a result, some water--core contact must occur at or near the supposed boundary between the two regions. The extent of that water--hydrocarbon contact will be determined by the surface area occupied by each head group and the radius of the core. It seems clear from a conceptual viewpoint that the relative ratio between the micellar core volume and surface area must play an important role in controlling the thermodynamics and architecture of the association process. Equally important is the need to understand the constraints that such molecular geometry places upon the ability of surfactants to pack during the aggregation process to produce micelles, microemulsions, vesicles, and bilayers.

Israelachvili et al[1] have shown that the geometric factors that control the packing of surfactants and lipids into association structures can be conveniently given by what they term the "critical packing parameter", $v/a_o l_c$, where v is the volume of the hydrophobic portion of the molecule, a_o is the optimum head group area, and l_c is the critical length of the hydrophobic tail, effectively the maximum extent to which the chain can be stretched out, subject to the restrictions noted earlier. The value of the packing parameter will determine the type of association structure formed in each case. A summary of the structures to be expected from molecules falling into various "critical packing" categories is given in Table 15.2.

While it is almost certain that examples will be found of materials that do not fit neatly into such a scheme, the general concepts are usually found to be valid. For surfactants and other amphiphiles that form bilayer structures, Israelachvili has offered several generalizations that make it easier to understand the geometric consequences of the structure of the amphiphile, including:

1. Molecules with relatively small head groups, and therefore large values for $v/a_o l_c$, will normally form extended bilayers, large (low-curvature) vesicles, or inverted micellar structures. Such events can also be brought about in anionic systems by changes in pH, high salt concentrations, or the addition of multivalent cations, especially.

Table 15.2. Expected aggregate characteristics in relation to surfactant critical packing parameter, $v/a_o l_c$.

Critical packing parameter	General surfactant type	Expected aggregate structure
< 0.33	Simple surfactants with single chains and relatively large head groups	Spherical or ellipsoidal micelles
0.33–0.5	Simple surfactants with relatively small head groups, or ionics in the presence of large amounts of electrolyte	Relatively large cylindrical or rod-shaped micelles
0.5–1.0	Double-chain surfactants with large head groups and flexible chains	Vesicles and flexible bilayer structures
1.0	Double-chain surfactants with small head groups or rigid, immobile chains	Planar extended bilayers
> 1.0	Double-chain surfactants with small head groups, very large and bulky hydrophobic groups	Reversed or inverted micelles

2. Molecules containing unsaturation, especially multiple cis double bonds, will have smaller values for l_c, and also will tend toward the formation of larger vesicles or inverted structures.

3. Multichained molecules held above the melting temperature of the hydrocarbon chain may undergo increased chain motion, allowing trans–gauche chain isomerization, reducing the effective value of l_c and resulting in changes in aggregate structures. This effect may be of particular importance in understanding the effects of temperature on biological membranes.

The above generalizations on bilayer assemblies of surfactants and other amphiphilic molecules offer a broad view of the types of structures that may be formed as a result of the self-assembly process. They consider only the fundamental relationships between structure and the geometric characteristics of the molecules involved. Not considered are any effects on the systems that may exist as a result of curvature or other distortions of the molecular packing. The unique characteristics of the bilayer and vesicle assemblies have attracted the attention of scientists in many disciplines for both theoretical and practical reasons.

The following brief discussion will only skim the surface of what is sure to become an even more interesting and important area of surfactant-related surface science.

Although it is convenient to visualize the micellar core as a bulk hydrocarbon phase, the density may not be equal to that of the analogous true bulk material. x-Ray evidence indicates that the molecular volumes of surfactants in micelles are essentially unchanged by the aggregation process. If a molecular volume for a hydrocarbon chain in the micellar core equal to that of a normal hydrocarbon is assumed, the core volume can be calculated from

$$V = n'(27.4 + 26.9 \, n_{c'}) \times 10^{-3} \ (\text{nm}^3) \qquad (15.15)$$

where, n' is an effective micellar aggregation number, and n_c' is the number of carbon atoms per chain in the core. In general, the value of n_c' will be one less than the total number of carbons in the hydrocarbon chain, n_c, since the first carbon after the head group is highly solvated and may be considered as a part of it. For normal surfactants with a single hydrocarbon tail, n' will be equal to the aggregation number, n, while for those that possess a double tail, $n' = 2n$.

If one assumes that the micellar core has no "hole" at its center, one dimension of the aggregate species will be limited by the length of the hydrocarbon chain when extended to its fullest. That maximum length can be calculated by assuming a distance of .253 nm between alternate carbon atoms of the extended chain and adding the value of the van der Waals radius of the terminal methyl group (= .21 nm) and half the bond distance between the first carbon in the core and that bonded to the head group ($\approx$ 0.06 nm). The maximum extended length l_{max} for a normal hydrocarbon chain with n_c' core carbon atoms, therefore, is given by

$$l_{max} = 0.15 + 0.1265 n_c' \ \text{nm} \qquad (15.16)$$

Since hydrocarbon chains in the liquid state are never fully extended, a dimension, l_{eff}, can be defined that gives the statistically most likely extension as calculated by the same procedure used for the calculation of polymer chain dimensions. For a chain with $n_c' = 11$, the ratio of l_{max} to l_{eff} will be approximately 0.75. In the micellar core, due to restrictions imposed by the attachment of the hydrocarbon tail to the head group bound at the surface, the mobility of the chains may be significantly limited relative to that of bulk hydrocarbon chains. The presence of "kinks" or gauche chain conformations, which may be imposed by packing considerations, will result in a calculated l_{max} amounting to only about 80% of the theoretical maximum.

Since hydrocarbon chains possess restricted bond angles as well as bond lengths, additional restrictions on the maximum extension of the chain arise beyond those mentioned previously. Chain segments located

Critical packing parameter	Approximate molecular shape	Expected aggregate structure
<0.33		
0.33 - 0.5		
0.5 - 1.0		
1.0		
>1.0		

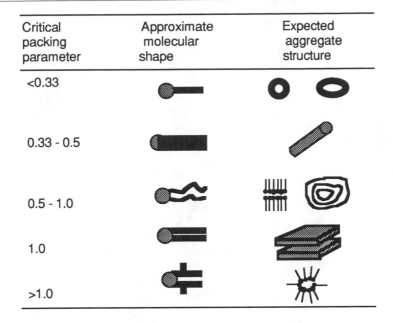

Figure 15.6. Summary of the expected relationship between surfactant molecular geometry and self-assembled Aggregate structures.

at the transition region from core to shell, for example, cannot assume arbitrary conformations in order to produce a perfectly smooth surface. The micellar surface, therefore, must be assumed to be somewhat rough or irregular, although the dynamic nature of the aggregate may obscure any practical effect of such roughness.

Israelachvili, et al,[1] and Tanford[2] have considered in detail many of the geometric restrictions that govern micelle sizes, shapes, and size dispersity. Their analysis of the geometrical and thermodynamic factors appears to allow for the prediction of most aspects of the aggregation of surface-active molecular species, including the CMC, average aggregation number, polydispersity of micelle sizes, and the most likely shape of the aggregated species (spherical, ellipsoidal, disk, or rod-shaped micelles, vesicles, extended bilayer, etc). A summary of the "rules" of association derived from their analyses of molecular structure is given in Figure 15.6.

While the geometric approach shows great promise, it has not worked its way into the general thinking on micelles. It is, however, finding wide acceptance in areas related to biological membranes and aggregates. As more experimental data is correlated with the predictions of geometric considerations, such an approach can be expected to gain ground as a useful basis for the design of surfactant molecules with specific desirable aggregation characteristics. With the preceding concepts in mind, we now turn to some of the experimental results that have, over the years, helped bring us to our current state of understanding of surfactant micellization.

Some Correlations Between Surfactant Structure, Environment, and Micellization

While the above model approaches to association phenomena are useful from a fundamental point of view, in practice, the association characteristics (the CMC, aggregation number, etc) of a surface-active material are very sensitive to such factors as the isomeric purity of the sample, the presence of contaminants, pH, electrolyte content, temperature, etc. A good working knowledge of micelle formation, therefore, must include some idea of how such factors will affect the behavior of the surfactant. The literature on those various topics is extensive and of varying quality; however, there have developed over the years a number of good generalizations that can be helpful in making "educated" extrapolations from ideality to reality. The following sections, then, will be devoted to the presentation of summaries and generalizations that illustrate many of the most significant effects of surfactant chemical structure and solution environment on the micellization process.

Aggregation Number

Aggregation numbers for many surfactants have been found to fall in the range of 50--100 molecules, although that can vary significantly according to structure and conditions. Some typical aggregation numbers for various surfactant types are given in Table 15.4.

Because the size and dispersity of micelles are sensitive to many internal (hydrophobic structure, head group type) and external (temperature, pressure, pH, electrolyte content) factors, it is sometimes difficult to place too much significance on reported values of n. However, some generalizations can be made that are usually found to be true. They include:

1. In aqueous solutions, it is generally observed that the longer the hydrophobic chain for an homologous series of surfactants, the larger will be the aggregation number.
2. A similar increase in n is seen when there is a decrease in the "hydrophilicity" of the head group --- for example, a higher degree of ion binding or a shorter polyoxyethylene chain.
3. External factors that result in a reduction in the hydrophilicity of the head group such as high electrolyte concentrations will also cause an apparent increase in n.
4. Temperature changes will affect nonionic and ionic surfactants differently. In general, higher temperatures will result in small decreases in aggregation numbers for ionic surfactants but significantly large increases for most nonionics.
5. The addition of small amounts of nonsurfactant organic

Table 15.3. Aggregation numbers for representative surfactants in water.[3]

Surfactant	Temperature T (°C)	Aggregation number, n
$C_{10}H_{21}SO_3^- Na^+$	30	40
$C_{12}H_{25}SO_3^- Na^+$	40	54
$(C_{12}H_{25}SO_3^-)_2Mg^{2+}$	60	107
$C_{12}H_{25}SO_4^- Na^+$	23	71
$C_{14}H_{29}SO_3^- Na^+$	60	80
$C_{12}H_{25}N(CH_3)_3^+Br^-$	23	50
$C_8H_{17}O(CH_2CH_2O)_6H$	30	41
$C_{10}H_{21}O(CH_2CH_2O)_6H$	35	260
$C_{12}H_{25}O(CH_2CH_2O)_6H$	15	140
$C_{12}H_{25}O(CH_2CH_2O)_6H$	25	400
$C_{12}H_{25}O(CH_2CH_2O)_6H$	35	1400
$C_{14}H_{29}O(CH_2CH_2O)_6H$	35	7500

materials of low water solubility will often produce an apparent increase in micelle size, although that may be more an effect of solubilization (see below) than an increase in the number of surfactant molecules present in the micelle.

6. The addition of a water-miscible organic material such as an alcohol will generally reduce the apparent aggregation number.

While the question of the size of micelles is of great theoretical interest, it is not usually very significant (as far as we know) in most surfactant applications, other than perhaps solubilization and microemulsion formation. Of more general importance is the concentration at which micelle formation occurs, the critical micelle concentration, since that is the time when many of the most useful surfactant properties come into play.

The Critical Micelle Concentration

Because there are many factors that have been shown to affect the observed critical micelle concentration strongly, the following discussion has been divided so as to isolate (as much as possible) the various important factors.

Any discussion of CMC data must be tempered with the knowledge that the reported values must not be taken to be absolute but reflect certain variable factors inherent in the procedures employed for their determination. The variations in CMC found in the literature for nominally identical materials under supposedly identical conditions must be accepted as minor "noise" that should not significantly affect the overall picture (assuming, of course, that good experimental technique has been employed).

The Hydrophobic Group: The "Tail"

The length of the chain of a hydrocarbon surfactant is a major factor determining the CMC. The CMC for a homologous series of surfactants decreases logarithmically as the number of carbons in the chain increases. For straight-chain hydrocarbon surfactants of about 16 carbon atoms or less bound to a single terminal head group, the CMC is usually reduced to approximately one-half of its previous value with the addition of each methylene group. For nonionic surfactants, the effect can be much larger, with a decrease by a factor of 10 following the addition of two carbons to the chain. The insertion of a phenyl and other linking groups, branching of the alkyl group, and the presence of polar substituent groups on the chain can produce different effects on the CMC.

The relationship between the hydrocarbon chain length and CMC for ionic surfactants generally fits the *Klevens equation* [4]

$$\log_{10} CMC = A - Bn_c \qquad (15.17)$$

where A and B are constants specific to the homologous series under constant conditions of temperature, pressure, etc, and n_c is the number of carbon atoms in the chain. Values of A and B for a wide variety of surfactant types have been determined, and some are listed in Table 15.4. It has generally been found that the value of A is approximately constant for a particular ionic head group, while B is constant and approximately equal to $\log_{10} 2$ for all paraffin chain salts having a single ionic head group. The value of B will change, however, in systems having two head groups, or for nonionic systems.

For nonionic surfactants, in which the mechanism of solubilization of the surfactant molecule is basically hydrogen bonding, the relative importance of the tail and head groups to the overall process changes. An empirical relationship between the CMC and the number of oxyethylene $(OE)_y$ groups present in several nonionic surfactant series has the form[5]

$$\ln CMC = A' + B'y \qquad (15.18)$$

where A' and B' are constants related to a given hydrophobic group. Examples of A' and B' for several commonly encountered hydrophobic groups are given in Table 15.5. In each case, the results are for one temperature and can be expected to vary significantly, given the sensitivity of such systems to changes in T.

For more complex surfactant structures, the following

Table 15.4. Klevens constants (eq. 15.17) for common surfactant classes.

Surfactant Class	Temperature (°C)	A	B
Carboxylate soaps (Na$^+$)	20	1.85	0.30
Carboxylate soaps (K$^+$)	25	1.92	0.29
n-Alkyl-1-sulfates (Na$^+$)	45	1.42	0.30
n-Alkyl-2-sulfates(Na$^+$)	55	1.28	0.27
n-Alkyl-1-sulfonates	40	1.59	0.29
p-n-Alkylbenzene sulfonates	55	1.68	0.29
n-Alkylammonium chlorides	25	1.25	0.27
n-Alkyltrimethylammonium bromides	25	1.72	0.30
n-Alkylpyridinium bromides	30	1.72	0.31

generalizations serve as a good guide:

(1) Ionic surfactants having two or three ionic groups at one end of the hydrocarbon tail such as α-sulfonated fatty acids and their esters, alkyl malonates or alkyl tricarboxylates, exhibit a linear relationship between CMC and chain length similar to eq. 15.17, although they usually have a lower Krafft temperature and a higher CMC than the corresponding singly charged molecule of the same.

(2) For surfactants having branched structures, with the head group attached at some point other than the terminal carbon, such as, sodium tetradecane 2-sulfonate $CH_3(CH_2)_{11}CH(SO_3^-Na^+)CH_3$, the additional carbon atoms off of the main chain contribute a factor equivalent to about one-half that for a main-chain carbons. Except for the lower members of a series, the relationship between carbon number and CMC follows a linear relationship similar to eq. 15.17.

(3) For surfactants that contain two hydrophobic chains, such as the sodium dialkylsulfosuccinates, it is generally found that the CMC values for the straight-chain esters follow the Klevens relationship, although the value of B is slightly smaller than that found for single-chain surfactants. The CMC's for the branched esters of equal carbon number occur at higher concentrations.

(4) In the alkylbenzene sulfonates, with various points of attachment of the alkyl group to the benzene ring, experimental data indicate that the aromatic ring has substantial hydrophilic character, with the benzene ring contribution being equivalent to about 3.5 carbon atoms.

(5) For surfactants that contain ethylenic unsaturation in the chain, one generally finds that the presence of a single double bond

Table 15.5. Empirical constants relating CMC and oxyethylene content for various hydrophobic groups in nonionic surfactants (eq. 15.18).

Hydrophobic Group	A'	B'
$C_{12}H_{25}OH$	3.60	0.048
$C_{13}H_{27}OH$	3.59	0.091
$C_{18}H_{35}OH$ (Oleyl)	3.67	0.015
$C_{18}H_{37}OH$ (Stearyl)	2.97	0.070
$C_9H_{19}C_6H_4OH$	3.49	0.065

increases the CMC by as much as a factor of 3-4 compared to the analogous saturated compound. In addition to the electronic presence of the double bond, the isomer configuration (cis or trans) will also have an effect, with the cis isomer usually having a higher CMC, presumably due to the more difficult packing requirements imposed by the isomer.

(6) The presence of polar atoms such as oxygen or nitrogen in the hydrophobic chain (but not associated with a head group), usually results is an increase in the CMC. The substitution of an ---OH for hydrogen, for example, reduces the effect of the carbon atoms between the substitution and the head group to half that expected in the absence of substitution. If the polar group and the head group are attached at the same carbon, that carbon atom appears to make little or no contribution to the hydrophobic character of the chain.

(7) A number of commercial surfactants are available in which all or most of the hydrophobic character is derived from the presence of polyoxypropylene groups. The observed effect of such substitution has been that each each propylene oxide group is equivalent to approximately 0.4 methylene carbon.

Two classes of materials that cannot easily be fitted into the known schemes for conventional hydrocarbons are the silicone-based surfactants and those in which hydrogens have been replaced by fluorine atoms. The hydrophobic unit of the silicone-based surfactants consists of low molecular weight polyorganosiloxane derivatives, usually polydimethylsiloxanes. Possibly because of their "nonclassical" nature they have received little attention in the general scientific literature, although their unique surface characteristics have proved useful in many technological applications, especially in nonaqueous solvent systems. The substitution of fluorine for hydrogen on the hydrophobic chain has produced several types of surfactants with extremely interesting and useful properties. The presence of the fluorine atoms results in large (ie, orders of magnitude) decreases in CMC's relative to the base hydrocarbon. Because of the electronic character of the carbon--fluorine bond, fluorinated materials have been found to have much lower surface energies and produce lower surface tensions than conventional materials. In general, a fully fluorinated surfactant with n carbon atoms will have a CMC roughly equal to that of a hydrocarbon material with $2n$ carbons.

The Hydrophilic Group

The effect of the hydrophilic head group on the CMC's of a series of surfactants with the same hydrocarbon chain will vary considerably, depending upon the nature of the change. In aqueous solution the difference in CMC for a C_{12} hydrocarbon with an ionic head group will lie in the range of .001 M, while a nonionic material with the same chain will have a CMC in the range of 0.0001 M. The exact nature of the ionic group has no dramatic effect, since the main driving force for micelle formation is the entropy gain on reduction of water--hydrocarbon interactions. The CMC's of several ionic surfactants are given in Table 15.6. Of the more common anionic head groups, the order of decreasing CMC values for a given hydrocarbon chain is found to be carboxylates (containing one more carbon atom) > sulfonates > sulfates. For cationic surfactants, one often finds that the CMC increases with methyl substitution on the nitrogen, probably due to increased steric requirements of the added methyl groups forcing an increase in ionization (ie, less ion pairing).

Counterion Effects

In ionic surfactants micelle formation is related to the interactions of solvent with the ionic head group. The degree of ionization, in terms of tight ion binding, solvent-separated ion pairing, or complete ionization, will therefore influence the value of the CMC and the aggregation number. Since electrostatic repulsions among the ionic groups would be greatest for complete ionization, one finds that the CMC of surfactants in aqueous solution decreases as the degree of ion binding increases.

From regular solution theory it is found that the extent of ion pairing in a system will increase as the polarizability and valence of the counterion increase. Conversely, a larger radius of hydration will result in greater ion separation. It has been found that, for a given hydrophobic tail and anionic head group, the CMC decreases in the order Li^+ > Na^+ > K^+ > Cs^+ > $N(CH_3)_4^+$ > $N(CH_2CH_3)_4^+$ > Ca^{2+} ≈ Mg^{2+}. In the case of cationic surfactants such as dodecyltrimethyl ammonium halides, the CMC's are found to decrease in the order F^- > Cl^- > Br^- > I^-.

Although within a given valency the size of the hydrated counterion will have some effect upon the micellization of an ionic surfactant, a more significant effect is produced by changes in valency. As the counterion is changed from monovalent to di- and trivalent, the

Table 15.6. The effect of the hydrophilic group on the CMC's of surfactants with common hydrophobes.

Hydrophobe	Hydrophile	Temperature (°C)	CMC (mM)
$C_{12}H_{25}$	COOK	25	12.5
"	SO_3K	25	9.0
"	SO_3Na	25	8.1
"	NH_3Cl	30	14
"	$N(CH_3)_3Cl$	30	20
"	$N(CH_3)_3Br$	25	16
$C_{16}H_{23}$	NH_3Cl	55	0.85
"	$N(CH_3)_3Cl$	30	1.3
"	$N(CH_3)_3Br$	60	1.0
C_8H_{17}	OCH_2CH_2OH	25	4.9
"	$(OCH_2CH_2)_2OH$	25	5.8
C_9H_{19}	$COO(CH_2CH_2O)_9CH_3$	27	1.0
"	$COO(CH_2CH_2O)_{16}CH_3$	27	1.8
$C_{10}H_{21}$	$O(CH_2CH_2O)_8CH_3$	30	0.6
"	$O(CH_2CH_2O)_{11}CH_3$	30	0.95
"	$O(CH_2CH_2O)_{12}CH_3$	29	1.1

CMC is found to decrease rapidly. The divalent and higher salts of carboxylic acid soaps generally have very low water solubility and are not useful as surfactants in aqueous solution. They do find use in nonaqueous solvents due to their increased solubility in those systems, especially in the preparation of water-in-oil emulsions.

The Effect of Additives

Many industrial applications of surfactants involve the presence in the solution of cosolutes and other additives that can potentially affect the micellization process through specific interactions with the surfactant molecules (thereby altering the effective activity of the surfactant in solution) or by altering the thermodynamics of the micellization process by changing the nature of the solvent or the various interactions leading to or opposing micelle formation.

Solution changes that might be expected to affect the association process include the presence of electrolytes, changes in pH, and the addition of organic materials that may be essentially water insoluble (eg, hydrocarbons), water-miscible (short-chain alcohols, acetone, dioxane, etc), or of low water solubility but containing polar groups that impart some surface activity although they are not classified formally as surfactants. The following generalizations are usually useful, although it must be remembered that each surfactant system can exhibit characteristics different from the general observations noted here.

Added Electrolyte

In aqueous solution the presence of electrolyte causes a decrease in the CMC of most surfactants, with the greatest effect being found for ionic materials. Nonionic and zwitterionic surfactants exhibit a much smaller effect. For ionic materials, the effect of addition of electrolyte can be empirically quantified with the relationship

$$\log_{10} CMC = - a \log_{10} c_i + b \qquad (15.19)$$

where a and b are constants for a given ionic head group at a particular temperature, and c_i is the total concentration of monovalent counterions in moles per liter. For nonionic and zwitterionic materials, the impact of added electrolyte is significantly less and the relationship in eq. 15.19 does not apply.

pH

For most modern, industrially important surfactants consisting of long alkyl chain salts of strong acids, solution pH has a relatively small effect, if any, on the CMC of the materials. Unlike the salts of strong acids, however, the carboxylate soap surfactants exhibit a significant sensitivity to pH. Since the carboxyl group is not fully ionized near or below the pK_a, pH changes may result in significant changes in the CMC as well as the Krafft temperature. A similar result will be observed for the cationic alkylammonium salts near and above the pK_b. Changes in pH will have little or no effect on the CMC of nonionic surfactants except, perhaps, at very low pH where it is possible that protonation of the ether oxygen of OE surfactants can occur. Such an event would, no doubt, alter the characteristics of the system. Little can be found in the literature pertaining to such effects, however.

A number of amphoteric surfactants have pH sensitivity related to the pK's of their substituent groups. The possibilities can be grouped in the following way:

1. Quaternary ammonium--strong acid salts will show little or no significant pH sensitivity.
2. Quaternary ammonium--weak acid combinations will be zwitterionic at high pH and cationic below the pK_a of the acid.
3. Amine--weak acid combinations will be anionic at high pH, cationic at low pH, and zwitterionic at some pH between the respective pK's.
4. Amine--strong acid combinations will be anionic at high pH and

zwitterionic below the pK_b of the amine.

Organic Materials

Organic materials that have low water solubility can be solubilized in micelles to produce systems with substantial organic content where no solubility would occur in the absence of micelles. More details on the phenomenon of solubilization in surfactant micelles will be presented below. In any case it is usually found that immiscible hydrophobic materials will have relatively little effect on CMC, although evidence for slight decreases has been reported.

Small amounts of organic additives with substantial water miscibility such as the lower alcohols, dioxane, acetone, glycol, and tetrahydrofuran have relatively minor effects on CMC. As the alkyl groups go beyond C_3, the inherent surface activity of the alcohol can begin to become significant. Otherwise, it will be only at high concentrations, where the additive may be considered a cosolvent, that major effects on CMC will be evident. In general, large amounts of water-miscible organics will increase the CMC by increasing the solubility of the tail, although the opposite effect may occur for highly ionized species, where the lower dielectric constant reduces head group repulsion.

The properties of a surfactant solution are found to change much more rapidly with the introduction of small amounts of long-chain alcohols, especially C > 3. Because so many classes of surfactants of importance academically and industrially are derived from raw materials containing alcoholic impurities, the recognition of the effects of such materials can be very important. Most of the observed effects can be attributed to the inherent surface activity of the longer alcohols.

The interactions between surfactants and alcohols have become of greater importance in recent years as a result of the intense interest in microemulsions and their potential application in various areas of technological importance. Some of the basic concepts in that area will be presented later.

Temperature

The effects of temperature changes on the CMC in aqueous solution have been found to be quite complex. It has been shown, for example, that the CMC of most ionic surfactants passes through a minimum as the temperature is varied from 0° through 60--70°C. Nonionic and zwitterionic materials are not quite so predictable, although it is has been found that some nonionics reach a CMC minimum around 50°C.

The temperature dependence of the CMC's of polyoxyethylene nonionic surfactants is especially important since the head group interaction is essentially totally hydrogen bonding in nature. Materials relying solely on hydrogen bonding for solubilization in aqueous solution are commonly found to exhibit an inverse temperature--solubility relationship. As already mentioned, major manifestation of such a relationship is the presence of the cloud point for many nonionic surfactants.

Micelle Formation in Mixed Surfactant Systems

When one discusses the solution behavior of many, if not most, industrially important surfactants, it is important to remember that experimental results must be interpreted in the context of a surfactant mixture rather than a pure homogeneous material. Studies of such systems are important both academically, assuming that the mixture can be properly analyzed, and practically, since most detergents and soaps contain homologs of higher or lower chain length than that of the primary component.

Determinations of the CMC of well-defined, binary mixtures of surfactants have shown that the greater the difference in the CMC between the components of the mixture, the greater is the effect of the chain length of the more hydrophobic member. The analysis of results for binary mixtures of an homologous series of surfactants must take into consideration the fact that at the CMC the mole fractions of the monomeric surfactants in solution are not equal to the stoichiometric mole fractions; each value must be decreased by the amounts of each mole fraction incorporated into the micellar phase. Interpretations may also be complicated by such effects as relatively small changes in the mole fraction of the smaller chain component due to preferential aggregation of the more hydrophobic material and the difficulty of inclusion of the longer chain into micelles of the shorter material. In some cases where the difference is very large, the component with the higher CMC may simply act as an added electrolyte, rather than becoming directly involved in the micellization process. When ternary surfactant mixtures are considered, it is usually found that the CMC of the mixture falls somewhere between the highest and lowest value determined for the individual components.

The presence of an ionic surfactant in mixture with a nonionic usually results in an increase in the cloud point of the nonionic component. In fact, the mixture may not show a cloud point, or the transition may occur over a broad temperature range, indicating the formation of mixed micelles. As a result of that effect, it is possible to formulate mixtures of ionic and nonionic surfactants for use at temperatures and under solvent conditions (electrolyte, etc) in which neither component alone is effective.

Many mixtures of surfactants, especially ionic with nonionic, exhibit surface properties significantly better than those obtained with either component alone. Such synergistic effects greatly improve many technological applications in areas such as emulsion formulations, emulsion polymerization, surface tension reduction, coating operations, personal care and cosmetics products, pharmaceuticals, and petroleum recovery, to name a few. The use of mixed surfactant systems should always be considered as a method for obtaining the optimal performance from any practical surfactant application.

Micelle Formation in Nonaqueous Media

The formation of micelle-like aggregates in nonaqueous solvents has received far less attention than the related phenomenon in water. In fact, there exists some controversy as to whether such a phenomenon in fact occurs in the same sense as in aqueous solutions. There can be no doubt that some chemical species, many surfactants included, do associate in hydrocarbon and other nonpolar solvents.

The changes involved in surfactant aggregation in nonaqueous solvents must differ considerably from those already discussed for water-based systems. The orientation of the surfactant relative to the bulk solvent will be the opposite to that in water (hence the term "reversed" micelle). In addition, the micelle, regardless of the nature of the surfactant, will be un-ionized in solvents of low dielectric constant, and so will have no significant electrical properties relative to the bulk solvent, although electrostatic interactions will play an important role in the aggregation process, but in an opposite sense to that in aqueous solution.

The primary driving force for the formation of micelles in aqueous solution is the gain in entropy resulting from a reduction of the unfavorable interactions between water and the hydrophobic tail of the surfactant. In nonaqueous solvents, there will be little significant change in the interactions between surfactant tail and solvent, even if one is hydrocarbon and the other aromatic. A more significant energetic consequence of nonaqueous micelle formation is the reduction of unfavorable interactions between the ionic head group of the surfactant and the nonpolar solvent molecules. Or even more likely, the gain of favorable interactions among polar or charged groups.

Unlike the situation for aqueous micelles in which interactions between the hydrophobic tails contribute little to the overall free energy of micelle formation, ionic, dipolar, or hydrogen bonding interactions between head groups in reversed micelles are one of the primary driving forces favoring aggregation. In the face of factors favoring aggregation, there seem to be few obvious factors tending to oppose the formation of nonaqueous micelles (unlike head group steric and electrostatic interactions in the aqueous case). The possible exception is unfavorable entropy losses as a result of fewer degrees of freedom for monomers in the micelle relative to those free in solution.

In contrast to aqueous surfactant solutions in which micellar size and shape may vary considerably, small spherical micelles appear to be the most favored, especially when the reduction of solvent--polar group interactions is important. Similar to water-based systems, geometric considerations often play an important role in determining micelle size and shape. Many materials that commonly form nonaqueous micellar solutions possess large, bulky hydrocarbon tails with a cross-sectional area significantly greater than that of the polar head group. Typical examples of such materials are sodium di-2-ethylhexylsulfosuccinate and sodium dinonylnaphthalene sulfonate.

Since unambiguous experimental data are much less available on micelle formation in nonaqueous solvents than for aqueous systems, it is far more difficult to identify trends and draw conclusions concerning the relationships between chemical structures, CMC, and aggregation numbers. However, some generalizations can be made.

In hydrocarbon solvents, the nature of the polar head group is extremely important in the aggregation process. It has generally been found that ionic surfactants form larger nonaqueous micelles than nonionic ones, with anionic sulfates surpassing the cationic ammonium salts. The aggregation number for an ionic surfactant in a given solvent will usually change little with changes in the counterion, indicating a lack of sensitivity to the nature of that species. The effect of the hydrocarbon tail length in a homologous series of surfactants is relatively small when compared to that in water. However, the aggregation number tends to decrease as the carbon number increases within a homologous series.

The presence of small amounts of water in a nonaqueous surfactant environment can have a significant effect on some systems. It can be presumed that the effects of water and other solubilized impurities on nonaqueous micelle formation stems from alterations in the dipolar interactions between head groups induced by the additive or impurity.

Although the vast majority of surfactants form micelles of some kind in aqueous solution, some materials, because of their special structure or composition, will not associate in the "normal" way described above. They will, however, take part in other association processes to for equally interesting and important association colloids, including especially vesicles and bilayer membranes.

VESICLES AND BILAYER MEMBRANES

As has been discussed previously, surfactants and related amphiphilic molecules, including the natural surfactants or *lipids* , tend to associate into a variety of structures in both aqueous and nonaqueous systems. In most cases, those assemblies can transform from one into the other as a result of sometimes subtle changes in the solution conditions (eg, concentration, solvent composition, added electrolyte, temperature changes, pH). The basic concepts that govern self-association into micelles also apply to the formation of the larger, more extended aggregate systems consisting of vesicles, bilayers, and membranes. An excellent discussion of the general field of molecular association structures can be found in the work of Israelachvili.[1] This section will be limited to a discussion of some of the general aspects of molecular association into such structures as vesicles and bilayers, and the presentation of a few of the possible practical applications of such systems currently under investigation.

Those amphiphilic materials that cannot conveniently pack into compact structures such as micelles generally associate to produce vesicles and extended bilayers. In general, such materials will have relatively small head groups or, as is more common, their hydrophobic groups will be too bulky to be packed in a manner necessary for normal micelle formation. Such a state of affairs is particularly common for molecules having more than one hydrocarbon chain, very highly branched chains, or structural units that produce molecular geometries incompatible with effective packing (eg, large, flat ring structures such as steroids).

Although extended (essentially infinite) planar bilayers are a thermodynamically favorable option for the association of some bulky surfactants or amphiphiles in aqueous solution, there are some conditions under which it is more favorable to form closed bilayer systems, leading to the existence of various types of membranes and vesicles. Such a situation can be seen to arise from two basic causes. First, even large, highly extended planar bilayers possess edges along which the hydrocarbon core of the structure must be exposed to an aqueous environment, resulting in an unfavorable energetic situation. Second, the formation of an infinitely extended structure is unfavorable from an entropic standpoint. The formation of vesicles, then, addresses both of those factors --- the edge effect is removed by the formation of a closed system, and the formation of structures of finite size reduces much of the entropy loss. As long as the curvature of the vesicle is gentle enough to allow the packed molecules to maintain close to their optimum area, vesicles will represent viable structures for the association of surfactants and related materials.

Vesicles

Many naturally occurring and synthetic surfactants and phospholipids that cannot undergo simple aggregation to form micelles will, when dispersed in water, spontaneously form closed bilayered structures referred to as liposomes or vesicles. They are constructed of alternating layers of lipid or surfactant bilayers spaced by aqueous layers or compartments arranged in approximately concentric circles (Figure 15.7a). If the spontaneously formed multilayer vesicles are subjected to ultrasound or other vigorous agitation, the complex multilayer structure may be disrupted to produce a single bilayer assembly consisting of a unilamellar vesicle in which a portion of the aqueous phase is encapsulated within the single bilayer membranes (15.7b). Typically, a vesicle so produced will have a diameter of between 30--100 nm, falling within the size range of classical colloidal systems.

Natural and synthetic amphiphiles that form vesicles are, by their nature, of limited solubility in aqueous systems, so that the exchange of individual molecules from the bilayer is often very slow. In addition, the

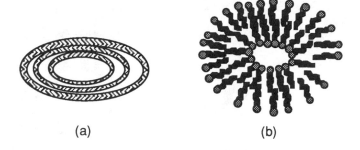

(a)   (b)

Figure 15.7. Schematic illustration of (a) multilamellar and (b) single lamellar vesicles.

bilayer structure has a significant degree of internal stability so that vesicles, once formed, can have a relatively long existence. Lifetimes of from a few days to several months have been reported. After extended periods, the unilamellar vesicles will begin to fuse to produce the more complex aggregate structures of the original systems.

One of the interesting and potentially useful characteristics of vesicles is their ability to entrap within the assembly a portion of the aqueous phase present at the time of their formation, along with associated solute. They therefore represent a unique microencapsulating technique, since residual solute located outside the vesicle can be removed by dialysis or some other related purification techniques. Oil-soluble materials can also be incorporated into vesicle systems, although they would then be located inside the hydrophobic portion of the membrane, much like materials solubilized in conventional surfactant micelles. The potential for the incorporation of both aqueous and nonaqueous additives into vesicles poses the interesting possibility of producing a system containing two active components, for example, a water- and an oil-soluble drug, for simultaneous delivery.

Other interesting and potentially useful physical characteristics of conventional vesicles include their activity as osmotic membranes, their ability to undergo phase transitions from liquid crystalline to a more fluid state, and their permeability to many small molecules and ions, especially protons and hydroxide. Because of their similarity to natural biological membranes, vesicles also have great potential as models for naturally occuring analogues that may be difficult to manipulate directly.

Polymerized Vesicles and Lipid Bilayers

Major barriers to the use of conventional vesicles in many applications include: (1) the inherent long-term instability of the systems, (2) their potential for interaction with enzymes and blood lipoproteins, and (3) their susceptibility to the actions of other surface-active materials. For such critical applications as controlled-release drug delivery, even the most stable systems with a lifetime of several months do not begin to

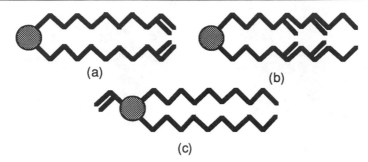

(a)   (b)

(c)

Figure 15.8. Typical structures of polymerizable vesicle-forming surfactants: (a) terminal unsaturation in the tail; (b) internal unsaturation in the tail; (c) unsaturation associated with the head group.

approach the shelf life requirements.

As a result of the potential utility and relatively low cost of vesicles, a great deal of effort has in recent years been applied to the development of polymerized surfactant and phospholipid systems. The ability to covalently crosslink the vesicle membrane after the encapsulation process should produce a system in which the basic nature of the vesicle as an encapsulating medium is retained while adding the structural integrity and increased stability of a crosslinked polymeric structure.

The general approach used to attain such structures has been the synthesis of conventional vesicle-forming amphiphilic materials containing polymerizable functionalities in the molecule, vesicle formation, and subsequent polymerization, preferably by some "nonintrusive" means such as irradiation. In principle, the polymerizable functionality can be located at the end of the hydrophobic tail, centrally within the tail, or in association with the ionic or polar head group (Figure 15.8). The choice of a preferred structure will probably be determined by the final needs of the system and the synthetic availability of the desired materials.

BIOLOGICAL MEMBRANES

In the last few years there has been a dramatic increase in interest in the molecular structure of biological membranes. While model systems composed of artificially prepared (or isolated) amphiphilic materials and associated colloids serve a very useful purpose, a better understanding of the reality of biological systems would be invaluable in many areas of biochemistry, medicine, pharmaceuticals, etc. While it is reasonably easy to determine the constituents of a biological membrane, elucidating just how the various components are put together, how they interact, and their exact function within the membrane represents a decidedly more difficult task.

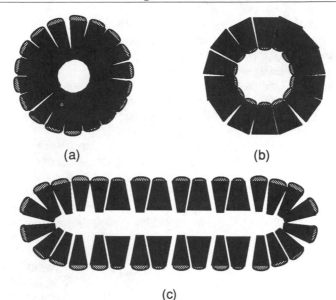

(a)   (b)

(c)

Figure 15.9. Illustration of the role of molecular geometry and the packing of lipids to form micelles, membranes, and complex cellular structures: (a) a "normal" micelle; (b) an inverted micelle; (c) mixed packing to form planar and cellular structures.

Membrane Surfactants or Lipids

The surface-active components of biological membranes are generally referred to as *lipids* , with the majority consisting of double-chained *phospholipids* or *glycolipids* . The hydrophobic tails normally contain chains of 16 to 18 carbons, with one generally being branched or unsaturated. The combination of those factors guarantees that the lipids will be very surface active and easily form self-assembled bilayer membranes that can encapsulate or isolate different regions and functions in an organism. In addition, the long chain lengths insure that the lipids will have a very low solubility in water (as the monomer), a low CMC, and therefore their assemblies will remain intact while contacting surrounding fluids. The presence of branching and unsaturation also guarantees that the membranes will remain fluid over a relatively wide temperature range, to insure the viability of the organism in the presence of varied environmental conditions.

The size, structure, and fluidity of membrane lipids are also important because those aspects of the molecules make it possible for them to pack efficiently into a variety of bilayer membrane structures with various degrees of curvature and flexibility. That flexibility, makes possible the inclusion of the various other important components of the cell wall including proteins, and cholesterol. In terms of the geometric concepts discussed previously (see Figure 15.6), one can visualize where one class of lipid will have a critical packing factor, F ($= v/a_o l_c$) < 1, which will produce a truncated cone shape, while another will have F > 1 for an inverted truncated cone. Combinations of the two can then accomodate the inclusion of, for example, proteins and cholesterol, while maintaining an overall planar structure (or a given degree of curvature), or increase curvature to produce a smaller associated unit. The situation is shown schematically in Figure 15.9.

Biological membranes are, like micelles and vesicles, dynamic structures in which the component lipids and proteins can move about relatively freely, even though the structure as a whole remains intact. In order to carry out its biological function, the cell membrane will also have heterogeneous regions of lipids, proteins, or other materials which may serve as specific binding sites, transport "channels," etc. The components of the entire structure, however, must all have one thing in common. They must be able to associate spontaneously to form the necessary assembly of molecules to do the job, even when all of the components (eg, cholesterol) will not form such structures alone. It appears that an organism can "sense" the specific lipid structures needed in a given situation to produce the membrane structure, fluidity, etc, called for. When conditions such as temperature change, the organism synthesizes the new molecules (more or less saturated fatty acid chains, for example) to fit the new conditions. Clearly, the creation and functioning of biological membranes cannot be a haphazard process of trial and error in selecting the proper lipids for a given cell structure. There must exist some feedback mechanism through which the organism can "know" what material is needed under given conditions so that it can be provided when and where called for.

Other aspects of the interactions of lipids and bilayer structures in biological systems can be understood in the context of molecular geometry, association phenomena, and general interfacial interactions. Unfortunately, those topics are too broad to be included here. It will be interesting to see how future research in molecular biology is able to incorporate the fundamentals of surface and colloid science into a better understanding of the function of membranes, cells, and entire organisms.

CHAPTER 16

SOLUBILIZATION, MICELLAR CATALYSIS, AND MICROEMULSIONS

In addition to being a fundamental consequence of the nature of amphiphilic molecules, micelle formation also plays a significant part in the practical application of surfactants in various areas. Because they represent what might be considered a second liquid "phase" in solution, micelles are often found to facilitate the production of apparently stable, isotropic "solutions" of mixtures of immiscible liquids (and sometimes solids), quite distinct from the obviously two-phase emulsions and sols previously discussed. Depending on the system (and the observer) such "solutions" are said to result from either *solubilization* of a material in the continuous phase or from the formation of *microemulsions* . In addition, the unique character of the micelle makes it a potentially useful "transition zone" between phases in which the unique environment may facilitate (ie, catalyze) chemical reactions difficult to achieve under normal two-phase conditions. The ability of a surfactants to carry out such functions is of great potential importance and warrants some closer attention.

SOLUBILIZATION

The increased solubility of organic materials in aqueous surfactant solutions is a phenomenon that has found application in many scientific and technological areas. It is only recently that a good understanding of the structural requirements for optimum solubilization has begun to develop as a result of extensive experimental and theoretical work.

The early work in this century addressing the mechanisms of micellar solubilization was, unfortunately, usually performed with surfactants of questionable purity. More recently, closer attention has been paid to using the purest or best characterized surfactant systems available, so that more confidence can be placed in the validity and interpretation of experimental results. That is not to say, however, that the pioneering work of the first half of this century is without merit. To the contrary, modern experimental techniques have done much to confirm the work of that era. Considering the relatively limited resources of the early investigators (compared to the modern chemical laboratory), one can only regard their results and interpretations with the highest respect.

There is some disagreement within the surfactant literature as to the exact definition of "solubilization," particularly as the ratio of surfactant to additive decreases, and one approaches the nebulous frontier between swollen micellar systems and the micro- and macroemulsion regimes. For present purposes, *solubilization* will be defined as the preparation of a thermodynamically stable, isotropic solution of a substance (the "additive") normally insoluble or only slightly soluble in a given solvent by the addition of one or more amphiphilic compounds at or above their critical micelle concentration. By the use of such a definition, a broad area can be covered that includes both dilute and concentrated surfactant solutions, aqueous and nonaqueous solvents, all classes of surfactants and additives, and the effects of complex interactions such as mixed micelle formation and hydrotopes. It does not, however, limit the phenomenon to any single mechanism of action.

One problem with that definition is that it says nothing about the relative amounts of surfactant and additive in the system. That question will arise again in the context of microemulsions. For present purposes, we will say that in solubilization, the ratio of additive to surfactant will generally be less than two. The reasons for that limitation will be discussed a bit more later.

For a specified solvent system, water or aqueous solutions for example, there are two variables that must be considered in the solubilization process: (1) the molecular nature, purity, and homogeneity of the surfactant and (2) the chemical nature of the additive. From a technological viewpoint, it is important to understand exactly what surfactant structural features serve to maximize the desired solubilizing effect, and the best way to achieve that understanding is through a fundamental knowledge of the molecular and thermodynamic processes involved. In addition, since most technological applications of solubilization involve complex multicomponent systems, such factors as temperature, electrolyte content, and the presence of polymeric species and other solutes must be examined.

The "Geography" of Solubilization

In order to better understand the "why's and wherefore's" of solubilization, it is helpful to understand the "geography" of solubilization - that is, the possible positions or loci in (or on) the micelle that can serve as host sites for the additive molecules, and the factors that determine where solubilization will occur.

It is well established that the location of a solubilized molecule in a micelle relative to the different structural components of the surfactant will be determined primarily by the chemical structure of the additive (Figure 16.1). In aqueous solutions, nonpolar additives such as hydrocarbons are intimately associated with the core of the micelle (Figure 16.1a),

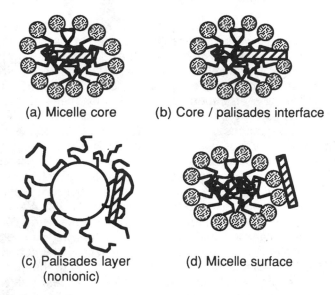

    (a) Micelle core   (b) Core / palisades interface

    (c) Palisades layer   (d) Micelle surface
       (nonionic)

Figure 16.1. The principle loci of solubilization in micellar systems.

while slightly polar materials, such as fatty acids, alcohols, and esters, will usually be located in what is termed the palisades layer --- the interfacial region between the core and surface head groups (Figure 16.1b). The orientation of such molecules is probably more or less radial with the hydrocarbon tail remaining closely associated with the micellar core. In some cases, that orientation can potentially have a significant effect on the energetics of the system, as will be discussed in the section on microemulsions. Other structural factors, such as the charge on the surfactant head group can significantly affect the locus of solubilization. Materials containing aromatic rings, for example, may be solubilized in or near the core of anionic systems but in the palisades layer of cationics due to electronic interactions between the ring and the cationic head group.

     In addition to the solubilization of additives in the micellar core and the core--palisades boundary region, they may also be found entirely in the palisades region (Figure 16.1c) and on the micellar surface (Figure 16.1d). The nature of the polar head group of nonionic surfactants, especially the polyoxyethylenes (POE), is such that a relatively large fraction of the micelle volume is in the palisades region. Because of the bulky nature of the POE chain and its attendant solvent molecules, it has been suggested that the hydrophilic chain is arranged in a spiral from the micellar core outward into the solution. As a result, areas of the palisades near the core will be sterically crowded with the POE chains, with relatively little room left for waters of hydration, or casual water molecules. As the distance from the core increases, the palisades layer becomes more hydrophilic, acquiring more characteristics of an aqueous solution. The net effect is that, deep in the palisades layer, the chemical environment will closely approximate that of a polyether, so that materials soluble in such solvents will be preferentially located in that region.

Even though chemical structures may dictate the preferred location for the additive, solubilized systems are dynamic, as are the parent micelles, and the location of specific molecules changes rapidly with time. It will always be important to remember that while a given region of the micelle may be preferred by an additive on chemical grounds, there is no guarantee that all phenomena related to the system (catalysis, for example) will be associated with that region.

In surfactant--nonpolar solvent systems where the "sense" of the micelle is reversed, the polar interactions of the head groups not only provide a driving force for the aggregation process, but also provide an opportune location for the solubilization of polar additives. Water is, of course, one of the most important potential polar additives to nonaqueous systems, and it is primarily located in the core. The nature of such solubilized water is not fixed, however. The initial water added likely becomes closely associated with the polar head group of the surfactant (as waters of hydration), while subsequent additions appear to have the character of free bulk water. Other polar additives, such as carboxylic acids, which may have some solubility in the organic phase, are probably associated with the micelle in a manner analogous to that for similar materials in aqueous systems.

The effects of solubilized additives on the micellar properties of nonaqueous surfactant systems vary according to the structures of the components. Such changes, however, are often greater than those found in aqueous solutions, so that due care must be exercised in evaluating the effects of even small additions on the aggregation characteristics of surfactants in nonaqueous solvents.

The Solubilization Process

Just as molecular structure is important to such surfactant characteristics as the critical micelle concentration, aggregation number, micellar shape, etc, it also controls the ability of a surfactant to solubilize a third component. Conversely, the presence of a third component in a surfactant solution can often affect its aggregation characteristics. Whether micelles formed in the presence of a third component are the same as those formed in its absence is a subject of some controversy. It has been shown that micellar activity may be induced in surfactant solutions below the "normal" CMC in the presence of small amounts of solubilized additives. In some cases such effects have been attributed to additive-induced micellization. In others, effects have been seen at concentrations several orders of magnitude below the CMC, suggesting the presence in solution of submicellar species possessing some properties of the fully aggregated system. It has been suggested that many, if not most, surfactants in dilute solution undergo a low level of molecular aggregation at concentrations well below their CMC, during which dimers, tetramers, and other "premicellar" aggregates are formed.

That may be especially true for surfactants having unusually large or bulky hydrophobic groups such as the bile acids and tetraalkylammonium halides.

Studies of micelle formation indicate that surfactant properties such as the CMC and aggregation number can be reasonably well correlated with the size and nature of the hydrophobic group. Unfortunately, comparably convenient relationships are not always so apparent in terms of surfactant structure and solubilizing power, probably because the structure of the additive can play such an important role in the overall aggregation process. Nevertheless, many of the factors discussed previously that cause an increase in micelle size might also be expected to increase the solubilizing power of the system.

Generalizations on Surfactant Structure and Solubilizing Power

An increase in the length of the hydrocarbon chain in a surfactant, for example, leads to a lower CMC and larger aggregation number so that more of a nonpolar additive can be incorporated into the micellar core per mole of surfactant in the system. Branching of the hydrocarbon chain of the surfactant usually results in a decrease in the solubilizing power of the system relative to that of the analogous straight-chain material. That is presumably due to geometric and packing constraints, which limit the ability of the micellar core to accommodate the added bulk of the solubilized molecules. The addition of ethylenic unsaturation and aromatic groups also tends to decrease the maximum amount of additive that can be fitted into the core packing arrangement.

In the case of nonionic surfactants, the amount of aliphatic hydrocarbon that can be solubilized generally increases as the length of the hydrophobic tail increases and decreases as that of the POE chain increases. Those results parallel changes in the CMC's and aggregation numbers of the respective materials. Divalent salts of alkyl sulfates quite often exhibit a greater solubilizing capacity than the corresponding monovalent salt for materials included in the micellar core, presumably reflecting the increased packing density attainable due to decreased head group repulsion.

If one considers the relative solubilizing powers of the different types of surfactant with a given hydrophobic tail, it is usually found that they can be ordered as nonionics > cationics > anionics. The rationale for such a result is usually related to the supposed looser packing of the surfactant molecules in the micelles of the nonionic materials, making available more space for the incorporation of additive molecules without greatly disrupting the basic structure.

The solubilizing power of amphoteric surfactants has not been as widely studied, or at least as widely reported, as that for the simpler ionics and nonionics. However, the available data indicate that they lie somewhere between the extremes in solubilizing capacity, the exact results being probably more sensitive to the nature of the additive than those for the other classes of surfactants.

Solubilization and the Nature of the Additive

The quantity of a substance that can be solubilized in surfactant micelles will depend on many factors, some of which have already been discussed. From the standpoint of the additive itself, such factors as molecular size and shape, polarity, branching, and the electronegativity of constituent atoms have all been found to be of some significance, depending on the exact system. One extensively explored factor relating the chemical structure of the additive to its solubilization is the relationship between the molar volume of the additive and the maximum amount of material that can be incorporated in a given surfactant solution. In general, one finds an inverse relationship between the molecular volume of the additive and the amount of material solubilized.

In general, increasing the chain length of an *n*- alkane or *n* - alkyl-substituted benzene reduces its solubility in a given surfactant solution. While the presence of unsaturation or cyclic structures tends to increase solubility, branching appears to have little or no effect. More complicated additive structures fail to behave in such an orderly fashion. The addition of a benzene ring, for example, tends to increase solubility while a second, fused ring, such as in naphthalene derivatives, produces the opposite effect.

In summary, the relationship between the chemical structure of the additive and its ability to be incorporated into a surfactant solution is quite complex and has so far not lent itself to simple analysis and structural correlation. Perhaps, as our understanding of the geometric packing requirements of molecules in the micellar core and palisades layer improves, a more rational scheme for predicting solubilization results will emerge.

The Effect of Temperature on Solubilization

When one considers the effects of changes in temperature on the solubilization process, two areas of concern must be addressed. First, the ability of a given surfactant to solubilize an additive is intimately related to the characteristics of the micelle --- size, shape, ionic nature, etc. Since changes in temperature are known to affect some of those characteristics, it should not be surprising to find alterations in the solubilizing properties of surfactants as a result of modifications in micellar structure. Second, changes in temperature can affect the intermolecular interactions between solvent and solutes (hydrogen bonding, for example), so that the overall solvent properties of the liquid for surfactant and additive may be significantly altered. In general, one can expect temperature changes that lower the CMC or increase the aggregation number of a surfactant toimprove its solubilizing capacity.

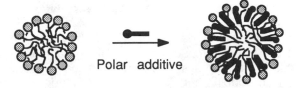

Polar additive

Figure 16.2. Schematic illustration of the proposed role of polar cosolutes in enhancing the solubilizing capacity of a micellar system by improving the "efficiency" of packing at the micelle surface.

Nonelectrolyte Co-Solutes

Nonelectrolyte cosolutes that are not part of the primary solubilized system (solvent-surfactant-additive) can have a significant effect on the solubilizing power of micellar solutions as a result of their effects on CMC's and aggregation numbers. It has become especially obvious that the addition of polar solutes such as phenols and long-chain alcohols and amines can greatly increase the solubility of nonpolar additives in ionic surfactant solutions. The mechanism for such enhancement likely involves the insertion of polar additive molecules between adjacent surfactant molecules in the micelle (Figure 16.2). As a result of the "isolation" of the ionic groups, repulsive interactions and unfavorable contact between the aqueous phase and exposed hydrocarbon in the core can be reduced. Those two modifications of the micellar surface would allow a decrease in surface curvature of the micelle and a subsequent increase in the capacity of the core to accommodate solubilized nonpolar additives. Since such additives may also act as cosurfactants for microemulsion formation (see below), it is possible that their function in each case is related.

Unlike polar cosolutes with relatively large hydrophobic tails, short-chain alcohols such as ethanol can significantly reduce the solubilizing power of a surfactant. In the earlier discussion of the effects of such materials on the micellization process, it was shown that the addition of significant quantities of short-chain alcohols, acetone, dioxane, etc, could result in profound changes in the CMC and aggregation number of surfactants, even to the point of completely inhibiting micelle formation. It is understandable, then, that such cosolutes would also adversely affect the solubilization capacity of a surfactant solution.

From the above, it seems clear that the effects of an added nonelectrolyte on the solubilizing capacity of a given surfactant system may be quite complex and may not lend itself to easy analysis. It can be assumed, however, that the fundamental relationships that exist between the cosolute and the micellization characteristics of the surfactant, in the absence of the solubilized additive, can be used to good advantage in predicting what may reasonably be expected in the four-component system.

The Effects of Added Electrolyte

For ionic micelles, the effect of addition of electrolyte is to decrease the CMC and increase the aggregation number. Such changes are predictable in micellar systems and might be expected to produce parallel effects on solubilization. The results, however, are not always so clear cut. At surfactant concentrations near the CMC, it is usually found that the solubilizing power of a system will increase with the addition of electrolyte, as a result of the greater number and larger size of micelles available in the system. At surfactant concentrations well above the CMC, however, the simplicity of the relationship may disappear due to more fundamental changes in the nature of the associated structures.

Such inconsistencies might also be related to the nature of the additive and its potential location in the micelle. For nonpolar additives or those lying deep in the palisades layer of the micelle, it seems reasonable to expect the increased volume of the micellar core produced by electrolytes to lead to a greater capacity for solubilization, as is generally the case. For more polar materials that might be incorporated less deeply in the micelle, added electrolyte results in a closer packing of ionic head groups, which could reduce the available space for solubilized molecules. Changes in micelle shape, from spheres to rods, for example, would also result in less surface volume available in the palisades layer as a result of closer packing of the head groups.

In the case of nonionic surfactants, the effects of added electrolytes seem to parallel their effects on the micellization process. When such addition produces an increase in micellar aggregation number, an increase in solubilizing capacity for hydrocarbon additives is also found. The results for the solubilization of polar materials is, again, less clear cut.

Miscellaneous Factors Affecting Solubilization

Other factors that can affect the ability of a particular surfactant system to solubilize materials include pH and pressure. The effects of such factors, however, have not been as extensively reported in the literature as the factors discussed above, and they are often very specific to each surfactant system. Obviously, surfactants that show extreme sensitivity to pH such as the carboxylate salts can also be expected to exhibit significant changes in solubilization with changes in that factor. In addition, changes in pH can affect the nature of the additive itself, producing dramatic changes in its interactions with the micelle, including the locus of solubilization. Such effects can be especially important in many applications of solubilization, such as in the pharmaceutical field.

The effects of such a variable as pressure on micelle formation and solubilization is a relatively new field of investigation. It can be assumed that significant effects will be observed once sufficient pressure levels have been attained. However, such levels lie outside the normally available range of experimental conditions and are of little practical concern. The exception being highly pressurized products such as fire fighting foams, shaving creams, whipped toppings, etc.

MICELLAR CATALYSIS

It is well recognized in all branches of chemistry that the rate of a chemical reaction can be very sensitive to the nature of the reaction environment. Reactions involving polar or ionic transition states can be especially sensitive to the polarity of the reaction medium. It should not be too surprising, then, that many chemical reactions, especially those in which one reactant may be soluble in water and the other in oil, can exhibit a significant enhancement in rate when carried out in the presence of surfactant micelles. The presence of the micellar species can provide a beneficial effect through two possible mechanisms. (1) The palisades region of the micelle represents a transition zone between a polar aqueous environment, which may be either the bulk phase or the micellar core, and a nonpolar hydrophobic region. Such a gradient in polarity can serve as a convenient area of intermediate polarity suitable for increased reactant interaction or for optimizing the energetics of transition state formation. (2) The potential for the micelle to solubilize a reactant that would not normally have significant solubility in the reaction medium means that it can serve as a ready reservoir of reactant, in effect increasing the available concentrations of reactants. The rate enhancements that have been reported range as high as 10^5, which makes such systems very attractive for potential practical applications.

Catalysis in Aqueous Solvent

In aqueous media a micellar system can serve as a catalyst for organic reactions, but it is also possible for it to retard such reactions. Possible mechanisms for catalytic action were suggested in Figure 16.3; inhibitory actions may arise from unfavorable electrostatic interactions between reactants and changes in the distribution of reactants between the bulk and micellar phases. In the case of electrostatic inhibition, the presence of a charge on the micelle surface can have two effects on a reaction involving a charged species. In the base hydrolysis of water-insoluble esters, if the micelle charge is negative, the transport of hydroxide ion into or through the palisades layer will be retarded by charge repulsion. If the micelle is positively charged, the inclusion of the oppositely charged species will be facilitated. For nonionic and zwitterionic surfactants, there will be little or no effect as a result of such electrostatic interactions.

Although such a model of electrostatic effects is extremely simplified, it has generally been supported by experiment. The basic hydrolysis of esters, for example, is catalyzed by cationic and inhibited by anionic surfactants, while the opposite is true for the acid hydrolysis of orthoesters.

The ability of a micellar system to solubilize a reactant may affect

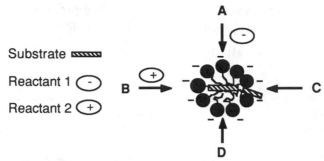

A - Reactant of same sign impeded in approach.
B - Reactant of opposite sign aided in approach.
C - Solubilization in palisades produces intermediate
     solvent environment.
D - Increased substrate availibility due to solubilization.

Figure 16.3. Schematic illustration of the predominant mechanisms of micellar catalysis.

its action as a catalyst or inhibitor in a reaction. When a surfactant system serves as a reservoir for increasing the availability of one reactant, any change that increases the solubilizing capacity of the micelle should also increase its effectiveness as a catalysis. If, on the other hand, the reaction must occur in the bulk phase, increased solubilizing power may remove reactant from the reaction medium and therefore decrease catalytic or increase inhibitor efficiency.

In aqueous solution, the effectiveness of micellar systems as catalysts is quite often found to increase with the length of the alkyl chain. For example, the rate of acid hydrolysis of methyl orthobenzoate in the presence of sodium alkyl sulfates increased in the order octyl < decyl < dodecyl < tetradecyl < hexadecyl. Such a result may be attributed to either electrostatic or solubilizing effects, or both. It might be expected that any effects due to electrostatic interactions would also increase. Alternatively (or additionally), the larger aggregation number may result in a significant increase in the solubilizing power of the system. The importance of each mechanism will depend upon the specifics of the reaction.

In addition to the effects noted for increases in the charge density on the micelle, the charge on the individual surfactant molecules can also be important. One generally finds, for example, that cationic surfactants containing two charge groups are significantly better at increasing the rate of nucleophilic aromatic substitutions than analogous singly charged materials. Similar results have been reported for singly versus doubly charged anionic surfactants.

As might be expected, the structure of the reactive substrate can have as much influence on micelle-assisted rate enhancement as that of the surfactant. Since the catalytic effectiveness of the micelle can be related to the location and orientation of the substrate in the micellar structure, the more hydrophobic the substrate (and the surfactant), the more significant may be the catalytic effect.

When nonsurfactant cosolutes (electrolytes, etc) are added to the micellar reaction mixture the results can be quite unpredictable. It is often found that the presence of excess surfactant counterions retards the catalytic activity of the micelle, with larger ions being more effective in that respect. In contrast, the addition of neutral electrolyte has been found to enhance micellar catalysis in some instances. It has been proposed that the retardation effect of excess common counterions is due to a competition between the excess ions and the reactive substrate most closely associated with the micelle for the available positions or "binding sites" on or in the micelle. The enhancing effect, however, has been attributed to the more general effects of added electrolyte on the properties of micelles, that is, lowering of the CMC, increasing the aggregation number, etc, all of which often tend to increase catalytic activity.

Catalysis in Nonaqueous Solvents

Interactions between polar head groups in nonaqueous solvents provide the primary driving force for the formation of micellar aggregates in such media. The nature of such reversed micellar cores is such that they provide a unique location for the solubilization of polar substrates. While keeping in mind the potentially dramatic effects of additives on the properties of micellar solutions, it is obvious that such nonaqueous systems hold great potential from a catalytic standpoint. They are especially of interest as models of enzymatic reactions.

The fundamental principles controlling activity in nonaqueous systems are the same as those for aqueous solutions, except that the specificity of the micellar core for the solubilization of polar substrates is much greater than for the aqueous situation. The popularity of reversed micelles as models for enzyme catalysis stems from the fact that the micellar core is capable of binding substrates in concentrations and orientations that can be very specific to certain functionalities, much as an enzyme would do. As a result, reaction rate enhancements can be obtained comparable (with luck) to those of the natural systems, and far in excess of what can be explained on the basis of partitioning or availability of substrate.

Work in the area of micellar catalysis in both aqueous and nonaqueous solvent systems is certain to continue to grow in importance as a tool for better understanding the chemistry and mechanics of enzymatic catalysis, as a probe for studying the mechanistic aspects of many reactions, and as a route to improved yields in reactions of academic interest. Of more practical significance, however, may be the expanding use of micellar catalysis in industrial applications as a method for obtaining maximum production with minimum input of time, energy, and materials.

MICROEMULSIONS

Microemulsions are composed of two mutually immiscible liquid phases, one spontaneously dispersed in the other with the assistance of one or more surfactants and cosurfactants. While microemulsions of two nonaqueous liquids are theoretically possible (fluorocarbon--hydrocarbon systems, for example), almost all of the reported work is concerned with at least one aqueous phase. The systems may be water continuous (O/W) or oil continuous (W/O), the result being determined by the variables such as the surfactant systems employed, temperature, electrolyte levels, the chemical nature of the oil phase, and the relative ratios of the components.

Most microemulsions, especially those employing an ionic surfactant, require the addition of a cosurfactant to attain the required interfacial properties leading to the spontaneous dispersion of one phase in the other. The structural character of the cosurfactant (chain length, head group, etc) and its relationship to the primary surfactant have been a major focus of much of the research in the area. It is commonly found that microemulsions prepared using ionic surfactants require a relatively high ratio of surfactant to dispersed phase, while some nonionic surfactants have been found to produce microemulsions with much lower levels of surfactant and without the addition of a cosurfactant.

For a given oil phase, it is usually possible to optimize the oil--water ratio by varying the structure of the surfactant and cosurfactant, if needed. An aromatic oil phase, for example, may produce microemulsions in a given composition range using a short-chain alcohol as cosurfactant while a hydrocarbon oil may require a longer chain material to achieve comparable results. Although the process requires a great deal of tedious laboratory work, three-dimensional phase diagrams (oil--water--surfactant--cosurfactant) can be quite useful in determining the optimum composition for the production of a microemulsion. To simplify matters somewhat, it is possible to fix one pair of variables, say the surfactant--cosurfactant ratio, and then prepare a simpler triangular phase diagram.

Microemulsions, like macroemulsions, can exist with either the oil or the water being the continuous phase. The characteristics of the system will, of course, be different in each case. The proper use of the phase diagram approach allows one to determine not only the component ratios necessary to produce a microemulsion, but also the component forming the continuous phase.

Micelle, Microemulsion, or Macroemulsion?

The status of the systems commonly referred to as *microemulsions* among surface and colloid chemists is somewhat uncertain. Various experimental approaches have been used in an attempt to ascertain the details of their structural and thermodynamic characteristics. As a result, new theories of the formation and stability of these interesting but quite complex systems are appearing. Although a great deal has been learned about microemulsions, there is much more to be learned about the requirements for their preparation and the relationships among the chemical structure of the oil phase, the composition of the aqueous phase, and the structures of the surfactant and the cosurfactant, where needed.

The distinction between microemulsions and conventional emulsions or macroemulsions is fairly clear. Although macroemulsions may be kinetically stable for long periods of time, in the end they will suffer the same fate --- phase separation to attain a minimum in interfacial free energy. The actions of surfactants, polymers, and other stabilizing aids may shift the rate of droplet coalescence to extremely long times through decreased kinetic rate constants, but the thermodynamic driving force to minimize the interfacial area between immiscible phases remains unchanged. On the other hand, the microemulsions appear to be thermodynamically stable compositions with essentially infinite lifetimes, assuming no change in such factors as composition, temperature, and pressure.

In addition to the thermodynamic distinction usually drawn between macro- and microemulsions, the two classes of colloids differ in several other more tangible characteristics, including the size of droplets formed and the mechanical requirements for their preparation. As far as droplet size is concerned, the conventional macroemulsions are generally found to have minimum diameters of 100--200 nm, meaning that such systems are usually quite turbid or opaque. Microemulsions, however, normally have droplet diameters of 100 nm or less, with many being only slightly larger than simple micellar systems. Because those particles are much smaller than the wavelength of visible light, they are normally transparent or slightly bluish.

The energy requirements for the formation of macroemulsions can be quite substantial. The formation of small droplets requires that the system overcome both the adverse positive interfacial free energy between the two immiscible phases working toward drop coalescence and bulk properties of the dispersed phase such as viscosity. Microemulsions, on the other hand, form spontaneously or with very gentle agitation when the proper composition is reached.

When one compares microemulsions and micelles, the demarcation line can become quite blurred and, in some cases, does not exist. There is some controversy as to the true definition of clear, isotropic solutions of oil, water, and surfactant (and cosurfactant if needed) as microemulsions rather than swollen micelles. Although the differences between the two systems may appear to many to be more semantic than factual, several arguments have been presented that strongly support a differentiation of the two systems.

If one constructs a "spectrum" of the possible situations for the dispersion of one liquid phase in another, the possible sizes of the dispersed phase units range from the molecularly dispersed true solution where "droplet" sizes are on the order of a few nanometers, to

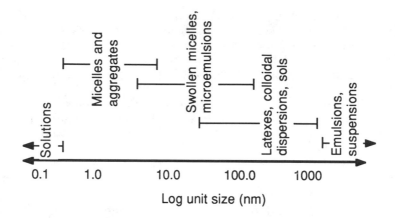

Figure 16.4. The spectrum of unit sizes found in various dispersed systems, ranging from molecularly dispersed solutions to coarse emulsions and suspensions.

macroemulsions with droplet diameters of hundreds or thousands of nanometers (Figure 16.4). Lying between the extremes are simple micelles (a few tens of nanometers), macromolecular solutions (tens to hundreds of nanometers), and colloids of several hundred to several thousand nanometers. The systems typically referred to as microemulsions will normally have unit sizes between 5 and 100 nm, generally well beyond the range of spherical micelles in dilute solution, but overlapping significantly with larger assemblies, eg, rod-shaped micelles, liquid crystals. The physical differences encountered among most of the different groups are sufficient to obviate any controversy as to their classification.

The question of differentiation between micelles and microemulsions is less simply answered. While it is undoubtedly true that, in the smaller size ranges especially, many systems classed as microemulsions are indistinguishable from swollen micelles, it is equally true that the larger microemulsion systems far exceed the solubilizing capacity of micelles as previously defined. Micelles will form under many circumstances, although the specifics of CMC, aggregation number, etc, may change with the environmental conditions. The formation of microemulsions, on the other hand, has been shown to have very specific compositional requirements. It is primarily due to those specific demands on the composition of the system and the chemical structures of the various components that the nomenclature for this separate class of dispersed species has developed.

An additional argument for a distinction between micelles and microemulsions is that in all of the literature on the solubilization of hydrocarbons, dyes, etc, in micellar solutions, the ratio of solubilized molecules to surfactant molecules very rarely exceeds, or even approaches, 2. Many microemulsion systems, on the other hand, have been described in which the dispersed phase-to-surfactant (and cosurfactant) ratio exceeds 100! Because of the relatively low ratios of additive to surfactant obtainable in micellar systems, it is clear that there can exist no oil phase that can be considered separate from the body of the micelle. In many microemulsions, however, the size of the droplet and the high additive-to-surfactant ratio requires that there be a core of dispersed material that will be essentially equivalent to a separate phase of that material. The seemingly obvious conclusion is that microemulsion systems (in the latter case, at least) possess an interfacial region composed primarily of surfactant (and cosurfactant), analogous to that encountered in macroemulsions.

Both theory and careful experimental work seem to indicate that the driving force for the spontaneous formation of microemulsion systems is the existence of a transient negative interfacial tension between the oil and aqueous phases, resulting in a rapid transfer of one of the two phases through the interface, producing the optimum droplet size for the given composition. It must be emphasized that the negative interfacial tension is a transient phenomenon, and at equilibrium must be zero or slightly positive.

The spontaneous formation of microemulsion systems has an especially important application in enhanced oil recovery processes. Oil deposits that cannot be practically recovered by conventional primary or secondary techniques (flooding, for example) can sometimes be treated with an excess of surfactant solution, leading to the formation of a microemulsion of the oil in the aqueous phase. The emulsified system can then be removed from the rock formation and the oil separated.

Like many other topics covered here, microemulsions hold a great deal of promise in many practical applications. To date, most research in the area has been closely associated with the formation and destruction of microemulsions in the context of secondary and tertiary petroleum recovery. While microemulsions are very attractive for use in many other areas, their sensitivity to composition makes their application much more difficult, since in many cases (ie, drug delivery) one or more component (which may be somewhat surface active) may be fixed, thereby reducing the options for the formulator.

CHAPTER 17

WETTING AND SPREADING

The wetting of a surface by a liquid and the ultimate extent of spreading of that liquid are very important aspects of practical surface chemistry. Many of the phenomenological aspects of the wetting processes have been recognized and quantified since early in the history of observation of such processes. However, the microscopic details of what is occurring at the various interfaces and lines of contact among phases has been more a subject of conjecture and theory than of known facts until the latter part of this century when quantum electrodynamics and elegant analytical procedures began to provide a great deal of new insight into events at the molecular level. Even with all of the new information of the last twenty 25 years, however, there still remains a great deal to learn about the mechanisms of movement of a liquid across a surface.

There are (as usual) two aspects of the system that must be considered: equilibrium thermodynamics and kinetics. The former topic has been fairly well in hand for many years. The latter has received less general attention but has begun to enter the spotlight more in recent years. The following discussion is intended to introduce the fundamental concepts underlying modern theories of wetting and spreading processes. Many of the topics discussed have already been introduced but will be reviewed where it seems useful. Other topics, especially kinetic aspects, which remain somewhat conjectural, will be presented as such. Before beginning the discussion of wetting, however, it will be useful to review somewhat a topic introduced in Chapter 6 in the context of capillary phenomena: the contact angle.

THE CONTACT ANGLE

One of the primary characteristics of any immiscible, two- or three-phase system containing two condensed phases, at least one of which is a liquid, is the *contact angle* of the liquid on the other condensed phases. The contact angle of one liquid on another, while being of theoretical interest, is normally of little practical importance. An

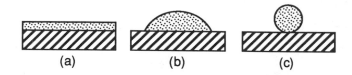

(a)   (b)   (c)

Figure 17.1. Schematic illustration of the various degrees of wetting: (a) complete wetting; (b) partial wetting; (c) complete nonwetting.

exception to that being in certain multilayer coating processes, in which several liquid layers are coated simultaneously on a solid surface. In such processes, it is important that the wetting properties of each layer on the one below be such that smooth, uniform coverage is obtained. Such processes are especially important in the photographic and graphics arts industries, where small coating flaws will make a coated material useless. Of more practical and widespread importance is the contact angle of a liquid directly on a solid. For liquids on solids, the contact angle can be considered a material property of the system, assuming that certain precautions are taken in the collection and interpretation of data, and that liquid absorption is taken into solution.

When a drop of liquid is placed on a solid surface, the liquid will either spread across the surface to form a thin, approximately uniform film (Figure 17.1a) or it will spread to a limited extent but remain as a discrete drop on the surface. The final condition of the applied liquid on the surface is taken as an indication of the *wettability* of the surface by the liquid or the *wetting ability* of the liquid on the surface, depending on your point of view. The quantitative measure of the wetting process is taken to be the contact angle, θ, which the drop makes with the solid as measured through the liquid in question (Figure 17.1b).

In the case of a liquid that forms a uniform film (ie, where $\theta = 0°$), the solid is said to be *completely wetted* by the liquid, or that the liquid *wets* the solid. Where a nonzero angle is formed, there exists some controversy as to how to describe the system. If a finite contact angle is formed ($\theta > 0°$), some investigators describe the system as being *partially wetted*. Others like to make a distinction based on the size of the contact angle. For example, a given worker may define as "wetting," any liquid that produces a contact angle of 30° or less on a solid. Between 30° and 89° the system would be "partially wetting." And 90° and above *nonwetting*. Alternatively, any system with $0° < \theta < 180°$ would be partially wetting, and only for $\theta = 180°$ would the nonwetting term be applied. The terminology one prefers will often depend on individual circumstances and is of relatively minor importance. The important thing is to know what descriptive system is being employed in a given set of circumstances.

While the contact angle of a liquid on a solid may be considered a characteristic of the system, that will be true only if the angle is measured under specified conditions of equilibrium, time, temperature, component purity, etc. Contact angles are very easy measurements to make (with a little practice) and can be very informative; but if the proper precautions are not taken, they can be very misleading.

The contact angle may be geometrically defined as the angle formed by the intersection of the two planes tangent to the liquid and solid surfaces at the perimeter of contact between the two phases and the third surrounding phase. Typically, the third phase will be air or vapor, although systems in which it is a second liquid essentially immiscible with the first are of great practical importance. The perimeter of contact among the three phases is commonly referred to as the *three-phase contact line* or the *wetting line* . On a macroscopic scale, such terminology is meaningful and useful; however, when one begins to zero in on the situation in the region of three-phase contact, it must be remembered that one is talking about a transition zone of finite (even though very small) dimensions in which three phases merge and the characteristics of one phase change to those of another. It is conceptually and mathematically convenient to think in terms of "lines," but for a more complete understanding of the situation the real facts of it must be kept in mind.

The great utility of contact angle measurements stems from their interpretation based on equilibrium thermodynamic considerations. As a result, most studies are conducted on essentially static systems in which the liquid drop has (presumably) been allowed to come to its final equilibrium value under controlled conditions. In many practical situations, however, its is just as important, or perhaps more important, to know how fast wetting and spreading occurs as to know what the final equilibrium situation will be. That will especially be true in situations where the process in question requires that wetting bring about the displacement of one phase by the wetting liquid. Typical examples would be: detergency in which a liquid or solid soil is displaced by the wash liquid; petroleum recovery, in which the liquid petroleum is displaced by an aqueous fluid; and textile processing, in which air must be displaced by a treatment solution (dyeing or waterproofing treatments, for example) in order to obtain a uniform treatment. Because of the economic importance of these and other processes, some emphasis will be placed on the dynamic aspects of the wetting processes.

As already mentioned, the interpretation of data on static contact angles must be done with the understanding that the system in question has been sufficiently well controlled so that the angle measured is the "true" angle and not a reflection of some contaminant on the solid surface or in the liquid phase of interest. Contact angles, for example, can be extremely useful as a spot test of the cleanliness of sensitive surfaces such as glass or silicon wafers for microelectronic fabrications. Both surfaces are "high energy" (see below) and are completely wetted by pure water. If the surface is contaminated by something such as an oil that interferes with the processing of the material (eg, the coating of a photoresist), a drop of water will have a nonzero contact angle, and the contamination will be immediately apparent.

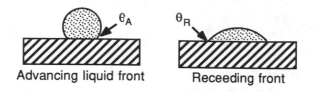

Advancing liquid front   Receeding front

Figure 17.2. Illustration of the phenomenon of advancing and receding contact angle.

For systems that have "true" nonzero contact angles, the situation may be further complicated by the existence of contact angle *hysteresis* . That is, the contact angle one observes may vary depending on whether the liquid is *advancing* across fresh surface (the *advancing* contact angle, θ_A) or receding from an already wetted surface (the *receding* contact angle, θ_R) (Figure 17.2). As an operational convenience, many, if not most, static contact angles measured and reported are in fact advancing angles. For a given system, it will be found that $\theta_A \geq \theta \geq \theta_R$. In practice, very few systems exhibit a complete lack of hysteresis, so that the problem can be operational as well as philosophical. Some of the primary causes of hysteresis will be discussed further below. For now, it is sufficient to keep in mind that when discussing contact angle data, one must always be aware of how the angle has been measured in order to interpret its significance properly.

In dynamic contact angle studies, additional complications arise because the movement of the wetting line is not always a steady, continuous process. It is often observed that the movement is "jerky," with the drop or liquid front holding a position for a time and then jumping to a new configuration. This phenomenon is often referred to as a *stick--slip* process and is not fully understood as yet. It has also been observed that in dynamic systems, the values of θ_A and θ_R will vary as a function of the velocity of wetting line movement, with θ_A increasing with velocity and θ_R decreasing.

Obviously, contact angle measurements and their interpretation are not without their hidden pitfalls and blind alleys. However, because of the ease of making such measurements, the low cost of the necessary apparatus, and the potential utility of the concept, they should be seriously considered as a rapid diagnostic tool for any process in which wetting phenomena play a role.

Contact Angle Measurement Techniques

There exist a variety of simple and inexpensive techniques for measuring contact angles, most of which are described in detail in various texts and publications and will only be mentioned here. The most common direct methods (Figure17.3) include the sessile drop (a), the

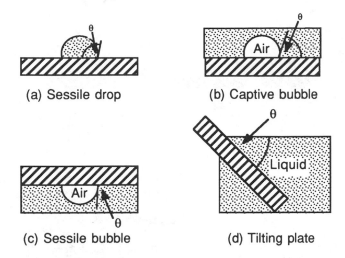

(a) Sessile drop   (b) Captive bubble

(c) Sessile bubble   (d) Tilting plate

Figure 17.3. Illustration of various laboratory methods for measuring contact angles.

captive bubble (b), the sessile bubble (c), and the tilting plate (d). Indirect methods include tensiometry and geometric analysis of the shape of a meniscus. For solids for which the above methods are not applicable, such as powders and porous materials, methods based on capillary pressures, sedimentation rates, wetting times, imbibition rates, etc, have been developed.

Contact Angle Hysteresis

When used with Young's equation and the other such relationships, the contact angle provides a relatively simple yet sensitive insight into the general chemical nature of a surface through such thermodynamic quantities as the work of adhesion. Unfortunately, as already mentioned, contact angles often exhibit hysteresis and cannot be defined unambiguously by experiment. It is always important to know as much as possible about the cleanliness, topography, homogeniety, etc, of a solid surface, as well as the purity and composition of the liquid employed, when attempting to interpret contact angle data.

Although the existence of contact angle hysteresis has been recognized for at least 100 years, the root of the "evil" has not always been understood. In addition to the physicochemical adsorption process already mentioned, which leads to differences in advancing and receeding contact angles, it is recognized that several physical and kinetic factors also contribute to the overall problem. Two of the most important, from a practical point of view, are discussed below.

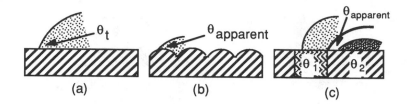

Figure 17.4. Illustration of various physical factors affecting the apparent contact angle of a liquid on solid surfaces: (a) "true" contact angle on a flat, homogeneous surface (θ_t); (b) apparent contact angle ($\theta_{apparent}$) on a rough surface of (a): $\theta_t < 90°$; (c) apparent contact angle on a composite surface where $\theta_1 > \theta_{apparent} > \theta_2$.

The Effects of Surface Roughness on Contact Angles and Wetting

The theoretical discussion of contact angle and wetting to this point has assumed implicitly that the solid surface in question is a smooth, ideal plane. In fact, of course, very few solid surfaces even begin to approach such a state. The finest polished glass surface, for example, will usually have asperities of 5 nm or more. Commonly encountered surfaces, will be much worse. The earliest quantitative attempt ot correlate the observed contact angle of a liquid on a solid with the surface roughness was that of Wenzel,[1] who proposed a thermodynamic relationship such that

$$\sigma_{12} \cos \theta = R_w (\sigma_{s2} - \sigma_{s1}) \tag{17.1}$$

where R_w is defined as the *surface roughness factor*, the ratio of the true and apparent surface areas of the solid (Figure 17.4). Defining the apparent contact angle as θ' and substituting for from eq. 17.1 yields

$$\cos \theta = R_w \cos \theta' \tag{17.2}$$

or

$$R_w = \cos \theta / \cos \theta' \tag{17.3}$$

Equation 17.3 may be taken as the fundamental definition of the effect of surface roughness on wetting and spreading phenomena.

Since surfaces having $R_w = 1$ are exceedingly rare, and highly polished surfaces usually have $R_w \geq 1.5$, the above relationships obviously has a great deal of relevence for exacting experimental work. Although the actual geometry of the surface roughness has no theoretical significance from a thermodynamic standpoint, in practice, the type of topography present may carry with it certain practical consequences. For example, if the surface, rather than being randomly rough, has pores, crevices, capillaries, or other structures that have their own characteristic wetting and penetration properties, the apparent contact angle will be affected by the thermodynamics and kinetics associated with such structures.

There have been developed a number of more detailed and sophisticated treatments of the effect of surface roughness on wetting phenomena; however, the utility of such approaches is limited by the necessity of having good data concerning the roughness of the surface in question, data that are not always easy or convenient to obtain. If there exists some consistency in surface roughness (as for some manufacturing process, for example) the ideas described above can still be employed with little concern for the details of surface topography, since R_w will presumably be a constant.

As a final note on the effects of surface roughness, examination of eq. 17.3 leads to a useful rule of thumb for some important applications of wetting and spreading phenomena; that is, if the "true" contact angle of a liquid (an adhesive, say) is less than 90° on the smooth surface, the angle will be even smaller on a rough surface. For a true contact angle > 90°, roughness will increase the apparent angle. Mathematically the situation can be described as

$$\text{for } \theta < 90°, \; \theta' < \theta$$

$$\text{for } \theta > 90°, \; \theta' > \theta$$

Practically, the above relationships indicate that if a liquid partially wets a surface, better wetting may be obtained if the surface is roughened in some way. Conversely, if wetting is not desired and a contact angle > 90° can be attained, the situation can be further improved by roughening.

Heterogeneous Surfaces

Roughness represents just one aspect of the effects of the nature of the solid surface on contact angles and wetting phenomena. A second potentially important factor is that of the chemical heterogeneity of the surface. Working from the basic approach of Wenzel, Cassie and Baxter[2] developed the following relationship relating apparent contact angle to the chemical composition of a surface

$$\cos \theta' = f_1 \cos \theta_1 + f_2 \cos \theta_2 \qquad (17.4)$$

where f_1 and f_2 are the fractions of the surface having inherent contact angles θ_1 and θ_2. Since $f_2 = 1 - f_1$, eq. 17.4 can be written in terms of one component. Theoretically, if the inherent contact angles of a test liquid on the homogeneous surfaces are known, then the composition of a composite surface can be determined from a simple contact angle measurement. Obviously, such an approach must be accepted as being rather qualitative, considering the pitfalls inherent in contact angle data. However, experiments employing specially prepared composite surfaces have shown that contact angle data can give results that agree reasonably well (± 15%) with more sophisticated surface composition data obtained using x-ray photoelectron spectroscopy.

Kinetic Aspects of Hysteresis

According to its definition, contact angle hysteresis should be concerned only with thermodynamic equilibrium situations and not with nonequilibrium kinetic events. However, because of experimental limitations, such as the time required for measurements, certain kinetic effects are almost unavoidable. Theoretically, such effects should fall under the heading of *kinetic contact angles* but are often reflected in "equilibrium" measurements.

Because the measurement of a contact angle must involve some movement of the wetting line, it is possible, or even probable, that the act of spreading of the liquid will displace certain surface equilibria that will not be reestablished over the time frame of the experiment. For example, the displacement of the second fluid may result in the establishment of a nonequilibrium situation in terms of the adsorption of the various components at the solid--liquid, solid--fluid 2, and liquid--fluid 2 interfaces. Time will be required for adsorption equilibrium to be attained, and it may not be attained during the time of the contact angle measurement if the transport and adsorption--desorption phenomena involved are slow. The kinetic effect may be especially significant for solutions containing surfactants, polymers, or other dissolved adsorbates.

A second potential kinetic effect may result from bulk interactions between the surface and the liquid. For example, if the liquid can penetrate the surface (eg, if the liquid can be absorbed as opposed to adsorbed), the rate of penetration may be so slow that the measured contact angle will not reflect the true equilibrium situation. Likewise, if the liquid swells the surface, the wetting line may lie on a ridge of swollen surface rather than on a flat surface, resulting in an error in θ' (Figure 17.5).

To summarize, while contact angle measurements represent a potentially powerful and practical tool for characterizing the nature and wettability of solid surfaces, variability leading to errors in interpretation can arise from various sources. That means that proper attention must be paid to experimental conditions, equilibria, solid and liquid purity, etc, to ensure the best possible data. Even then, interpretation must be done with the above-mentioned caveats in mind. Nevertheless, contact angle data should never be excluded from studies or processes in which

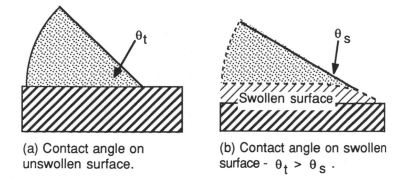

(a) Contact angle on
unswollen surface.

(b) Contact angle on swollen
surface - $\theta_t > \theta_s$.

Figure 17.5. Schematic illustration of the effect of surface swelling on
the apparent contact angle (θ_s).

wetting and spreading are involved.

THE THERMODYNAMICS OF WETTING

The basic framework for the application of contact angles and
wetting phenomena lies in the field of thermodynamics. However, in
practical applications it is often difficult to make a direct correlation
between observed phenomena and basic thermodynamic principles.
Nevertheless, the fundamental validity of the analysis of contact angle
data and wetting phenomena helps to instill confidence in its application
to nonideal situations.

Young's Equation

If one considers the three-phase system depicted in Figure 17.6, in
which the liquid drop is designated fluid 1, the surrounding medium fluid
2, and the solid surface S, then at equilibrium the contact angle θ will be
given by Young's equation as

$$\sigma_{12} \cos \theta^o = \sigma_{S2} - \sigma_{S1} \qquad (17.5)$$

where σ_{12}, σ_{S1}, and σ_{S2} are the interfacial tensions at the respective
interfaces. Although eq. 17.5 was originally proposed based upon a
mechanical analysis of the resultant forces at the three-phase contact
line it has since been derived rigorously on the basis of fundamental
thermodynamic principles.

While Young's equation provides a thermodynamic definition of
the contact angle, its experimental verification is prevented by the fact
that the values of σ_{S1} and σ_{S2} cannot be directly determined
experimentally. In this sense, the contact angle of a liquid on a solid
differs from that of a liquid on a second liquid since in the latter case all

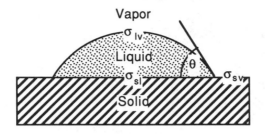

Figure 17.6. Schematic illustration of the mechanical equilibrium of surface forces leading to contact angle formation as given by Young's equation.

three interfacial tensions can be determined independently and the relationship can therefore be verified directly.

When the liquid of interest (1) is found to spread completely over the solid surface, $\theta = 0$ and eq. 17.5 reduces to

$$\sigma_{12} = \sigma_{S2} - \sigma_{S1} \qquad (17.6)$$

In terms of the mechanical (hydrostatic) equilibrium derivation of Young's equation, eq. 17.6 appears nonsensical, since the wetting line for which it describes the equilibrium does not exist. That is, there is no three-phase contact line at which the two fluid phases contact each other and the solid surface. Thermodynamically, however, one can show that the equation is eaxctly obeyed when spreading occurs.

If eq. 17.6 is rewritten to take into account the microscopic mechanism of spreading, the situation can be clarified. For example, spreading proceeds first by a molecular level spreading of liquid on the solid surface. If it is assumed that the second fluid (2) is air and the only component adsorbed at the S--2 interface is the vapor of liquid 1, eq. 17.6 can be written

$$\sigma_{12} = \sigma_S - \pi_{S2,1} - \sigma_{S1} \qquad (17.7)$$

where σ_S is the surface tension (energy) of the solid surface with no adsorbed molecules and $\pi_{S2,1}$ is the surface pressure of adsorbed liquid 1 at the S--2 (see also Chapter 8) interface so that

$$\sigma_S - \pi_{S2,1} = \sigma_{S2} \qquad (17.8)$$

If spreading occurs, the spreading pressure $\pi_{S2,1}$ will increase, reducing σ_{S2} until eq. 17.8 is exactly satisfied. At that point the thickness of the adsorbed film will be such that only the S--1 and 1--2 interfaces exist. Further thickening of the spread film may then occur by spreading of

Figure 17.7. Illustration of the autophobic effect: (a) initial spreading; (b) final spreading after orientation of adsorbed molecules at the interface.

liquid 1 on itself.

An interesting sidelight to the above concept is the phenomenon generally referred to as *autophobicity*. In that situation, as the spreading liquid is adsorbed onto the "bare" solid surface in the usual way (Figure 17.7). However, the molecules in this case are adsorbed with a specific molecular orientation relative to the solid and the spreading fluid. As a result, when a complete monolayer of adsorbed molecules is formed, the spreading liquid no longer "sees" a surface tension σ_S or σ_{S2}, but rather a new surface tension $\sigma_{1,O}$ (liquid 1, oriented). If the spreading liquid is a polar organic material such as an alcohol or carboxylic acid and the solid surface one that can interact strongly with the polar group (glass, mica, metals and oxides, etc) the orientation process will expose an essentially hydrocarbon surface (mostly ---CH_3 groups) with $\sigma_{1,O} \ll \sigma_S$ (or σ_{S2}). In order for eq. 17.5 to be satisfied under those conditions, cos θ must be < 1. That is, the spreading liquid will retract from the surface to produce a finite contact angle.

The Spreading Coefficient

Young's equation is usually found to be a very useful and adequate means of describing wetting equilibria in most circumstances. However, it is sometimes found useful to define another term that indicates from a thermodynamic point of view whether a given liquid--solid system will be wetting (θ = 0°) or nonwetting (θ > 0°). Such a term is the *spreading coefficient*, S.

For a spontaneous process (such as spreading) to occur, the free energy of the process must be negative. In terms of surface free energies, then, one can write the relationship

$$S_{S12} = \sigma_S - \sigma_{S1} - \sigma_{S2} \qquad (17.9)$$

where the subscript S12 refers to the *initial spreading coefficient* for the spreading of liquid 1 over the solid S in the presence of (or displacing) fluid 2. At first sight, the autophobic phenomenon mentioned above would seem to be exceptions to eq. 17.9. However, if a relationship like that in eq. 17.8 is invoked, then

$$S_{S12} = \sigma_{S2} + \pi_{S2,1} - \sigma_{S1} - \sigma_{12} \qquad (17.10)$$

and the autophobicity produced by the oriented monolayer adsorption of liquid 1 is easily accommodated.

An alternative to the relationship in eq. 17.10 is the so-called *equilibrium spreading coefficient*

$$S_{S12}{}^\circ = \sigma_{S2} - \sigma_{S1} - \sigma_{12} \qquad (17.11)$$

in which the solid surface tension σ_{S2} is now that of the solid with saturated adsorbed layer of fluid 2.

Classification of Wetting Processes

While the term *wetting* may conjure up a fairly simple image of a liquid covering a surface, from a surface chemical standpoint the situation is somewhat less clear cut. Classically, there are three types of wetting phenomena of importance: *adhesional*, *spreading*, and *immersional* wetting. The distinctions may seem subtle, but they can be significant from a thermodynamic and phenomenological point of view.

Adhesional wetting refers to the situation in which a solid, previously in contact with a vapor, is brought into contact with a liquid phase. During the process, a specific area of solid--vapor interface, A, is replaced with an equal area of solid--liquid interface (Figure 17.8a). The free energy change for the process is given by

$$-\Delta G = A \left(\sigma_{SV} + \sigma_{LV} - \sigma_{SL} \right) = W_a \qquad (17.12)$$

where the σ's refer to the solid--vapor (SV), liquid--vapor (LV), and solid--liquid (SL) interfacial tensions. The quantity in parentheses is the thermodynamic work of adhesion, W_a. From eq. 17.12, it is clear that any decrease in the solid--liquid interfacial energy σ_{SL} will produce an increase in the work of adhesion (and a greater energy decrease), while an increase in σ_{SV} or σ_{LV} will reduce the energy gain from the process.

Spreading wetting applies to the situation in which a liquid (L1) and the solid are already in contact and the liquid spreads to displace a second fluid (L2, usually air) as illustrated in Figure 17.8b. During the spreading process, the interfacial area between solid and L2 is decreased by an amount A, while that between the solid and L1 increases by an equal amount. The interfacial area between L1 and L2 also increases during the process. The change in interfacial area in each case will be the same, so that the total decrease in the energy of the system (for a spontaneous process), will be

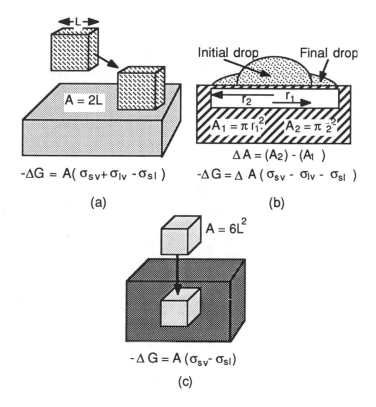

$-\Delta G = A(\sigma_{sv} + \sigma_{lv} - \sigma_{sl})$

(a)

$-\Delta G = \Delta A(\sigma_{sv} - \sigma_{lv} - \sigma_{sl})$

(b)

$-\Delta G = A(\sigma_{sv} - \sigma_{sl})$

(c)

Figure 17.8. Schematic illustration of important wetting processes: (a) adhesional, (b) spreading, and (c) immersional wetting.

$$-\Delta G = A(\sigma_{SL2} - \sigma_{SL1} - \sigma_{12}) \tag{17.13}$$

where σ_{12} is the interfacial tension between fluids 1 and 2. If the term in parentheses, defined as the spreading coefficient S, is positive, then L1 will spontaneously displace L2 and spread completely over the surface (or to the greatest extent possible). If S is negative, the spreading process as written will not proceed spontaneously. Rewriting Young's equation in the form

$$\cos\theta = (\sigma_{SL2} - \sigma_{SL1})/\sigma_{12} \tag{17.14}$$

and combining with that for the spreading coefficient S gives

$$S = \sigma_{12}(\cos\theta - 1) \tag{17.15}$$

It is clear from eq. 17.15 that for $\theta > 0$, S cannot be positive or zero, and spontaneous spreading will not occur.

The third type of wetting, *immersional wetting* , results when a solid substrate not previously in contact with a liquid is completely immersed in liquid L1, displacing all of the solid--L2 interface (Figure 17.8c). In this case, the free energy change at equilibrium is determined by two factors, a component related to the solid--air interface $A \sigma_{SL2}$ and that of the new solid--liquid interface $A \sigma_{SL1}$, where A is the total surface area of the solid. The free energy change is then given by

$$- \Delta G = A (\sigma_{SL2} - \sigma_{SL1}) \qquad (17.16)$$

From the above relationships for wetting processes, it is clear that the interfacial energies between a solid and any contacting liquid, and the interfacial tension between the liquid and the second fluid (usually air), control the manner in which the system will ultimately perform. The ability to alter one or several of those surface energy components makes it possible to manipulate the system to attain the wetting properties desired for a given system. It is generally through the action of surfactants at any or all of those interfaces that such manipulation is usually achieved. A more specific discussion of the role of surfactants in the alteration and control of the wetting process is given below.

Additional Useful Thermodynamic Relationships for Wetting

Although the mathematical relationships encountered in wetting phenomena are usually quite simple, they are found to be very useful in many practical applications. Their combinations and variations have given rise to still more relationships, which further expand their utility without expanding the amount of information necessary for their application.Two thermodynamic relationships that can be useful in the analysis of wetting and spreading phenomena are the works of cohesion and adhesion.

The *work of cohesion* , W_c, has been defined as the reversible work required to separate two surfaces of unit area of a material with surface tension σ, given simply as

$$W_c = 2\sigma \qquad (17.17)$$

The related quantity the *work of adhesion* , W_a, was similarly defined as the work required to separate unit area of interface between two different materials or phases to leave a "bare" solid surface.

$$W_{aS12} = \sigma_S + \sigma_{12} - \sigma_{S1} \qquad (17.18)$$

The subscript S12 is employed here to emphasize the fact that, as the two surfaces (S and liquid 2) are separated, two new interfaces are formed --- S--2 and 1--2. If the idea of the spreading pressure $\pi_{S2,1}$ is included, eq. 17.18 becomes

$$W_{aS12} = \sigma_{S2} + \pi_{S2,1} + \sigma_{12} - \sigma_{S1} \qquad (17.19)$$

which takes into account the energetic effect of molecules of the spreading liquid being adsorbed on the solid surface ahead of the moving three-phase boundary.

Working from eq. 17.19, one can define a new quantity, the *reversible work of adhesion*, which includes $\pi_{S2,1}$

$$W^{\circ}_{aS12} = \sigma_{S2} + \sigma_{12} - \sigma_{S1} \qquad (17.20)$$

Comparing eqs. 17.19 and 17.20 one can see that $W_{aS12} = W_{aS12}{}^{\circ}$ only when $\pi_{S2,1} = 0$. In strongly interacting systems in which $\pi_{S2,1}$ is significant, $W_{aS12} \gg W_{aS12}{}^{\circ}$. Equations 17.18 and 17.20 can be useful in situations in which fluid 2 is a liquid that may also wet the solid (eg, *competitive wetting*) or when it contains some component (a surfactant, for example) that can be adsorbed on the solid.

Some additional useful relationships based on the concepts already introduced include:

$$S_{S12} = \sigma_{12} (\cos \theta^{\circ} - 1) + \pi_{S2,1} \qquad (17.21)$$

$$S_{S12}{}^{\circ} = \sigma_{12} (\cos \theta^{\circ} - 1) \qquad (17.22)$$

$$W_{aS12} = \sigma_{12} (1 + \cos \theta^{\circ}) + \pi_{S2,1} \qquad (17.23)$$

$$W_{aS12}{}^{\circ} = \sigma_{12} (1 + \cos \theta^{\circ}) \qquad (17.24)$$

and

$$\cos \theta^{\circ} = -1 + 2 (W^{\circ}_{aS12}/W_{c12}) \qquad (17.25)$$

Analysis of eq. 17.25 shows that if $W_{c12} = W_{aS12}{}^{\circ}$, $\theta = 0^{\circ}$; if $W_{c12} = 2 W_{aS12}{}^{\circ}$, $\theta = 90^{\circ}$; and finally, if $W_{c12} > 2 W_{aS12}{}^{\circ}$, $\theta > 90^{\circ}$. Phenomenologically, this says says that the contact angle observed for a given liquid--solid system will reflect the competition between the drive of the liquid to "stick to its own kind," and the pull of the solid surface. It also says that it is impossible for a liquid--solid--vapor system to have a 180° contact angle since that would require that ($W_{aS12}{}^{\circ}/W_{c12}) = 0$, which is physically impossible. Such a situation would require, according to the definition of $W_{aS12}°$, that there be no net attractive interaction between liquid 1 and the solid in the presence of a noncondensed third phase, a very difficult event to imagine given the nature of dispersion forces.

CONTACT ANGLES AND THE CALCULATION OF SOLID SURFACE ENERGIES

It is generally a rather simple task to measure the surface tension of a liquid. However, from the standpoint of acquiring a better theoretical handle on surface and interfacial process (including wetting) it would be useful if one could calculate from first principles the surface tension of a system based on an understanding of the intermolecular forces giving rise to it. If such a calculation could be made for a pure liquid (σ_{LV}), it could conceivably be carried out for more complex liquid--liquid (σ_{12}), and solid--liquid (σ_{SL}) systems, opening the way for analysis of wetting systems requiring terms that are not experimentally accessible.

Using a statistical treatment of the variation of local intermolecular forces as the interface is traversed from the liquid to the vapor phase, it is possible to calculate the surface tension of a simple liquid (eg, argon) that agrees well with experiment. However, such exact methods become quite complex or impossible for calculating surface tensions for practical systems.

As is usually the case when theoretical and experimental science meet, it is necessary to make some simplifying assumptions in order to apply theories to practical systems. Good and Girifalco[3] proposed an empirical approach to the problem based on the Berthelot principle that the interaction constant between two different surfaces or particles will be the geometric mean of the interaction constant for the individual surfaces or particles, an approach already introduced in Chapter 4. Good and Girifalco suggested that the work of adhesion between two different liquids could be expressed as a similar function

$$W_{a12} = \Phi(W_{c1} W_{c2})^{1/2} \qquad (17.26)$$

where the constant Φ takes into account differences in the molecular size and polar content of the two materials involved. Combination of eqs. 17.17, 17.18, and 17.26 gives

$$\sigma_{12} = \sigma_1 + \sigma_2 - 2\Phi(\sigma_1\sigma_2)^{1/2} \qquad (17.27)$$

For nonpolar liquids with $\Phi \approx 1$, results agree reasonably well with experiment. For dissimilar substances, such as water and alkanes or water and mercury, values of Φ between 0.32 and 1.15 must be used to obtain agreement.

The Good and Girifalco approach has been extended to the use of contact angles in the computation of surface tension values for solid--liquid interfaces. Considering a system where the fluid 2 is either vapor or air (in which case it can be ignored), and combining with Young's equation, one obtains the expression

$$\cos \theta = -1 + 2\Phi(\sigma_S/\sigma_1)^{1/2} - \pi_{S,1}/\sigma_1 \qquad (17.28)$$

Equation 17.28 is found to give reasonable values of σ_S for nonpolar solids and represents a potentially useful tool for characterizing solid surfaces, empirically, at least.

The logical next step in the process of extending the utility of theory to practical systems is to include polar molecular interactions. For this step, Fowkes[4] suggested that the intermolecular forces contributing to surface and interfacial tensions, and subsequent phenomena such as wetting, could be broken down into independent and additive terms. For example, a polar molecule such as an ester would have two terms making up its surface tension --- dispersion forces (d) and dipolar interactions (p) --- so that

$$\sigma = \sigma^d + \sigma^p \qquad (17.29)$$

where σ^d and σ^p are the dispersion and dipolar contributions to the total surface tension. Application of the principle of eq. 17.26 to eq. 17.29 produces a reasonable approximation for the work of adhesion for interactions involving only dispersion and dipole forces.

$$W_{a12} = 2(\sigma_1\sigma_2)^{1/2} [(d_1d_2)^{1/2} + (p_1p_2)^{1/2}] \qquad (17.30)$$

where d and p represent the respective fractions of dispersion and polar components contributing to the energy densities of the adjacent phases in which $p = (1 - d)$.

For nonpolar materials involving only dispersion forces such as simple alkanes $d = 1$ and $\sigma = \sigma^d$. Fowkes developed an expression for the dispersion force contribution to the total work of adhesion of the form

$$W_{a12}{}^d = 2(\sigma_1{}^d\sigma_2{}^d)^{1/2} \qquad (17.31)$$

If one combines the suggestion of Good and Girifalco with the modifications proposed by Fowkes one may rewrite eq. 17.30 in the form

$$W_{a12} = W_{a12}{}^d + W_{a12}{}^p \qquad (17.32)$$

in which the work of adhesion is divided into dispersive and polar components. The identification of the separate components of surface tension (σ^d and σ^P) with the corresponding works of adhesion has important theoretical and practical consequences.

Equations similar to 17.27 and 17.28 can be written for solid--liquid interfaces and, when combined with Young's equation, give an experimentally accessible relationship between the contact angle a liquid will make on a solid surface and the attractive components of the surface tensions of the two phases.

$$\cos \theta = -1 + 2[(\sigma_s{}^d\sigma_1{}^d)^{1/2} / \sigma_1] - (\pi_{s,1}/\sigma_1) \qquad (17.33)$$

The great potential utility of eq. 17.33 lies in the fact that, since all of the variables are accessible through reasonably simple experiments, a solid interacting by dispersion forces alone (eg, a pure hydrocarbon wax or polymer) can be used to determine the dispersion component of the surface tension of a liquid. Alternatively, a standard nonpolar liquid can be used to characterize the surface tension of a solid, or if the "clean" solid has been previously characterized, the contact angle can be used to assist in determining the nature of a surface contaminant.

When using eq. 17.33, it must be kept in mind that it was derived primarily for dispersion and dipolar interactions. It has been used with some success for strongly hydrogen bonding systems (a special class of dipolar interaction), but with greater variability than found for simpler systems. Other types of intermolecular interactions are also possible, including ionic interactions and metallic forces. Whether such forces can be reasonably incorporated into the Good-Girifalco-Fowkes theory via the geometric mean approach remains an active question. More complex equations have been proposed to include such interactions. Such extended equations have been used to investigate hydrophilic and biological surfaces with some success. Because of the relationship between intermolecular interactions, cohesive energy densities, and solubility parameters, solubility parameters have also been used in the context of wetting phenomena. For most practical systems, the Good-Girifalco-Fowkes approach remains one of the best tools for estimating the interactions at solid--liquid interfaces. A workable alternative, however, is that of the *critical surface tension* discussed in the next section.

The Critical Surface Tension of Wetting

Before the introduction of the theory of Good and Girifalco, Zisman and coworkers developed a useful and practical systematic method for characterizing the "wettability" of solid surfaces.[5] The system is based on the observation that for solid surfaces having a surface tension (σ_s) < 100 mN m^{-1} (generally classed as "low-energy surfaces"), the contact angle formed by a drop of liquid on the solid surface will be primarily a function of the surface tension of the liquid, σ_{12} (where phase 2 is air saturated with the vapor of liquid 1). They found, in particular, that for a given solid

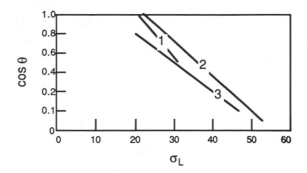

Figure 17.9. The relationship between contact angle and surface tension for several classes of liquids on teflon: curve 1 --- nonpolar liquids; curve 2--- alkyl halides; curve 3 --- miscellaneous polar liquids.

surface and a homologous series of related liquids (eg, alkanes, dialkyl ethers, alkyl halides,) cos θ was an approximately linear function of σ_{12}. The relationship for several liquid types is illustrated in Figure 17.9. For nonpolar liquids, the relationship holds very well, while for high surface tension, polar liquids, the correlation begins to break down somewhat and the line begins to curve.

The general limitation of the technique to the so-called low-energy surfaces must be made because such materials as metals, metal oxides, or ionic solids, which have surface free energies in the hundreds and thousands, are almost always covered with an adsorbed layer of a low-energy substance such as water or oils from the atmosphere. Under rigorously controlled experimental conditions, the technique may be applied to such materials, but interpretation of the results can be difficult .

From a plot of cos θ vs σ_{12} one can obtain the value of the liquid surface tension at which cos θ = 1, a value which has been termed the *critical surface tension of wetting* , σ_c. It is defined as the surface tension of a liquid that would just spread on the surface of the solid to give complete wetting. In other words, if $\sigma_{12} \le \sigma_c$, the liquid will spread; if $\sigma_{12} \ge \sigma_c$, the liquid will form a drop with a nonzero contact angle. Typical values of σ_c for commonly encountered materials are given in Table 17.1.

While σ_c is an empirically determined value, depending to some extent on the nature of the liquids used for its determination, attempts have been made to identify it with such theoretical terms as σ_s or $\sigma_s{}^d$. According to the Good-Girifalco equation (eq. 17.27) , for $\pi_{s,1} = 0$ and cos θ = 1,

$$\sigma_{12} = \sigma_1 = \Phi^2\sigma_s = \sigma_c \qquad (17.34)$$

Table 17.1. Values of the critical surface tension of wetting (σ_c, mN cm^{-1}) for a variety of materials.

Solid	σ_c	Solid	σ_c
Teflon	18	Polytrifluoroethylene	22
Polyvinylidene Fluoride	25	Polyvinylidene chloride	40
Polyvinyl fluoride	28	Polyvinyl chloride	39
Polyvinyl alcohol	37	Polyethylene	31
Polystyrene	33	Nylon 6,6	46
Polyethyleneterephthalate	43		

For nonpolar liquids and solids, $\Phi \approx 1$, therefore $\sigma_s = \sigma_c$. Starting with the Fowkes postulate

$$\sigma_{12} = \sigma_1 = (\sigma_s{}^d \sigma_1{}^d)^{1/2} = \sigma_c \qquad (17.35)$$

Therefore, when $\sigma_1{}^d = \sigma_{12}$, then $\sigma_c = \sigma_s{}^d$. For nonpolar solids, for which only dispersion interactions occur, the same might be expected to hold regardless of the nature of the liquid employed. However, experiment has shown that the condition of $\pi_{s,1} = 0$ is unlikely to hold for systems in which θ is near or equal to zero.

Young's equation can also be employed to test the theoretical significance of the σ_c concept. If $\cos \theta = 1$, then

$$\sigma_c = \sigma_{12} = \sigma_s - \pi_{s,1} - \sigma_{s1} \qquad (17.36)$$

From eq. 17.36 it is clear that $\sigma_c = \sigma_s$ only if both σ_{s1} and $\pi_{s,1} = 0$. It must be concluded, then, that the value of σ_c is not a characteristic property of the solid alone but of the solid--liquid combination. That does not, however, greatly diminish the practical utility of the concept as a method for characterizing the wettability of a surface, which was, after all, its purpose all along.

While the critical surface tension concept is attractive as a practical tool for characterizing solid surfaces, it does have several drawbacks. Of particular significance is the fact that the procedure requires the use of several liquids of different surface tensions in which $\sigma_{12} > \sigma_c$. That means that a relatively large number of measurements are required for each solid surface. In addition, it is sometimes difficult to obtain a sufficient quantity of pure liquids with the required range of σ_{12}, leading many investigators to employ liquid mixtures (eg, alcohols or glycols in water) or solutions of surfactants as their test liquids. While such a procedure is attractive, making available a wide range of surface tensions, from 72 mN m^{-1} for water to about 20 mN m^{-1} for solutions of some fluorinated surfactant solutions, it carries with it its own pitfalls.

While the solution method may give results in agreement with pure hydrocarbon liquids for surfaces such as paraffin or polytetrafluoroethylene (PTFE), the agreement is not general. For example, while solutions of hydrocarbon surfactants on PTFE and polyethylene give good agreement, solutions of fluorocarbon surfactants give good agreement for PTFE, but deviate significantly on hydrocarbon surfaces. Low values of σ_c may also be obtained for solutions on slightly polar surfaces such as polyesters, polyacrylates, or polystyrenes. The generally accepted reason for the problems with surfactant solutions is that surfactant molecules will adsorb at the σ_{s1} and σ_{s2} interfaces and alter the nature of the surface being measured. The general topic of the effects of surfactants on the wetting of solid surfaces is treated in a following section.

From a practical applications point of view, both the critical surface tension approach and the use of contact angles with the Good-Girifalco-Fowkes equation represent handy tools for the characterization of the wettability, and therefore something of the chemical nature, of solid surfaces. The choice of technique is basically one of preference and convenience.

THE KINETICS OF WETTING

So far the discussion of wetting and contact angles has been essentially limited to "equilibrium" situations. In many practical applications, the wetting phenomena of interest are "dynamic" in nature, involving a moving wetting line at which equilibrium is never attained. The contact angle of a moving wetting line is generally called a *dynamic contact angle* . It is generally found that the dynamic contact angle for a given system will differ from the equilibrium value, even for very slow rates of movement, with the difference usually being velocity dependent.

Although exceptions have been reported, it is generally true that dynamic contact angles are dependent on both the direction and speed of movement of the wetting line; that is, the velocity of the movement. Experimentally, is it observed that advancing contact angles increase and receding angles decrease as the velocity of the wetting line increases. When thought of in terms of the contact angle hysteresis discussed previously, it indicates that the dynamic contact angle is a function of thermodynamically irreversible processes. In the case of the static angle, hysteresis may at times be attributed to the movement of the wetting line between the so-called metastable states. In the dynamic case, it must be assumed that the wetting line never reaches even a metastable "equilibrium" and irreversibility results from the constant drive toward an equilibrium that can never be reached.

A "typical" curve shape for the relationship between dynamic contact angle and wetting velocity v is shown in Figure 17.10. It is

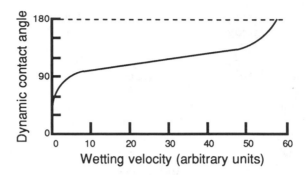

Figure 17.10. Hypothetical curve illustrating the dependence of the dynamic contact angle on the velocity of the wetting line.

generally found that at low wetting rates, the contact angle θ_d is a rapidly changing function of velocity, v. As the velocity increases, the slope of the curve decreases until, at still higher values of v, it again changes rapidly to approach 180°. The velocity at which $\theta_d = 180°$ is often denoted as v_{180}. At still higher velocities, it is observed that the wetting line becomes irregular, often taking on a regular sawtooth shape (Figure 17.11a). Air becomes entrained at the trailing vertices leading to a loss of contact between wetting liquid and solid surface. In a practical coating process, such an event obviously leads to the production of a spotty, irregular coating which is unlikely to be of use. A similar effect results in dewetting operations in which the fluid being displaced from the solid surface stops being displaced in a smooth process but begins to be entrained by the displacing fluid as drops or rivulets (Figure 17.11b).

The practical aspects of dynamic contact angle phenomena, then, center around determining the maximum wetting rates that can be attained before entrainment or wetting failure occurs, and how a system can be modified to increase that velocity, since in many if not most cases, speed is money. Liquid coating operations are obviously impacted by the dynamic restraints of a system. If one can increase the maximum speed of coating a substrate from, say, 200 to 400 m min^{-1}, productivity gains will be impressive. Less obviously, petroleum recovery by various flooding techniques is affected because, if the flooding fluid is injected at a velocity that surpasses v_{180} for the system, petroleum may be trapped by the flooding fluid resulting in a loss in recovery efficiency and ultimately total reservoir yield.

Factors Affecting Dynamic Wetting Phenomena

It is generally observed that the main factors affecting v_{180} for a given solid substrate are the viscosity and surface tension of the coating liquid. The dynamic contact angle at a given value of v is found to increase with increasing viscosity, η, and decreasing surface tension, σ.

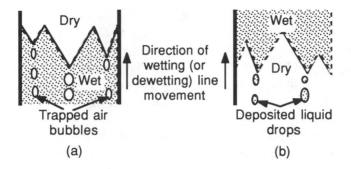

Figure 17.11. Schematic representation of wetting line shapes at wetting velocities greater than a critical value: (a) sawtooth wetting line with the inclusion of air bubbles in the coating; (b) dewetting with the deposition of residual liquid drops in the dewetted area.

In quantitative terms, the relationship between θ_d and those two values is correlated with the *capillary number*, C_a

$$C_a = (\eta v / \sigma) \tag{17.37}$$

so that θ_d always increases with C_a.

The effect of viscosity can be rationalized somewhat in that viscous forces would be expected to work against wetting by slowing the rate at which the wetting line approaches its equilibrium (or metastable) position. The effect can be seen, for example, in the fact that even for "static" contact angles, the rate at which the contact angle of viscous liquids such as molten polymers change is a function of viscosity. In general, then, viscous forces tend to oppose wetting, so that higher viscosities lead to an increased velocity dependence of θ_d and a smaller v_{180}.

The observed effect of surface tension is less obvious. For static or equilibrium wetting, it is usually found that a lower liquid surface tension will improve wetting (ie, reduce $\theta°$). In the dynamic case, however, the velocity dependence of θ_d, and therefore v_{180}, for liquids of the same viscosity but different surface tensions was in the direction of greater dependence on v with lower σ.

It has been suggested that the role of surface tension forces in the dynamic wetting process may be represented by

$$F_w = \sigma_{12} (\cos \theta° - \cos \theta_d) \tag{17.38}$$

where F_w is the nonequilibrium surface tension force acting at the wetting line in the direction of wetting (Figure 17.11), which must be a constant for a given set of circumstances. The work performed by F_w, then, is the irreversible work done by surface forces leading to wetting at some finite velocity, as opposed to infinitely slowly. According to eq. 17.38, if σ_{12} is decreased, θ_d must increase to maintain a constant value of F_w. On the other hand, reducing σ_{12} may also alter $\theta°$, so that the overall change may be more complicated than the simple relationship given by eq. 17.38.

If the reduction in surface tension for the liquid is brought about by the addition of a surfactant, the net affect on θ_d will depend of levels and rates of adsorption of the surfactant molecules at the various interfaces involved --- 1--2, S--1, and S--2. Obviously, an analysis of such a complex situation becomes very involved (or impossible), and experiment and experience become the best tools of the trade.

COMPETITIVE WETTING

The above discussion of wetting phenomena was restricted to the situation in which fluid 2 was air. In many practical situations, ie, detergency, petroleum recovery, the second fluid will also be a liquid. Not surprisingly, such systems may exhibit even more complex behavior than those having one vapor phase. Qualitatively, one can analyze the situation as follows: at the interface between water and a nonpolar solid in the presence of a nonpolar liquid (SW2), one expects that the work of adhesion between water and solid, $W_{aSw2}°$ will be significantly smaller than that between the nonpolar liquid and the solid, $W_{aS2w}°$. The reason for that, of course, is that the only attractive interactions possible between water and solid are the dispersion forces, which will be of similar magnitude for each side of the system (≈ 22 mN m^{-1} for water and 24--28 mN m^{-1} for a nonpolar hydrocarbon solid). In comparison, the internal binding in water, including hydrogen bonding, etc, will be much greater ($W_{cw} = 2\sigma_w \approx 144$ mN m^{-1}). As a result of this balance (or more accurately, imbalance) of forces, $\theta_w°$ will be quite large, values of well over 110° being common.

Beginning with the basic definition of the reversible work of adhesion (eq. 17.20) and the Fowkes relationship (eq. 17.31) one can place the above situation on a more quantitative footing, yielding the relationship

$$W_{aSw2}° = 2[\sigma_2 - (\sigma_w{}^d \sigma_2)^{1/2} - (\sigma_s{}^d \sigma_2)^{1/2} + (\sigma_s{}^d \sigma_w{}^d)^{1/2}] \quad (17.39)$$

If eq. 17.39 is evaluated for the system water-tetradecane-PTFE, a value of $W_{aSw2}° = 0.4$ mJ m^{-2} is obtained. For the same system, $W°_{aS2w} = 99.8$ mJ m^{-2}. Clearly, any wetting process that requires the displacement of an oil from a nonpolar solid surface by an aqueous solution must work against a considerable thermodynamic barrier.

Of particular practical importance in that sense is the primary mechanism of detergency for oily soils. In that case, the main role of the detergent solution is to displace or "roll up" the oily soil from the solid surface so that it can be more easily and completely removed from the surface by mechanical action. Obviously, for a nonpolar surface, such action must be limited by the above balance of adhesive and cohesive forces. Due to the complexity of the situation it is difficult to generalize as to how or even if, the addition of a surfactant will improve a given situation. (That point is discussed in the following section.) However, one can venture the following propositions:

1. If significant surfactant adsorption occurs at the oil--solid interface, displacement of the oil by aqueous solution will be hindered.
2. If little or no adsorption occurs at the solid--water interface, detergency will be adversely affected.
3. The most effective surfactants will be those which lower σ_{12} the most, while not coming into conflict with statements 1 and 2.

In other words, the best surfactant from the point of view of detergency should be one that adsorbs well at the solid-water and water--oil interfaces, but not at all at the solid--oil interface.

THE EFFECT OF SURFACTANTS ON WETTING PROCESSES

Having mentioned several times the use of surfactant solutions in wetting studies, we now consider specifically some of the effects their presence can have on contact angles and wetting. The action of surfactants derives from their adsorption at the various interfaces and the resultant modification of interfacial tensions. In terms of the Gibbs equation, the relationship between the specific adsorption of a solute, Γ, and surface tension is given by

$$(d\sigma / d \ln c)_{T,P} = -RT\ \Gamma \tag{17.40}$$

where T, P, R, and c have their usual significance. Looking again at Young's equation

$$\cos \theta^{\circ} = (\sigma_{S2} - \sigma_{S1}) / \sigma_{12} \tag{17.41}$$

one can see that θ° will decrease if either σ_{S1} or σ_{12} or both are reduced and σ_{S2} remains essentially unchanged. The effect of such changes will be greater if σ_{S2} is larger, that is, if the second fluid in the system is air. The contact angle will increase with surfactant addition only if the surfactant is adsorbed at the S--2 interface. Such a situation requires some sort of transport mechanism for carrying surfactant from the solution to that interface. For most surfactants of low volatility, such a mechanism is not readily available when fluid 2 is a vapor. If the second fluid is a liquid, transport of surfactant from liquid 1 through liquid 2 can result in significant adsorption at the S--2 interface. For more mobile surface-active materials such as alcohols, molecular diffusion may be sufficient. That is, the mechanism already mentioned leading to the phenomenon of autophobicity.

A general relationship between contact angle, surfactant concentration, and specific adsorption can be obtained by differentiating Young's equation with respect to $\ln c$

$$d(\sigma_{12} \cos \theta°)/d \ln c] = (d\sigma_{S2}/d \ln c) - (d\sigma_{S1}/d \ln c) \quad (17.42)$$

Combining with the Gibbs equation gives

$$\sigma_{12} \sin \theta° (d\theta°/d \ln c) = RT (\Gamma_{S2} - \Gamma_{S1} - \Gamma_{12} \cos \theta°) \quad (17.43)$$

In eq. 17.43, $\sigma_{12} \sin \theta°$ will always be positive, so that $(d\theta°/d \ln c)$ must always have the same sign as the right-hand side of the equation. Using that relationship, one can define three situations for changes in contact angle and wetting:

1. The addition of surfactant lowers $\theta°$ and improves wetting. This situation corresponds to the inequality $\Gamma_{S2} < \Gamma_{S1} - \Gamma_{12} \cos \theta°$.
2. The addition of surfactant increases $\theta°$ and dewetting occurs. In that case it must be that $\Gamma_{S2} > \Gamma_{S1} - \Gamma_{12} \cos \theta°$.
3. If $\Gamma_{S2} = \Gamma_{S1} - \Gamma_{12} \cos \theta°$, the addition of surfactant has no net effect on $\theta°$ and wetting is unaffected.

In some practical situations it is found that the effect of surfactant addition on wetting is variable, with behavior 1 being observed in one concentration range and 2 in another. Such an example will be discussed below. In general, however, one finds that situations 1 and 3 are most commonly encountered in systems where the solid substrate is a low-energy, nonpolar material. Situation 2 is usually observed only with higher energy, more polar substrates.

In some cases, one finds that solutes that adsorb strongly at solid--liquid interfaces, are not as strongly adsorbed at the liquid--fluid (1--2) interface. Such materials, including many polymers, will affect wetting depending on how the adsorbed layer interacts with liquid 1. If the adsorbing polymer presents a lower energy surface to the liquid, dewetting will be observed; if a higher energy surface is developed, improved wetting will result. In practice, it is often found that in order to coat a low energy polymeric substrate (a polyester, for example) effectively with an aqueous solution, it is necessary first to apply a very thin layer --- a primer or *mordant* coating --- of a more polar polymer (eg, gelatin or a lightly carboxylated vinyl polymer) to improve wetting and adhesion of the coating. Alternatively, one can modify the surface by chemical etching or corona discharge to improve wetting and adhesion.

Surfactant Effects on Nonpolar Surfaces

When one considers the effects of low molecular weight surfactants on wetting, it is helpful to divide the subject into two regimes: the effect on nonpolar surfaces and that on polar surfaces, the reason being that, for nonpolar surfaces it is often possible to assume that adsorption at the S--2 interface will be negligible. If surfactant adsorption at the S--2 interface is small (eg, $\Gamma_2 \approx 0$), as is usually observed for the situation where fluid 2 is air, eqs. 17.41 and 17.42 indicate that

$$d\sigma_{12} \cos \theta^\circ/d \ln c) = - (d\sigma_{S1} / d \ln c) = RT \, \Gamma_{S1} \qquad (17.44)$$

Theoretically, one can determine the adsorption at the S--1 interface from the change in θ° with surfactant concentration. Using the Gibbs equation, it is possible to determine the adsorption of surfactant at the 1--2 interface (from $d\sigma_{12}/d \ln c$). It is found experimentally that for most hydrocarbon surfactants on nonpolar surfaces, the adsorption at S--1 is the same as that at the 1--2 interface, or that $\Gamma_{S1} \approx \Gamma_{12}$.

A more quantitative picture of the situation can be obtained by looking at the effect of variations in σ_{12} on the reversible work of adhesion, W_{aS12}°. Differentiating eq. 17.20 by σ_{12} yields

$$(dW_{aS12}^\circ/d\sigma_{12}) = (d\sigma_{s2}/d\sigma_{12}) + 1 - (d\sigma_{s1}/d\sigma_{12}) \qquad (17.45)$$

Assuming that adsorption at the S--2 interface is zero, and setting $(d\sigma_{S2}/d\sigma_{12}) = 0$ and $(d\sigma_{S1}/d\sigma_{12}) = 1$, the result is

$$(dW_{aS12}^\circ/d\sigma_{12}) = 0 \qquad (17.46)$$

That is, the reversible work of adhesion will be independent of surfactant concentration. Based on the above assumptions, a number of interesting relationships can be derived relating the effects of surfactant concentration on wetting phenomena.

It should be kept in mind that the above ideas are based on the assumption that $\Gamma_{S2} = 0$, which is probably not true in some cases. However, it can be shown that if Γ_{S2} is significant, but still proportional to σ_{12} then useful linear relationships between, say, W_{aS12}° and σ_{12} can be obtained.

A variation of eq. 17.44 useful for predicting whether the addition of a surfactant will improve wetting is to differentiate Young's equation and combine with the Gibbs equation to yield

$$(d\sigma_{12} \cos \theta°/d\sigma_{12}) = (\Gamma_{S2} - \Gamma_{S1})/\Gamma_{12} \qquad (17.47)$$

If $\Gamma_{S2} = 0$ then a plot of $\sigma_{12} \cos \theta°$ vs σ_{12} will have a slope of $-(\Gamma_{S1}/\Gamma_{12})$. From eq. 17.47, it can be seen that if $\Gamma_{S1}/\Gamma_{12} = 1$, complete wetting of a surface should occur if the surfactant is capable of lowering σ_{12} to σ_c or below. If $\Gamma_{S1}/\Gamma_{12} = 0$ (eg, no surfactant adsorption at the S--1 interface) then complete wetting cannot be obtained for any value of σ_{12}. Finally, if $\Gamma_{12}/\Gamma_{S1} = 0$ (no adsorption at the 1--2 interface) complete wetting can occur with no change in σ_{12}. Obviously, in order for wetting improvement to be obtained with surfactant addition, it is necessary for adsorption to occur at the S--1 interface.

Experimentally, it is found that few systems adhere completely to the "pure" behavior described by eq.17. 47. As pointed out in Chapter 9, many surfactant--solid combinations can show a complex pattern of adsorption, resulting in equally complex patterns of wetting. In any case, with a little thought, the qualitative concepts related to adsorption and wetting can serve as useful guides to understanding the possible role of surfactants in controlling wetting processes.

Surfactants and Wetting on Polar Surfaces

When one extends the above discussion to include more polar solid surfaces, several things can be expected to change as a result of the possibility of additional interactions between the aqueous solution and the polar surface. For present purposes, one can include as polar surfaces not only ceramics and minerals, but also polar organic solids. The importance of being able to modify the wetting properties of such solids is seen in many important industrial processes including detergency, petroleum recovery, mineral ore flotation, the wetting of powders and pigments prior to dispersion, and the wetting of stone by road tars.

Inorganic polar solids (eg, minerals, ceramics, metals and metal oxides) generally have relatively high surface energies, ranging from a few hundred to several thousand mJ m^{-2}. Organic polar solids, on the other hand, normally have surface energies of 30--50 mJ m^{-2}. As a result, the inorganic materials generally are completely wetted by high surface tension liquids such as water, while the organics are only partially wetted. The situation is complicated by the fact that strongly polar surfaces can undergo specific interactions with polar liquids which can alter their wetting characteristics once liquid--solid contact has taken place. A good example of such a situation is that of the autophobic effect mentioned previously.

When surfactants are added into the equation, wetting can become even more complicated due to the many specific interactions that can occur between surfactant and solid, surfactant and water, and surfactant and oil. The exact effect of a given surfactant on a system will be determined by the degree and mode (ie, orientation) of its adsorption at the various interfaces, and the reversibility of that adsorption. Some of the factors that affect those variables include:

1. The degree and nature of the polarity of the surface (dipole moment, polarizability, etc)
2. The presence and nature of surface charges (charge density, degree of ionization, nature of charge-determining ions, etc)
3. The pH and ionic strength of the aqueous solution, both of which may affect charge density, ionization of the surface, ionization of the surfactant, etc
4. The presence of ions that may specifically interact with the solid surface or surfactant molecules, resulting in the formation of chelates or insoluble salts, such as Al^{3+}, Ca^{2+}, Cu^{2+}
5. The degree of surface hydration

While surfactant adsorption on weakly polar surfaces such as polyesters and polymethylmethacrylate is often sufficiently nonspecific to allow the use of models based on nonpolar soilds, interactions with more polar ionic surfaces tend to be more complicated. Even those cases, however, can be successively analyzed in terms of the above concepts, so that the modification of wetting characteristics by surfactant adsorption can be predicted with reasonable confidence, possibly saving a great deal of time (= money) in various processes.

Of particular interest is the case of surfactant adsorption onto

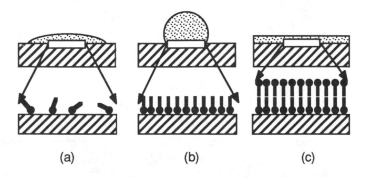

(a)   (b)   (c)

Figure 17.12. Illustration of the relationship between surfactant adsorption on a surface of opposite charge and wetting: (a) stage 1, low level adsorption --- some dewetting; (b) stage 2, complete monolayer formation --- maximum contact angle; (c) stage 3, formation of a second layer of opposite orientation --- wetting behavior returns.

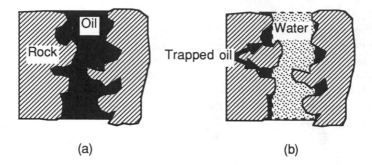

(a)   (b)

Figure 17.13. Schematic illustration of the effect of poor wetting by aqueous displacing fluids on petroleum recovery: (a) oilbearing rock formation wetted by petroleum; (b) petroleum is only partially displaced by flooding liquid leaving significant quantities of oil trapped in pores and crevices.

surfaces of opposite charge, in which a complex relationship between surfactant concentration and wetting is often encountered (Figure 17.12). At low surfactant concentrations (stage 1), for example, the strong electrostatic attraction between the surface and surfactant will produce a significant level of adsorption at S--1 (ie, a significant increase in σ_{s1} due to the orientation of the adsorbed molecules) while there will be only limited adsorption at the 1--2 interface (ie, little change in σ_{12}). The reduced attraction between the aqueous solution and solid surface (as reflected in a "new" $W_{aS1}°$) results in an increase in θ and a retraction of the wetting line (eg, "dewetting").

As the surfactant concentration increases, adsorption at the S--1 interface approaches completion while that at the 1--2 interface continues to increase, reducing σ_{12}. Eventually, θ will reach a maximum value, usually corresponding closely to the zero point of charge for the surface (stage 2). As the surfactant concentration continues to increase, θ will often begin to decline due to the formation of a second adsorbed monolayer in which the heads of the surfactant molecules are now oriented outward to the aqueous phase (stage 3). This results in a reduction of σ_{S1} and another "new" value for $W_{aS1}°$. In some surfactant--solid systems, especially for surfactants with long, straight-chain hydrocarbon tails, a cycle of complete wetting to nonwetting ($\theta > 90°$) to complete wetting can be obtained.

From a practical standpoint, the phenomenon just described can have significant impact on various processes. For mineral ore flotation, for example, it is generally desirable to obtain a large contact angle so that air bubbles will adhere to the ore particles and "float" them to the surface. If the concentration of the surfactant used as the flotation agent is too low or too high, optimum results will not be obtained. Likewise, if the charge and adsorption characteristics of the various minerals present in an ore are properly evaluated, it becomes easier to formulate the surfactant system that will give the maximum separation of ores.

For processes such as petroleum recovery, on the other hand, the maximum degree of wetting is desired so that oil attached to the rock formations can be displaced more efficiently. If poor wetting by the aqueous fluid occurs, significant amounts of oil will be left "stranded" as the plug of displacing fluid passes (Figure 17.13).

While wetting is obviously a complex process, careful evaluation of the examples and guidelines presented above, and the application of a bit of intuition and common sense, can help one arrive at a pretty good analysis of most situations.

CHAPTER 18

FRICTION, LUBRICATION, AND WEAR

One of the most important natural phenomena we must constantly battle in our modern mechanized world Is that of friction and wear between the multitude of moving systems that function to maintain our lifestyles. While most aspects of the three related areas --- *friction, lubrication,* and *wear* --- lie outside the true domain of surface and colloid chemistry, a few of the most important practical aspects of how to attack the problems they produce do fall into that area. For that reason, the subjects have been included here in the form of a brief descriptive discussion. Each of the three fields has developed its own basis of theory and practice along with an extensive literature. The following discussion is intended only as a general introduction, with the goal of trying to integrate those aspects most directly concerned with surface chemistry so that the novice can begin to see how some knowledge of interfacial phenomena may help clarify problems, and hopefully solutions, in practical applications.

FRICTION

When two solid objects are in contact under a normal load W, a certain finite amount of force will be required to initiate and maintain tangential movement with respect to one another. When at rest, no recoverable energy is stored at the interface between the two, so that when force is applied and work is done, most of that work is dissipated as heat. The force which must be overcome in order to make the two objects move is known generally as *friction* . In general, one finds that two frictional forces will be involved in such a process: the force necessary to initiate movement or that to overcome *static friction* , and that necessary to maintain movement or *kinetic friction* .

The general nature of frictional forces was recognized as early as the time of Leonardo da Vinci (and probably earlier, but not recorded). Since then they have been rediscovered several times and formulated into "laws" of friction that have served well, even though they are

(a)   (b)

Figure 18.1. Illustration of the difference between true and apparent areas of contact between solid surfaces: (a) for two perfectly smooth surfaces, the true and apparent areas are equal; (b) for "real" surfaces, the true area of contact will be less that the apparent area due to spaces created by surface irregularities.

generally found to be limited in their application. The three "laws" of friction, generally known as Amontons' law, can be stated as follows: (1) the frictional force will be proportional to the load, W; (2) the force will be independent of the geometric area of contact between the two objects; and (3) the kinetic frictional force for a system will be approximately one-third of the value of the normal load. The first two statements are generally found to be true over a relatively wide (and useful) range of conditions. The third is much more limited in application but, in the absence of other data, may be found useful.

The most general modern model used to describe frictional phenomena assumes that the friction between two unlubricated surfaces arises from two sources. The first and generally most important is that of adhesion between points of actual contact between the surfaces. We have seen on various occasions that real solid surfaces are almost never smooth. A very smooth surface will normally have asperities of between 50 and 100 Å so that the "true" area of contact between surfaces will be less that the apparent area (Figure 18.1). At those areas of contact, the two surfaces will be bound by a certain adhesional force arising from the interaction between the materials at the molecular level --- the same basic forces we have encountered before plus, in some cases, more physical interactions due to mixing or interpenetration. In order for the two surfaces to move tangentially, the points or areas of adhesion, welds, or junctions must be sheared or broken. If the real area of contact is A and the *shear strength* of the welds is s, then the frictional force due to adhesion will be

$$F_{ad} = As \qquad (18.1)$$

The adhesional friction, then, can be considered to be truly a "surface" as well as a bulk phenomenon.

The second contribution to the total friction (F_{def}) is more of a bulk physical contribution. It arises as a result of deformation, cracking, or plowing caused by penetration of the asperities of one surface into that of the other. Without going into detail about the exact processes involved, the deformation process may be summed up as a nonspecific term, P. If it is assumed that the two terms act independently, then the total frictional force for a system, F, may be written as

$$F = F_{ad} + F_{def} = As + P \qquad (18.2)$$

It can be seen that, in order to understand friction, the important unknowns include the real area of contact between surfaces, A, the shear strength of the points of contact, s, and the deformation component, P. If the various unknowns do not operate independently, their mutual interrelationships obviously becomes important.

If a normal force W is applied to the system, it may be expected that the added pressure on the asperities will cause some deformation leading to an increase in the real area of contact. If the materials respond to the added pressure by *plastic deformation* (ie, a permanent change in the shape of the asperities brought about by the application of the mechanical force), the real area of contact can be written as

$$A = W / p_0 \qquad (18.3)$$

where p_0 is the *yield strength* of the weld. For two objects of the same material, the *coefficient of friction*, μ_f, can be defined

$$\mu_f = F / W = s / p_0 \qquad (18.4)$$

For metals, it is usually found that the yield strength, p_0, is about five times the shear strength, s, which helps explain the fact that the coefficient of friction between unlubricated metal surfaces is quite often found to be about 0.2.

If a material responds to the applied load by *elastic deformation* (ie, temporary, reversible deformation under the load), the contact area will vary as $W^{2/3}$; if response is viscoelastic (ie, the deformation is reversible, but time dependent), the exponent of W will be between 2/3 and 1 and will have a time component so that μ_f will vary with the speed of the tangential movement.

If the two surfaces are of different materials, with different hardness, asperities on the hard surface will tend to plow into the soft surface, forming grooves. The force required for such a process will depend on the cross section of the groove as well as the yield strength of the soft material. The cross section will depend on the geometry of the asperity and the depth it penetrates into the soft surface. If the asperity is assumed to be a cone with an apical angle of 2θ (Figure 18.2), the plowing contribution to the coefficient of friction, μ_p, will be

$$\mu_p = (2 / \pi) \cot \theta \qquad (18.5)$$

The total coefficient of friction will, of course, be a combination of adhesional and plowing contributions. However, where both are found to

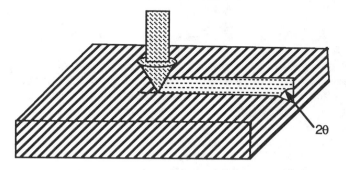

Figure 18.2. Schematic illustration of the plowing mechanism in friction.

be important, it is usually difficult to separate the relative roles of each. That is to say, the two are not necessarily directly additive in their contribution. If the asperities on a surface are known to be relatively blunt, it can usually be assumed that they will not make a significant contribution to μ_f in terms of the plowing mechanism.

FRICTION AND THE NATURE OF THE SURFACE

When trying to understand friction between two surfaces, it is necessary to know specific details about those surfaces. As stated above, the frictional force between surfaces will depend on the forces of interaction between them (ie, their chemical nature) and the hardness and strength of each material (their physical nature, so to say). In a general way, one can divide commonly encountered surfaces into four types: metals and their oxides, relatively isotropic crystalline materials, layered crystals, and polymeric (eg, amorphous) materials. Each type of surface may have its own unique frictional characteristics related to the chemical and physical nature of the relevant interactions.

Metals and Metal Oxides

From a practical standpoint, metal--metal friction is probably the most heavily studied system. In fact, the study of metal friction is in reality the study of two topics --- metal friction and metal oxide friction --- because except in very special (and usually not very practical) systems, a truely clean metal--metal contact is never encountered. In the "best" of situations, a normal metal surface will be covered with at least a monolayer of adsorbed gas molecules or other contaminants. As a result, frictional forces will be less than would be expected in their absence. For truly clean metal surfaces that have, for example, been treated by electron bombardment at elevated temperatures and ultra high vacuum, coefficients of friction in the range of 3 to 6 have been recorded. Sometimes, when good contact is achieved between clean surfaces of the same metal, the two surfaces will in fact weld or seize, to the point that the union is as strong as the bulk metal. For dissimilar metals, a similar result may be encountered if there exists a degree of mutual miscibility. If the two metals are mutually immiscible, there may or may not be such seizure.

The behavior of metal surfaces in the presence of air will be quite different from the clean surfaces. When oxygen is present, all but the most noble of the metals will rapidly develop a layer of metal oxide on the surface. The oxide, then, will have significantly different frictional properties to the metal. It is found, for example, that a clean copper surface has a coefficient of friction of about 6.8. After exposure to air, the coefficient falls to the range of 0.8.

In the case of friction between oxide layers, several patterns of behavior may be observed, depending primarily on the strength of the oxide and its bonding to the bulk metal surface. At very light loads and/or low sliding velocity, an oxide layer may completely separate the metal surfaces, resulting in a μ_f in the range of 0.6 --1. As the load is increased, the oxide layer, if it is relatively weak, will begin to break or detatch, allowing for more direct metal--metal contact and a significant increase in μ_f. The exact effect of load, in that case, will depend on various factors. In aluminum, for example, the oxide layer is much harder than the metal. The frictional interaction therefore easily deforms (plastic deformation) the metal supporting the oxide, the latter, being a thin film with little inherent strength, then breaks under even small loads. Friction, then is relatively independent of load. Copper, on the other hand, generally shows a dependence of friction on load, indicating that the oxide layer is relatively strong and better matched to the characteristics of the underlying metal.

In practice, then, the friction between nonlubricated metal surfaces should probably be considered that between oxide layer or composite surfaces in which there are oxide--oxide, oxide--metal, and metal--metal components. If one considers only two of the three --- metal--metal and oxide--oxide friction --- one may estimate surface composition from the frictional force using the relationship

$$F = A[(\alpha \, s_m + (1 - \alpha) \, s_o]$$ (18.6)

where α is the fraction of metal-metal contact, and s_m and s_o are the respective shear strengths of the metal and the oxide. Clearly, the effect of an oxide layer, assuming that it does not chemically "bridge" the two metal surfaces, may be considered a special instance of boundary lubrication by adsorbed films, which will be discussed further below.

Crystals with Relatively Isotropic Structures

Examples of materials falling into this class include salts such as sodium chloride, diamond, sapphire and other similar minerals, and solid nonmetallic elements such as krypton. The softer members of the class

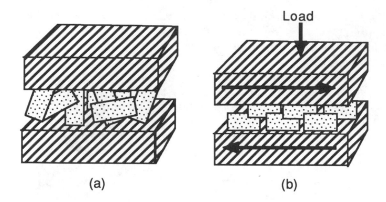

Figure 18.3. Schematic illustration of lubrication by anisotropic solids such as graphite: (a) in the absence of load, the solid lubricant may have a random orientation among particles; (b) under load, the particles reorient to align lower energy surfaces, reducing frictional forces between moving surfaces.

are generally found to obey Amontons' laws with frictional coefficients falling in the range of 0.5 to 1.0. The harder, more brittle substances such as sodium chloride tend to suffer extensive surface damage due to cracking but still hold more or less to "normal" behavior.

Diamond and sapphire differ in that they have lower than normal μ_f's, in the area of 0.1, and its value depends on the load, as might be expected for materials that deform elastically rather than plastically. Such materials also begin to show surface damage beyond a certain load. Under very clean conditions, μ_f for diamond has been found to rise to 0.6, suggesting that some mechanism such as the adsorption of a monomolecular water layer or slight surface oxide formation may act to lubricate the diamond surface naturally.

Anisotropic or Layered Crystalline Materials

Many commonly encountered solid lubricants are in fact highly anisotropic crystalline materials that can form layered structures. These include graphite, molybdenum disulfide, and talc, all of which are known as good lubricants under certain conditions. What might one expect to be the primary mechanism of lubricating action is such systems?

A schematic representation of layered crystalline materials is shown in Figure 18.3. Because the anisotropic structure is the "natural" scheme of things for graphite, MoS_2, etc, one must assume, based upon the discussion of surface energies and crystal faces given in Chapter 7, that the larger faces represent surfaces of lower surface energy. It would seem safe to assume, therefore, that the interatomic forces involved in the adhesional component of friction along the large, flat surfaces would be significantly less than those at the narrow edges. The result should be

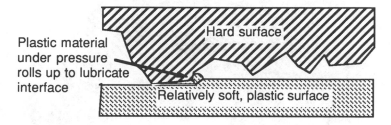

Figure 18.4. Schematic illustration of the "roll-up" mechanism of lubrication: (a) the softer, plastic surface material under pressure "rolls up" in front of areas of contact, producing a ball bearing effect and reducing friction.

a lower shear strength along those faces and a lower coefficient of friction. If it could be accurately measured, one would then expect to find a higher μ_f for edge-to-edge encounters and something intermediate for edge-to-face ones.

For molydenum disulfide, which has a structure like that of graphite, the intermolecular forces in the crystal are relatively weak, so that the above picture, along with some contribution by a more complete breakdown of the crystals, seems a reasonable explanation for its low μ_f, independent of such factors as the "cleanliness" of the system. Graphite, on the other hand, has a $\mu_f = 0.1$ in air, but in vacuum it rises to about 0.6. That indicates that for graphite, the adsorption of gases at the interface probably plays some role in weakening the adhesional forces between the parallel faces. Other mechanisms may also be involved that have yet to be fully explored, such as the "rollup" of graphite layers in front of the sliding face producing a "ball bearing effect" on the motion (Figure 18.4).

Polymeric (Amorphous) Materials

In the last few decades, the importance of friction between polymeric materials has rapidly increased. While the coefficients of friction for such materials are usually found to fall in the "normal" range, their behavior as a function of load indicates that the deformations occurring at the points of contact are elastic rather than plastic in nature.

For many polymers it is found that the static friction is significantly greater than the kinetic component, and that that value will be very load dependent. For an interface between two polymers, due to the relative mobility of polymer units at an interface, some small degree of chain interpenetration between surfaces may be expected to occur with time, depending on temperature, load, and the mutual miscibility of the two materials, leading to a relatively strong adhesive joint --- strong relative to a "fresh" joint joint, that is. The result would be the observed high static μ_f. Once sliding is initiated, it has been suggested that the relatively low shear strength of most polymers allows for the detatchment of a thin

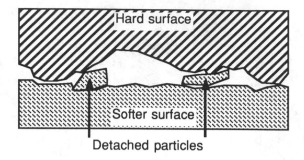

Detached particles

Figure 18.5. Lubrication by particle detachment: flecks of softer material removed from the surface facilitate movement, lowering the apparent kinetic coefficient of friction.

layer of polymer from the sliding surface (Figure 18.5) which then acts as a boundary layer lubricating film (see below).

The effect of such a transfer mechanism between two polymer surfaces might be expected to be small. For metal--polymer interfaces, on the other hand, the effect may be significant, since one is now going from a situation of metal--polymer sliding contact to one of polymer--polymer contact, which would be expected to have a much smaller inherent coefficient of friction. Because polymers are relatively soft materials, the questions of plowing contributions and elastic and viscoelastic work loss must be considered.

Obviously, the question of friction and its various contributing factors can become quite complicated, which explains the large volume of scientific and technical literature on the subject. So far, the discussion has been limited to unlubricated (formally, at least) surfaces. From a surface chemical point of view, the more interesting subject, perhaps, is not friction but how to combat it, or perhaps in a few cases increase it. With that in mind, we now turn to the subject of lubrication.

LUBRICATION

Functionally, one may define *lubrication* as the reduction of the friction between two surfaces by providing some mechanism(s) for the reduction of the various chemical (eg, adhesional) and physical (eg, plowing) interactions between them. A *lubricant* may be a solid, plastic, or liquid substance entrained between two sliding or rolling surfaces. The materials may be further classified as a *wetting lubricants* , which interact strongly with one or both surfaces (ie, adsorb to or "wet" them), leaving any intervening film relatively fluid; or strictly *hydrodynamic* , in which case there is no specific interaction between lubricant and the sliding surfaces, but the physical presence of the lubricant acts to separate the two surfaces and reduce friction.

Of the solid lubricants, the most common include graphite, molydenum disulfide, and talc, although as we shall see, spontaneously formed oxide layers also serve that purpose in some cases. "Plastic" lubricants include soaps and fatty acids, and petroleum residues (greases). Liquids include animal, vegetable, and mineral oils, although almost any liquid may serve the purpose in a given situation.

Mechanisms of Lubrication

Mechanisms for the reduction of friction between moving surfaces can be conviently broken down into four regimes: (1)*hydrodynamic* , in which an intervening, relatively thick layer of material physically prevents contact between sliding surfaces, thereby reducing friction; (2) *elastohydrodynamic* , in which the lubricating film has thinned to the point where the bulk properties of the lubricant begin to change to reflect special characteristics of systems only a few molecules thick; (3) *boundary layer* , in which the lubricating action is a result of the existence of an adsorbed monomolecular film of material that reduces the adhesional forces acting at points of contact; and (4) what may be called *chemical* lubrication, in which the lubricant effectively weakens or destroys the welds at points of contact by chemical attack. Although from the point of view of surface chemistry boundary lubrication is the most directly applicable, it is of interest to discuss each category briefly in order to better understand their overall significance.

Hydrodynamic Lubrication

Under certain operating conditions, it is possible for moving parts to operate with a continuous film of material separating them, such films being of sufficient thickness that the bulk properties of the lubricating material are maintained. In such conditions, the friction is a result of the work needed to overcome the viscosity of the lubricant, and no contact or wear between surfaces results (in principle, at least). These are the conditions for classical hydrodynamic lubrication.

While we normally think of such lubricating systems as containing some liquid material, in fact solids and gases can also act in that capacity. A simple example of hydrodynamic lubrication would be that pictured in Figure 18.6, in which an asymmetric cam rotates in a bearing. The lubricant forms a film between the cam and the bearing, but as the cam rotates, the thickness of the film varies from point to point due to the eccentricity of the cam movement. In the regions of closest approach between cam and bearing surfaces, the thickness of the lubricating film is reduced and the pressure between the surfaces (the frictional load) increases. The friction is found to be a linear function of the viscosity of the lubricant, η; the number of revolutions sec^{-1}, ω; and the nominal pressure between the surfaces, P :

$$\mu_f = f\,(\eta\omega/\,P)\qquad\qquad(18.7)$$

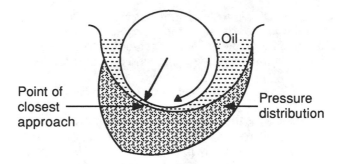

Figure 18.6. Schematic illustration of hydrodynamic lubrication of a cam in a half bearing.

For hydrodynamic lubrication, it is obviously best to maintain a relatively thick film of lubricant between the surfaces. In order to maintain such a situation it is necessary to keep the lubricant viscosity and relative speed of movement as high as possible, and the load as small as possible. The problems of load and speed are usually variables set by the operation of the device in question, so that their control is somewhat limited by use. That leaves viscosity as the primary handle on the friction problem.

The optimum viscosity for a lubricant will be determined by the configuration and running conditions of the device. However, as movement occurs and heat is generated, there is a tendency for the viscosity of most liquids to decrease, which could be dangerous in most systems. The temperature dependence of the viscosity of a liquid is given by a relationship such as

$$\eta = \eta_0 \, e^{Q/RT} \tag{18.8}$$

where Q is the activation energy for the viscous process. In order to counteract the natural tendency of viscosity to decrease with increases in temperature, critical lubricants usually contain materials termed viscosity "modifiers" or "improvers." Such materials are normally polymeric species that, under normal conditions are tightly coiled, thereby adding little to the viscosity of the system. As heating due to friction or other causes occurs the tight coils begin to expand (ie, the polymer becomes more "soluble"), thereby increasing, or in this case, maintaining, the viscosity of the system. Such is the case as long as the temperature does not reach a level at which chemical degradation of the system or some components begins.

In some situations it is found convenient to use a gas rather than a liquid as a lubricant. Gaseous lubricants are convenient because they are cheap (air in an "air bearing," for example), their viscosity naturally increases with temperature, so that additives are not required, and there is little chance of chemical degradation. Such systems are limited, however, by the fact that gas lubricants cannot support large load factors.

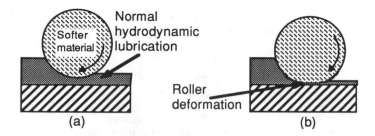

Figure 18.7. The effect of pressure on lubrication mechanism: (a) under normal load, hydrodynamic lubrication occurs; (b) under extreme load, the hydrodynamic layer thins, resulting in abnormally high pressures at the points of closest approach --- the softer material may then deform in order to spread the load over a larger area, thereby decreasing the localized pressure at the critical points.

In addition, for a gas lubricating system to function properly, the volume between the moving surfaces must be relatively small, calling for a very fine surface finish (ie, small asperities) and very good alignment of the parts.

Elastohydrodynamic Lubrication

With liquid lubricants, as the thickness of the lubricant film decreases, one may reach the point at which the properties of the lubricant are no longer those of the bulk material, but rather those of a special film only several molecules thick, which is penetrated by large surface asperities leading to excessive wear. Such a situation does not correspond to normal hydrodynamic lubrication, but neither does it indicate the onset of classical boundary lubrication (see below). This intermediate regime is referred to as *elastohydrodynamic lubrication*.

In a system in which at least one of the surfaces in question is relatively soft, such as a polymer, when the lubricating film thickness becomes very small, the pressure being applied to the oil film can become quite large, leading to elastic deformation of the softer surface. For a system of an optically smooth rubber roller on a glass surface, one can describe the situation as shown schematically in Figure 18.7. Under light loading conditions, the lubricating system operates according to normal hydrodynamic rules (Figure 18.7a). As the load increases, the film thickness decreases to the point where the rubber deforms or flattens, giving a configuration with greater load carrying capacity (Figure 18.7b). Finally as the localized load is removed the pressure drops suddenly and the rubber elastically rebounds to its original condition.

For two hard surfaces, the situation is somewhat different. In that case, elastic deformation as in the rubber example is no longer a significant factor. However, the pressure increases on the lubricant are still there. In a finely machined and aligned system, the pressures on the lubricant film may be on the order of 700--7000 kg cm^{-1}. The approximate relationship between the viscosity of a liquid and pressure is

$$\eta = \eta_o \, e^{\alpha P} \qquad\qquad (18.9)$$

where the constant α for most mineral oils is on the order of 2.8 x 10^{-3} cm^2 kg^{-1}. For the pressure range indicated above, the viscosity changes to be expected for the lubricant are something like the following:

$$P = 700 \text{ kg cm}^{-1} \;\Rightarrow\; \eta \approx 7\,\eta_o$$
$$= 1000 \text{ kg cm}^{-1} \;\Rightarrow\; \eta \approx 16\,\eta_o$$
$$= 3000 \text{ kg cm}^{-1} \;\Rightarrow\; \eta \approx 4450\,\eta_o$$
$$= 5000 \text{ kg cm}^{-1} \;\Rightarrow\; \eta \approx 1.2 \times 10^6\,\eta_o$$

Obviously, at very high contact pressures, the lubricating liquid between the two surfaces rapidly increases in viscosity until it must attain the consistency of a solid or wax rather than a liquid. In such a case, it is easy to see why some lubricating oils that exhibit such behavior show better performance than would be predicted for classical hydrodynamic theories. It also helps explain why other materials (eg, silicone oils), which have less dramatic viscosity increases with pressure, do not perform as well under those conditions. Other factors may be that in the viscosity range where elastohydrodynamic lubrication occurs, fluids may begin to exhibit non-Newtonian behavior.

When one considers the above mechanism of elastohydrodynamic lubrication, the question may arise, if the situation as described is true, why should one ever see the failure of a lubricant? While the answer has yet to be determined definitively, two possible explanations are: (1) even at the extremely high pressures involved, the extreme local temperatures that also exist surpass the "critical" point so that the lubricant is effectively evaporated away; or (2) shear forces near points of contact are sufficient to effectively break up the "solidified" lubricant film, leaving bare spots that can have direct contact, leading to excessive wear and possible seizure.

If a lubricant contains a small amount of a surface-active material, that is, one which has a specfic adsorptive interaction with the surfaces, it is often found that the conditions under which a system may operate without excessive wear or failure may be extended significantly. While it can be shown that the added material has no effect on the bulk or viscous properties of the lubricant, its presence provides an important degree of added protection not anticipated by the models of hydrodynamic or elastohydrodynamic lubrication. From the available evidence, it can be concluded that the added surface-active material must form a thin, probably monomolecular, film at the two surfaces which does not affect the viscosity of the lubricant, yet affords lubrication protection when the elestohydrodynamic film breaks down. Obviously, such action leads to another of the important mechanisms, that most directly related to surface chemistry, *boundary lubrication* .

Boundary Lubrication

The addition of small amounts of materials that adsorb specifically at the moving surfaces affords an added protection against excessive friction due to the presence of the adsorbed monomolecular film. That is, the lubricating action derives not from the bulk viscous properties of the oil, but from the specific surface interactions involved. In boundary lubrication, materials that are most effective are those which have relatively long hydrocarbon tails and strong specific interactions with the surfaces, especially those containing such groups as hydroxyl (---OH), amino (---NH$_2$), or carboxyl (---COOH). Groups containing phosphorous (eg, ---PO$_4$) and sulfur (---SH) are also found useful.

While boundary lubrication is usually encountered as the third in a series of lubricating mechanisms under increasingly harsh conditions, it can also be employed as a "stand alone" mechanism by the direct application of a monomolecular film to a surface in situations in which hydrodynamic or elastohydrodynamic processes are impractical or impossible. An example would be in the lubrication of magnetic tapes that must continuously pass over metal surfaces (recording and playback heads, guide posts, etc) but that cannot suffer much wear or abrasion without rapid loss of signal quality. Effective lubricating action can be attained in such a case by including in the oxide coating a surfactant that produces the required low coefficient of friction while being strongly adsorbed to the coating surface or directly to the oxide. Some polymeric materials, especially those containing silicones and fluorinated hydrocarbons, also serve well in that situation.

Because an adsorbed monomolecular film will have a thickness on the order of 2.5 nm, while the surface asperities present on all but the finest surfaces will seldom be less than 5--10 nm, it is important to have a clear picture of the mechanism of boundary lubrication at the molecular level. A "typical" situation is shown schematically in Figure 18.8. In the figure it can be seen that there are two types of contact between the two surfaces in the total contact area A : contact between the adsorbed lubricant films (area αA in the figure) and that between the actual surfaces where the adsorbed film has broken down (area βA). The total frictional force between the two will be the sum of each contribution

$$F = \alpha As_l + \beta As_s = \alpha As_l + (1 - \alpha) As_s \qquad (18.10)$$

where subscripts *l* and *s* refer to the lubricant and bare surface, respectively. Obviously, the greater the fraction of contact between adsorbed films (αA), the lower will be the friction.

The situation may be somewhat complicated when one surface is

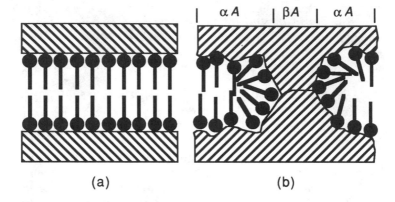

Figure 18.8. Schematic illustration of boundary layer lubrication: (a) ideal orientation of lubricant molecules on a smooth surface; (b) a "typical" situation on normal rough surfaces.

lubricated and the other is not. In that case, one often finds that with each pass of the unlubricated surface, some amount of lubricant film is transferred to the bare surface, leaving bare spots on the lubricated surface. With excessive transfer, the effectiveness of the lubrication will clearly decline. The rate of such transfer will depend on several factors, the most important of which are the applied load and the strength of adsorption of the lubricating film. The load factor is, of course, an operating parameter that lies outside the realm of surface chemistry; it is the adsorption process that is of most direct significance for our present purposes.

The are two general mechanisms for the adsorption of a monomolecular film, as we have seen from earlier chapters: *physical adsorption* and *chemisorption* . The simplest type of adsorption is the physical adsorption of materials such as hydrocarbons in which the only attractive factor is dispersion force interactions. If the solid surface is relatively hydrophobic and the hydrocarbon chain is relatively long, such adsorption can be quite strong. However, the adsorbed molecules will generally lie more or less horizontal along the surface (Figure 18.9a), which means that the thickness of the monolayer will be very small, about 0.25 nm. As a result, the effectiveness of such films will be limited and will decrease rapidly with repeated passing of the surfaces.

A more effective physically adsorbed film would involve interactions such as dipolar or hydrogen bonding interactions. Such would be the case for lubricating films of alcohols, amines, fatty acids, and fatty acid soaps, excluding the direct formation of metal soaps at the surface, which would fall under the category of chemisorption, discussed below. Because dipolar and hydrogen bonding interactions are more directional than dispersion forces, the adsorbed monolayer will tend to be oriented with respect to the surface, as illustrated in Figure 10b. In this case, the adsorption is stronger and the resulting film is much thicker, approximately 0.4--0.6 nm, depending on the chain length, resulting in a

(a)   (b)

Figure 18.9. Lubricant molecule orientation in boundary layer lubrication: (a) physically adsorbed --- random orientation and relatively low effectiveness; (b) chemisorbed --- highly oriented and highly effective.

significant increase in film effectiveness and durability. Comparative results for various systems are given in Table 18.1.

Clearly, the nature of the adsorbing species significantly affects its ability to function as an effective boundary lubricant. In general one can rely on the following "rules of thumb" for predicting the efficacy of various potential materials:

1. For optimum effectiveness the film should be in the close-packed, condensed state (see Chapter 8). It is generally found that as lubricant transfer occurs with repeated movement (ie, a reduction in the density of the packed film), or as the temperature of the system increases (ie, the condensed film "melts" to become "liquid condensed") the coefficient of friction increases.

2. The coefficient of friction decreases as the chain length of the hydrocarbon (or fluorocarbon) tail increases up to about 14 carbons, after which little change is observed.

3. Although the initial coefficient will be about the same for a given lubricant on various surfaces, its "durability" will vary with the strength of the adsorption; stronger adsorption results in greater durability.

4. Related to 3, while the initial coefficients for various lubricants of the same chain length will be similar regardless of the nature of the polar head group, durability will usually increase in the order ---NH_2 < ---OH < ---COOH, although some reversal of the amine--alcohol relationship may be seen in going from, say, a glass surface to a metal, depending on the reactivity of the metal.

Chemisorbed films will generally follow the above rules except that their interactions with the surface are generally stronger, resulting in the production of more durable films. The dividing line between physical adsorption and chemisorption is, in some cases, quite blurred, especially where acid groups such as carboxylates and phosphates are concerned. In cases in which salt or compound formation between surface and lubricant is possible, particularly effective boundary lubrication can result. For example, if a metal surface is basic, in the sense that it can react with a carboxylic acid to form a metal carboxylate, the resulting film will be especially durable. The same may be said about surfaces that are acidic and therefore react with amines or other basic groups.

Examples of films that overlap with the "chemical" lubrication to be discussed below, are encountered in systems in which the bare metal

Table 18.1. Effectiveness of physically adsorbed films as boundary layer lubricants under a normal load of 100 g.[1]

System	Lubricant	μ_f (sliding speed, cm sec^{-1})
Polyethyleneterephthalate	None	0.39 (.001)
"	Hexadecane	0.28 "
"	Oleic acid	0.25 "
"	Stearic acid	0.14 "
Steel on glass	Octanoic acid	0.18 (0.01)
"	Decanoic acid	0.13 "
"	Dodecanoic acid	0.09 "
"	Tridecanoic acid	0.06 "
"	Hexadecanoic acid	0.06 "
"	Octadecanoic acid	0.05 "
Cadmium on cadmium (load = 2 kg)	None	0.8 (0.1)
"	Cetane	0.06 "
"	Cetyl alcohol	0.4 "
"	Palmitic acid	0.07 "
"	Copper palmitate	0.05 "

is especially reactive. For example, as an asperity is worn away by friction, it may expose an area of bare metal that, in the absence of other alternatives such as oxidation by air or reaction with polar groups as discussed above, may react with other organic functionalities present, including especially elements of unsaturation (ie, double and triple bonds, carbonyl groups). It is found, for example, that aluminum surfaces that undergo sufficient wear to produce sigificant areas of "clean" metal surface are lubricated much better by unsaturated hydrocarbons than by the saturated analogues. Cetene ($C_{15}H_{30}=CH_2$), for example, is found to be more effective than cetane ($C_{15}H_{31}$---CH_3). While the details of such reactions are not completely clear, there is some evidence that the clean metal surface acts as a polymerization catalyst, producing a more viscous, strongly adsorbed boundary layer film.

Similar results have been reported under conditions in which the hydrocarbon may be "cracked," producing free radicals, which then react to form polymers or react with oxygen to produce peroxides, hydroxyls, carboxylates, etc, which consequently improve the efficacy of the system. In addition, it has been observed that boundary lubricants such as carboxylic acids will sometimes be much more efficient and durable lubricants for metal surfaces when the surface has a covering of at least a monomolecular layer of metal oxide. If the oxide is removed, lubrication effectiveness falls significantly.

Obviously, under extreme conditions, any number of chemical transformations may occur in the lucricant leading to alterations in its lubricating ability. The chemistry involved is complex and difficult to study, but its existence must always be kept in mind. Other chemistries can also be very important under some circumstances, which leads to the fourth lubricating mechanism, what will be referred to as *chemical lubrication* .

Chemical Lubrication Under Extreme Loads

So far the lubrication mechanisms discussed have involved ever thinner layers of lubricating liquid. The obvious limit to that progression is the complete absence of an external lubricant. When devices operate under extreme conditions of load, speed, temperature, etc, conventional lubricants will usually begin to break down and drastic mechanisms must be employed to prevent complete seizure and failure of the machinery. One way to approach that problem has been the development of *sacrificial lubricants* , which, under extreme conditions, react with fresh metal surfaces formed by wear to produce a new inorganic chemical layer that can then be more easily sheared, thereby preventing seizure.

The conventional lubricants in this class of materials are compounds containing reactive chlorine, sulfur, or phosphorus groups, which react to produce inorganic metal chlorides, sulfides or phosphates. Such compounds are generally included as additives in more conventional lubricants and are therefore passive until extreme conditions are reached at which time they begin to "do their thing."

The chemistry of the sacrificial lubricants can be quite complex, especially if oxygen is present. For example, systems that nominally produce metal sulfide layers may produce mixed sulfide--oxide layers with oxygen present. That may or may not be detrimental to the operating device but must be considered. Chloride forming chemical lubricants were once popular because the resulting surface layers generally have very low shear strength, leading to better lubricating properties. However, under certain conditions of temperature and in the presence of moisture, they were found to decompose to produce hydrochloric acid, which is corrosive and certainly undesirable. Such materials have lost much of their popularity for that reason.

Because the chemical lubricants are generally included as additives in more conventional systems, certain precautions must be taken with respect to the overall composition of the mixture. Obviously, there must be no significant chemical reaction between the various components of the lubricant as formulated. In addition, one must be certain that the primary lubricant in the mixture will not dissolve or otherwise attack the chemically formed inorganic layer. If it does, both the chemical lubricant and the surface being lubricated will be rapidly depleted, leading to excessive wear of the surfaces.

While chemical lubrication is a viable, mechanism for the lubrication of machinery which must operate under extreme conditions, its beneficial effects are obtained at the price of significant loss of material from the lubricated surfaces --- that is, one pays for the operation of the device in terms of increased wear to the lubricated surfaces.

Some Final Comments on Lubrication

Before closing the book on lubrication, there are a few additional points of interest that should be mentioned briefly. Two potentially important ones from a practical standpoint are the so-called *Rehbinder effect* and *weeping lubrication* . The Rehbinder effect relates to the effects of the adsorption of surface-active materials on the strength of materials. While there exists some uncertainty on the matter, there is significant evidence that the adsorption of surfactants or other materials onto surfaces, especially in cracks and surface flaws, can reduce the mechanical strength of the material.

It has been stated several times in other chapters that the driving force for the adsorption of surfactants is the reduction of surface energy. In solids, areas of surface flaws and cracks usually represent areas of higher than normal surface energies, implying that adsorption should occur in those areas more rapidly than in the more "normal" areas of the surface. If the phenomenon occurs as described, it can have a number of practical consequences --- some bad, some good. On the detrimental side, the weakening of a component by an adsorbed lubricant film could obviously lead to device failure. On the positive side, however, the effect may play an important role in several processes related to lubrication. For example, the shear strength of points of contact between surfaces or oxide or other inorganic layers may be reduced, making movement of the surfaces easier and thereby reducing friction. In lubricants for metal cutting operations, the adsorbed lubricant film may weaken the metal being worked sufficiently to reduce mechanical wear of the cutting surface. And finally, in drilling operations, especially in oil exploration, the Rehbinder effect may aid significantly in reducing wear on expensive drill bits by weakening the rock structure being penetrated.

The second potentially important point to be mentioned is that of weeping lubrication, in which a lubricant is trapped inside a porous material, in surface cracks and flaws, or *absorbed* by a polymeric material, to be slowly extruded as needed when subjected to pressures and temperatures as a result of friction. Such a "weeping" action can provide a mechanism for the continuous renewal of a lubricating layer under conditions in which confinement of a normal lubricant would be impractical or impossible.

Weeping lubrication has been suggested as one of the possible mechanisms by which body joints are lubricated by the synovial fluid. In that case, the fluid would be trapped within the bone or cartilage structure and released "on demand." If the supply of lubricant is cut off or the porous structure of the bone and cartilage changed by some medical condition, lubrication would be lost, leading to obvious problems.

A second application of weeping lubrication is seen in the working of certain metal surfaces. If a metal is lightly abraded and lubricated prior to working, the lubricant can penetrate the flaws and cracks and be held in reserve for use during subsequent processing. For example, metals so treated can often be subjected to significant deformation by rolling and drawing without the formation of gross surface defects or scuffing. The lubricant stored in the initially abraded surface is squeezed out during processing, providing continuously renewed lubrication and preventing excessive metal--metal contact.

Lubrication is clearly an important practical topic that is ever changing and improving to meet the new demands of technology. The above discussion is obviously superficial (pardon the pun) but does introduce many of the more fundamental aspects of related problems and perhaps some solutions. While there exists a great deal of uncertainty and some disagreement on details, there are some general conclusions that can be drawn and that seem to have broad applicability.

For most conditions, the best effect is obtained when there exists a thick film of fluid between the moving surfaces affording efficient aerodynamic, hydrodynamic, or elastohydrodynamic lubrication. If the lubricating film is thinned or broken by operating conditions or system failure, additional protection is afforded by adsorbed films through boundary lubrication. Finally, under extreme conditions, protection against seizure and complete failure may be obtained as a result of chemical processes which produce "weak" oxide, sulfide, phosphate, etc, surface layers that can be more easily sheared that direct metal--metal contacts For hydrodynamic and elastohydrodynamic lubrication, careful attention to the viscous properties of the lubricant, including the addition of viscosity improvers or modifiers, can significantly extend the practical operating range for many liquids.

In the case of boundary layer lubrication, in which the adsorption of monomolecular films is required, the best protection is provided by materials such as fatty acids and soaps which can adsorb strongly at the surface to form a solid condensed film. Less durable, but effective protection can be obtained with polar groups such as alcohols, thiols, or amines. The least effective protection is obtained with simple hydrocarbons which adsorb more or less randomly and through dispersion forces alone. For adsorbed monomolecular films, best results are obtained when the hydrocarbon tail has at least 14 carbons. In some cases fluorinated carboxylic acids and silicones may provide a lower initial coefficient of friction, but their weaker lateral interaction sometimes results in a less durable surface film that "melts" at a lower temperature resulting ultimately in less overall protection. If a polar lubricant can form a direct chemical bond to the surface, as in the formation of metal soaps, even better results can be expected.

At elevated temperatures and pressures, even strongly adsorbed films may be desorbed, leading to excessive surface contact. In such cases, some form of chemical lubrication must be used, because simple organic liquids may begin to degrade, oxidize, or polymerize, rapidly losing their lubricating properties. Obviously, the optimum system where extreme conditions may be expected is one in which all of the bases are covered: a good hydrodynamic lubricant with viscosity modifiers, long-chain fatty acids for adsorption and boundary layer protection, and extreme pressure sacrificial additives as a last resort. Closing the subject of lubrication leads directly to the consequences of friction or the lack of lubrication --- wear.

WEAR

Wear between moving surfaces is not directly a surface chemical problem. However, if the mechanisms of friction described above, namely adhesion between points of contact followed by shearing and material transfer between surfaces, are correct, it is a direct consequence of friction, which is a surface problem. It is therefore of interest to include a few comments on the subject to close this chapter.

Unlike many practical problems, wear seems to be a complex phenomenon that has not lent itself to the formulation of useful generalizations or the generation of "laws of wear." The safest things one can say about the subject is that wear increases with time of operation, with severity of operating conditions (ie, load, speed, temperature), and appears to be more severe with soft and brittle materials than with hard surfaces. Unfortunately, there are many exceptions to those three generalizations. In addition, the existence of chemical wear, the Rehbinder effect, etc, complicate matters since surfaces that suffer less physical wear, may suffer more from chemical attack.

Physical Wear

Physical wear may be loosely defined as the removal of surface material due to adhension--shear cycles. The process may be divided into three general regimes --- light, medium or mild, and severe wear, depending on the extent of material transfer and the general effect of the process on the nature of the surface produced.

Light wear can conveniently be termed the incidental transfer of small amounts of material from one surface to another, or to the lubricating fluid, as a result of contact between large asperities. Light wear is an inevitable consequence of the nature of the solid state and results, in some cases, in a beneficial effect on the overall operation of a device. The "breaking in" of new machinery, for example, is partly a process of allowing the new surfaces of gears, bearings, etc, to wear off large asperities to produce a smoother surface and better "fit" between moving parts.

Medium or mild wear is usually assumed to involve the interaction of thin films, primarily oxides for metal parts, on the sliding surfaces. In this case, it is generally found that the oxide has a lower shear strength than the metal and its transfer occurs more easily (eg, at lighter loads) than would be the case for direct metal--metal contact. Even when metal contact does occur under conditions of mild wear, those areas of contact will be small and will generally produce a reactive fresh metal surface which will rapidly oxidize, effectively returning the system to a condition of oxide contact. As a result the material transferred will be primarily oxide, which, because of its relatively soft, brittle nature, will cause few major problems due to plowing and other abrasive action.

Severe wear is obviously the condition of most practical importance. It is normally associated with conditions of adhesion and subsequent shearing of points of contact between primary bulk materials --- metals in most cases. As has been stated previously, it is observed that junctions or welds between metal surfaces often exhibit the same shear strength as the bulk material. In fact, in welds between the same materials adhesion will be strong and shear strength may be greater than that of the parent materials due to work hardening. In such a case, one can expect cleavage to occur at some other point in one of the two surfaces away from the original area of contact. As a result of random transfer from both of the surfaces, roughening will be relatively more severe and likely to produce operational problems more rapidly. As a result, it is generally not good practice to have critical machined parts made from the same material.

If the shear plane is at the original area of contact, little transfer of material will occur and the effect on the topography of the surfaces will be negligible. For the case for surfaces of two different materials, shearing usually occurs at a new location within the bulk of the softer of the two. In that way, unlike the case of the same material, one surface is attacked preferentially, somewhat simplifying the transfer situation.

When material transfer occurs during movement, its effect on the surfaces, the obvious wear involved, may not correspond directly to the actual amount of material transferred. For example, if material is transferred and adheres to the receiving surface, the net result may not alter significantly the topography of the surface and therefore not appear as wear. In fact, for some situations, repeated contact between surfaces may result in the passing back and forth (*backtransfer*) of material, obscuring the true rate of wear. It is only when the transferred materials become detached that general wear becomes readily obvious. The net rate of wear, then, will depend not only on the absolute rate of material transfer, but on the degree of backtransfer and complete detachment.

The relationship between the observed coefficient of friction, μ_f, and the production of wear fragments for a given system is quite complex and not subject to easy interpretation. Typical results are given in Table 18.2 for various materials interacting under a standard set of conditions. From the table one can see that although the coefficients of friction vary by a factor of 2--3, the wear rate, as defined by the fraction of contacts which result in the production of wear fragments, k , varies by a factor of 10^5. Interestingly, the system with the highest coefficient of friction, polyethylene on hard steel, showed the lowest wear rate, while brass on hard steel, with μ_f of about one third, exhibited a wear rate four orders of magnitude higher.

The reasons for such seemingly illogical results must lie in the different responses of the materials involved to the pressures and shearing forces acting in the wear process, and the mechanism by

Table 18.2. Coefficients of friction, μ_f, and wear rates, k, for representative materials sliding at 180 cm sec^{-1} under a normal load of 400 g.

Interacting surfaces	μ_f	k
Mild steel on mild steel[2]	0.6	10^{-2}
60/40 Leaded brass on hard steel[2]	0.24	10^{-3}
Polytetrafluoroethylene on hard steel[3]	0.18	2×10^{-5}
Stainless steel on hard steel[3]	0.5	2×10^{-5}
Polyethylene on hard steel[4]	0.65	10^{-7}

which detachment occurs. For hard materials such as steel, deformation at points of contact is expected to be plastic. With repeated passage of an area through loading and unloading cycles, it is reasonable to expect strain to be built up in the areas around and beneath the contact points, leading at some point to the situation in which surface forces (adhesion) are sufficient to break loose the fragment, or the strain will induce material fatigue leading to detachment. Brass, of course, is a rather soft, ductile material that can be gouged easily by the harder steel, with fragments being plucked off relatively frequently.

For polymers, deformation will be primarily elastic, so that strain and fatigue will be less of an immediate problem. In addition, any strains induced by contact may be annealed away by localized heating due to the relatively low glass transition temperatures of polymeric materials. The greater wear rate of fluorinated polymers is undoubtedly related to their inherently lower cohesive strength.

Abrasive Wear

Once wear fragments detach from sliding surfaces, they become free to move about within the lubricating film and inflict additional damage on the surfaces by plowing and other abrasive mechanisms. In order for significant abrasion to occur, the detached particles must be harder than the surfaces involved. That is not to say, however, that softer particles will not cause abrasive wear --- it will simply be at a slower rate and often of a different nature. One can often determine the nature of abrasive particles by careful examination of the abraded surface. Spherical particles, for example, characteristically produce relatively smooth grooves in soft surfaces or flaking in harder materials. Sharp, irregular particles, on the other hand, will cut the surface in a more or less random pattern and generally have a higher wear rate.

In some materials, abrasion can, over time, induce physical changes in the natures of the moving surfaces. For example, abrasion may induce changes in the structure and orientation of surface layers, changes in crystallization, etc, resulting in a significant alteration in the physical properties of the surface. Work hardening by abrasion, in fact, materials.

wear in many cases is
arts. That can often be
process, heat treatment,
is that should a fragment of
an inflict much greater damage
a fragment of the underlying bulk
hardness may be found to be
would be to increase the elasticity of
ue is reduced, thereby reducing the
agments. Such an approach might be
al industry where mechanical abrasion
ical attack on the surfaces. The use of an
olymeric coating is indicated in such cases.

ll, such as alloying
lly, small changes
etc) may lead to
wear, calling for
the information
tand friction,
r ideas to the
deas to the

Chemical Wear

that chemical attack on sliding surfaces can be
g distinct surface layers that serve as extreme
s. However, under normal operating conditions,
, especially oxide formation, is generally found to be
ile oxide films may be harder than the underlying metal,
sometimes more brittle and become easily detached. As
nts, the oxides can inflict significant abrasive damage on a
ading to early failure. Likewise, while the use of chlorine- or
ontaining extreme pressure lubricants may be justified under
operating conditions, care must be exercised to ensure that the
ve is not too reactive, since such a condition can lead to rapid
osion of the surfaces and failure just as rapidly as the processes it
was introduced to retard.

In the case of polymeric surfaces, chemical wear or degradation is a more complex problem because it involves (or may involve) a wide variety of bond-breaking and bond-making processes. Polymers, for example, may undergo oxidative degradation to produce polar surface groups (---OH, ---COOH, etc) which may increase adhesive forces at points of contact, thereby increasing friction and wear. Under repeated strain cycles, polymers may also undergo chain rupture to produce free radicals that can react further to "depolymerize" the surface reducing shear strength and increasing wear, or lead to crosslinking which may reduce the elasticity of the surface and, again, increase wear. Obviously, the inclusion of additives in the polymer (or the lubricant) that retard such degradative processes would probably be beneficial.

While the individual mechanisms of wear are fairly well understood, it is a fact of life that one mechanism seldom operates independently, and the combination of two or more will generally lead to a situation that is difficult or impossible to interpret unequivocally. While one mechanism may be entirely detrimental to the operation of a device, another may lead to an improved resistence to wear by work hardening

or some other process. Processes not mentioned at a
of the surfaces, may be important in some cases. Fina
in operating conditions (ie, load, speed, temperature,
significant alteration of the predominant mechanism of
changes in lubricant formulation.

Hopefully, even in the face of all of that uncertainty,
provided can assist the novice in the field to under
lubrication, and wear processes, and apply these and othe
solution of individual practical problems.

CHAPTER 19

ADHESION

Adhesion is an extremely important concept in both practical and theoretical terms. Unfortunately, there is no completely satisfactory definition of the term that fulfills the needs of both the theoretical surface chemist and the practicing technologist. So far in this book, the term "adhesion" has been encountered as applied in the ideal or theoretical sense --- referring to the reversible thermodynamic process of separating unit area of two phases which originally had a common interface. That aspect of the term was defined in Chapter 2 and will not be repeated here, in all of its detail. Some comments about the "reality" of that concept will be in order, however. A more practical definition of adhesion is a state in which two bodies (usually, but not necessarily dissimilar) are held together by intimate interfacial contact in such a way that mechanical force or work can be applied across the interface without causing the two bodies to separate. It is the latter definition that will be of most concern in the discussion which follows, although the two concepts are, in fact, inseparable. Before entering into that discussion, it will be useful to clarify a few terms commonly encountered in the field of adhesion and often misinterpreted by the novice.

TERMINOLOGY

Thermodynamic adhesion is the term that applies to the "ideal" adhesion already defined in terms of reversible work needed to separate two surfaces by overcoming the molecular interactions across the interface. *Chemical adhesion* is a term that may be applied to adhesion involving the formation of formal chemical bonds (covalent, electrostatic, or metallic) across an interface. *Mechanical adhesion* refers to the situation in which actual mechanical interlocking of microscopic asperities at the interface occurs over a significant portion of the contact area.

The term *practical adhesion* may be used with reference to the so-called failing load of a particluar joint, particularly as applied to the performance of glues, cements, etc. *Adherence* is a term sometimes used to describe the degree of practical adhesion, while *adhesive* applies to a material used to join two surfaces together in an adhesive *joint* or *weld* . The *adherend* is the phase (or phases) being joined by the adhesive. A *proper adhesive joint* is a utilitarian term used to describe an ideal situation in which an adhesive joint has been prepared with complete intimate contact between components (ie, no flaws) and in the absence of intervening contaminants (ie, moisture, dirt, oil) that might reduce the ideal strength of the system.

The following discussion will attempt to put into some degree of focus the interelationship between the various definitions as well as illustrate some of the more important aspects of the subject. The subject is quite broad, with an extensive literature, so the coverage will be limited in scope.

THERMODYNAMIC OR IDEAL ADHESION

We have already encountered the concept of thermodynamic adhesion and its related terms such as the work of adhesion. The term is applied to a defined model system and does not take into consideration conditions before or after the formation of the interface, the presence of random flaws or defects in the system,or the bulk physical properties of the components, all of which are of primary importance in the practical application of the concept of adhesion. It is related to molecular interactions such as van der Waals, dipolar, and electrostatic forces but does not consider mechanical or chemical interactions as defined above. It is therefore not a very useful device in terms of practical adhesion problems, but it serves as a good theoretical tool and to indicate a maximum force or work that a given interface may be expected to transmit before failure (ie, separation) occurs.

Because, in theory at least, the concept of ideal or thermodynamic adhesion applies equally well to liquid and solid phases, it is of interest to see how a calculation of such an ideal value compares with reality. The complete expression for the work of adhesion betwen two phases with each phase completely saturated by the other is[1]

$$W_{A(B)B(A)} = \sigma_{A(B)} + \sigma_{B(A)} - \sigma_{AB} = W_{AB} - \pi_{A(B)} - \pi_{B(A)} \quad (19.1)$$

where the phase in parenthesis saturates the one it follows. The maximum force required to separate unit area of interface, the ideal adhesive strength of the interface, Z^a, can be approximated by

$$Z^a = 1.03 \, W_{A(B)B(A)} / r_0 \quad (19.2)$$

where r_0 is the equilibrium distance of separation, usually on the order of a few molecular diameters (0.2--0.5 nm). Referring to the calculation for the system benzene-water given in Chapter 8, the work of adhesion is

$W_{A(B)B(A)}$ = 56 mJ m^{-2}. If one assumes that the major portion of that work will be done over a distance r_o = 0.4 nm, then the ideal force required to separate the two phases would be on the order of 1.44 x 10^5 kg m^{-2}, well above the actual strength of most adhesive joints. As a more practical example, for two polymers that interact by dispersion forces alone, a typical value for W_{AB} will be 100 mJ m^{-2} leading to a maximum ideal adhesive strength of about 2.6 x 10^5 kg m^{-2} which is several orders of magnitude greater than practically obtainable adhesive strengths. Similar calculations can be made for cohesive strengths. A few comparisons of ideal and practical cohesive strengths for some materials are given in Table 19.1.

Table 19.1. Ideal and real cohesive (tensile) strengths of some common materials (z_o = 0.4 nm).

Material	Cohesive (tensile) strength, MPa	
	Ideal	Real
Polyethylene (molded)[2]	180	38[2]
Polystyrene (molded)[2]	210	69[2]
Aramid yarn[3]	7,900	2,760[3]
Drawn steel[3]	9,800	1,960[3]
Graphite whisker[4]	100,000	24,000[4]

Based upon the ideal calculations similar to the above, it has been suggested that for practical adhesives, if good wetting of the surfaces to be joined can be achieved, dispersion forces alone should be sufficient to ensure a strong adhesive bond. Unfortunately, reality seems not to agree with that conclusion. The ideal calculations are, of course, based on the concept of thermodynamically reversible separation processes, while the fact of life is that such conditions are almost never attained. In fact, fracture processes are invariably accompanied by irreversible viscoelastic processes that dissipate energy and complicate the analysis of the situation. In addition, and perhaps more importantly, real adhesive joints will contain flaws that will greatly reduce the practical strength of the system. Some of those points will be discussed more fully below.

PRACTICAL ADHESION

The brief analysis of ideal adhesion given above indicates that, under the best of circumstances, one might expect to be able to attain very strong adhesive joints with materials interacting through the universal dispersion energy, with no need to invoke stronger molecular interactions, mechanical interlocking, or chemical bond formation. Experience, unfortunately, generally indicates otherwise. A simple analysis of the situation indicates that the adhesive strength between two

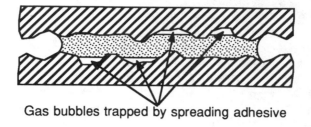

Gas bubbles trapped by spreading adhesive

Figure 19.1. Schematic illustration of the entrapment of gas bubbles by spreading liquid adhesive, resulting in loss of adhesive strength in the final joint.

dissimilar materials should be greater that the cohesive strength of the weaker of the two, leading to cohesive failure rather than adhesive failure. In some instances, particularly in the adhesive component of friction (see Chapter 18) that concept seems to be born out since shearing of the contact area between surfaces normally occurs in the weaker material. In adhesive joints, however, the area of contact between surfaces is several orders of magnitude greater than that in friction, and the question of the completeness of interfacial contact becomes important.

A liquid spreading over a rough surface (the normal situation) can easily trap air in depressions in the surfaces, leading to the formation of a composite surface (Figure 19.1). That will be the case whether the spreading liquid has a small or large contact angle, although a smaller θ_A will obviously improve matters. When a portion of the composite involves air--adhesive interfaces, the the actual area of adhesive contact is greatly reduced. In addition, the three phase boundaries thus formed represent excellent sites for the initiation of cracks and flaws in the system. The net result --- a significantly weakened joint. Similar effects can be seen if the entrapped material is water, oils, or other materials with significantly lower adhesional interactions or cohesion strength.

In practice, it is usually found that the actual adhesive strength of a joint with a "good" (ie, wetting) adhesive will be at least one order of magnitude less than the ideal value. Poorly wetting systems would be expected to perform correspondingly less effectively. The primary reason for the descrepency between ideal and real adhesive strength appears to lie in the almost invariable presence of bubbles, cracks, and flaws that are associated with the interfacial zone (Figure 19.2). When stress is applied to a joint, it tends to concentrate at such flaws, so that the "local" stress is significantly greater than the average value. When the local stress exceeds the local strength (already lessened by the presence of the composite interface, for example), failure occurs.

The fracture of practical adhesive joints involves two primary processes --- cohesive or adhesive failure at or near the joint and work (reversible and irreversible) involved in plastic, elastic, or viscoelastic deformation of one or all of the components of the joint --- one of the two

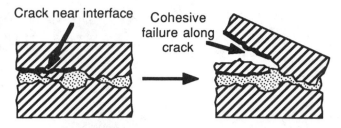

Figure 19.2. Illustration of the role of cracks and flaws near the interface in controlling the locus of joint failure.

solid surfaces or an adhesive, if present. As indicated in the preceding chapter on friction, cohesive failure of the weaker of two solids in contact is common. The same can be said for normal adhesive joints, in that actual *adhesive* failure (ie, exactly at the interface) is less common that *cohesive* failure, of for example, the adhesive material, near the interface. What, then are the necessary conditions for obtaining "good" adhesion between two surfaces?

SOME CONDITIONS FOR "GOOD" ADHESION

Any workable theory of adhesion must take into consideration all of the possible aspects of energy transfer across an adhesive joint, not only molecular forces or ideal adhesive strengths, but also the presence, number, and size of flaws, various dissipation processes, and irreversible fracture processes. It has already been demonstrated that in an ideal situation, forces of molecular interaction should be sufficient to produce a very strong adhesive joint, assuming perfect contact across the interface, but that reality lies far from the ideal.

For reference purposes, Table 19.2 lists the various attractive molecular forces that may operate across an interface along with the approximate range of strengths they cover. Since van der Waals forces, for example, fall off rapidly with distance of separation by r^{-3}, for such forces to be effective the interacting surfaces must be as close as possible, typically 0.2--0.5 nm. Beyond that distance, such interactions will be quite weak and the ability of the joint to transmit any applied stress will be accordingly reduced. Clearly, intimate molecular contact between phases is a necessary condition for good adhesion --- necessary but not sufficient!

It is a fact of life that in practical adhesion problems, materials properties are often as improtant as interfacial forces. It has been stated that a significant portion of the fracture energy of a joint is dissipated in various deformation processes. Obviously, the nature of the joint interface in terms of physical "mixing" is an important aspect of the overall problem.

The intermolecular forces listed in Table 19.2 are common to all liquids and solids, depending on chemical composition, yet liquids have significantly lower mechanical strength than comparable solids. In solids,

Table 19.2. Values of attractive molecular interactions at interfaces.

Type of interaction	Approximate energy range (kcal mol^{-1})
van der Waals	
Dispersion	5
Dipole--dipole	0--10
Dipole--induced dipole	0--0.5
Hydrogen bonding	0--40
Chemical bonds	
Covalent bonds	15--170
Ionic bonds	140--250
Metallic bonds	27--83

the intermolecular distances are generally smaller and are reinforced by the forces stemming from the more ordered structure (crystal lattice, etc). In polymers, which constitute the majority of adhesives, mechanical strength (eg, the ability to transmit stress without failure) is also a function of molecular weight. Below a certain value, strength falls off rapidly with molecular weight. In such a system, the intermolecular forces between chains are essentially unchanged, so that the loss in strength must be a result of some more physical phenomenon, in this case so-called *molecular entanglement* . That is, as the polymer molecules become longer, they become more "wrapped up" in their neighbors (Figure 19.3). For the polymer to fail under stress, the entangled chains must slide past one another and disentangle, a process that requires a significant amount of energy, thereby serving as an excellent process for the dissipation of the stress forces, and providing a good added mechanism to back up intermolecular forces for producing good adhesion.

Dissipation processes depend directly on the amount of chain entanglement and the nature of the forces acting between chains. In a bulk polymer, the degree of entanglement is a direct result of the nature of the polymer and its method of preparation (eg, cast from the melt, from solvent, spun). At an interface, entanglement is more problematical --- the process may be assisted by the use of a solvent which swells the adherend surface, allowing interpenetration; the application of heat, which increases the mobility of the polymer chains; the use of a monomeric system which is polymerized after applications; etc. In any case, if entanglement can be increased, the strength of the bond at the interface will be improved.

If entanglement does occur, it should be obvious that one can no longer talk about a sharp interface, in the classical sense, but must consider an interfacial zone the structure of which will be a primary determinant of the strength of the joint. For simplicity, one may consider two types of interfaces --- sharp, in which the primary component of strength derives from classical intermolecular attractive forces, and

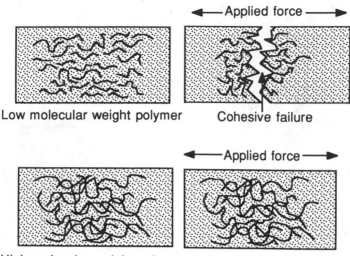

Figure 19.3. The role of polymer molecular weight in determining the strength of a polymer surface or adhesive: low molecular weight materials have relatively little chain entanglement or other interactions --- the free movement of adjacent chains results in inferior cohesive strength; for high molecular weight materials, chain entanglement and increased lateral interactions result in greater cohesive strength and failure at higher applied forces.

diffuse, in which entanglement plays a significant role. Those can be further classified by the relative strengths of the molecular interactions to give three general classes of adhesive behavior (Figure 19.4).

The first, and simplest, class can be described as having a sharp interface and weak intermolecular interactions. An example might be a joint between a nonpolar polymer (eg, polyethylene) and a polar polymer (eg, polyvinyl alcohol--polyvinyl acetate copolymer). In such a system, the only molecular interaction is that due to dispersion forces, with little tendency for chain entanglement due to the inherent incompatibility of the two polymers. The mechanical strength of the resulting joint derives solely from the dispersion forces, which will not be able to inhibit the movement or slippage of the joint significantly. A joint of very poor strength is the result.

The second class is a joint with a sharp interface, but with significant specific chemical interactions between the two phases. For example, a polymer containing groups capable of significant hydrogen bonding or acid--base interactions (eg, ---COOH) can interact strongly with, say, a metal or metal oxide surface (M^{n+} or ---M---O---M---), even though no significant entanglement is possible. The resulting joint would have significant mechanical strength because of the larger magnitude of the attractive forces, even without the assistance of entanglement.

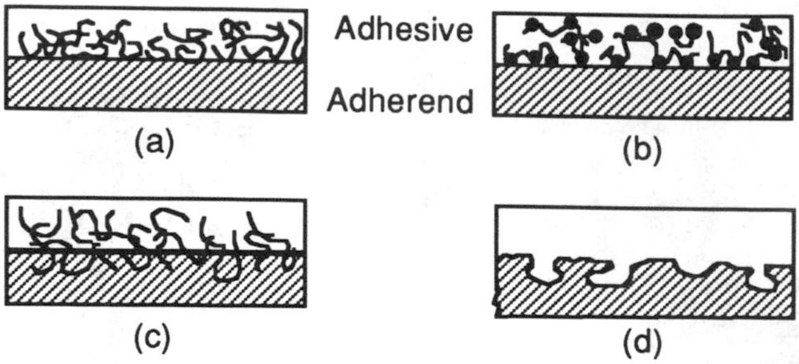

Figure 19.4. General classification of individual mechanism of adhesion: (a) physical adhesion at the interface; (b) specific interactions (chemisorption) at the interface (black circles); (c) interpenetration of adhesive molecules across the interface; (d) physical interlocking of the two phases at irregularities in the interface.

In the third class, there is significant interpenetration of the adhesive into the surfaces to be bound. The interpenetration may be at the molecular level, in the case that the adhesive polymer and polymeric adherend are mutually miscible, or more at the microscopic bulk level, as for porous substrates. If the interface is diffuse and significant entanglement occurs, a strong joint may be expected, regardless of the nature of the intermolecular forces acting between elements. In such a case, the entanglement of entire polymer molecules is not necessary for significant strength to be developed. Entanglement may be considered a sufficient condition for good adhesion, even in the absence of strong intermolecular interactions.

If one considers the action necessary to displace the two surfaces in a system with a sharp boundary, it can be seen that for the first case, the molecular forces being overcome are not only small (relative to specific interactions) but are essentially limited to a one-dimensional action across the interface. If entanglement occurs, the same small forces will be acting in three dimensions --- that is, each entangled molecule will have nearest neighbors on all sides, which will mean --- crudely speaking --- that the interactions will be multiplied many times, depending on the degree of interpenetration. The concept is illustrated schematically in Figure 19.4. If stronger specific interactions are present, or if chemical bonds are formed, so much the better! From the above discussion, then, it appears that the optimum conditions for good adhesion include a diffuse interfacial zone and/or strong specific intermolecular interactions between phases. One may add to that (in the opinion of some) the existence of a direct physical interlocking between surfaces. With all of the best designed into practical systems, however, adhesive joints never attain the strength one would predict based on theory. The following section addreses the question of why life can be so difficult.

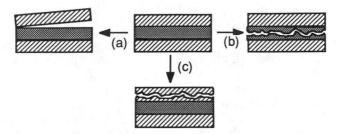

Figure 19.5. The principle loci of adhesive failure: (a) adhesive failure exactly at the interface; (b) cohesive failure in the adhesive; (c) cohesive failure in the adherend.

ADHESIVE FAILURE

When an adhesive bond fails under a small applied stress, it is commonly described as a "weak" bond or "poor" adhesion. In fact, such a description may be misleading since failure may have occurred at the interface, near the interface within one of the phases comprising the system, or well away from the interface. The situations are illustrated in Figure 19.5. It is only the first case that one can accurately describe as being a result of poor adhesion. The other two are more properly called failures of the bulk materials, that is, failure of cohesion, which is not the same thing. However, usage dictates that failure of any kind be termed adhesion failure. Of course, in an actual joint, failure may be due to a combination of all three processes.

When failure occurs exactly at the joint interface or well away from it (say, more than 100 nm) then identification of the locus of failure is generally a simple process. It is when the failure occurs within 10 --100 nm of the interface that identification becomes a problem.

Correct identification of the locus of failure can be of great practical and theoretical importance. If one can determine that failure occurred cohesively near the actual interface, it can be inferred that improvement in bond strength can be obtained by increasing the cohesive strength of the ruptured material without worrying about the nature of the interactions at the interface (ie, molecular attraction or entanglement). If the failure is found to occur at the interface, on the other hand, it will clearly be necessary to change the chemical nature of the components to increase the intermolecular attractive forces --- introduce more specific interactions, form chemical bonds, and/or increase interpenetration of the phases. If the locus of failure is not correctly identified, a great deal of time, energy, and money may be wasted solving problems that do not exist! All of that depends, of course, on the fact that the bond in question was actually between the phases expected (ie, a "proper" joint), and that some cohesively weak contaminant (ie, moisture, oil, air, dirt, etc) is not the primary cause of failure.

Based on the simple calculations of ideal adhesive bond strength given earlier, it has been suggested that bond failure in a "proper" adhesive joint will almost never occur at the interface. Instead failure will occur in a weak boundary layer near the true interface, or within the weaker of the two bonded phases. Modern experimental techniques and theoretical considerations, however, indicate that all three possibilities for failure do, in fact, occur, depending on the given situation.

For a system with a sharp interface and weak intermolecular interaction, as in the nonpolar--polar polymer system mentioned above, failure exactly at the interface is a distinct possibility (or even probability). That is, interfacial separation may be expected when interfacial strength is weaker than the bulk strength of the bonded materials. As we have seen, if the intermolecular interactions across the interface are more specific (including chemical bond formation) or if significant interpenetration of polymer chains occurs, rupture at the interface becomes less likely. In some cases, the locus of failure may depend on the rate at which stress is applied --- rapid application leading to cohesive failure and very slow application tending more toward true adhesive failure (since slow application of stress gives more time for the entangled molecules to "slide past" one another).

A great deal of theory has grown up around the subject of failure in adhesive joints. It would be prohibitive to attempt to cover the subject here; however, the practical importance of understanding the topic cannot be overemphasized. One has only to consider the large number of critical structural joints employed in modern construction (of airplanes, for instance) to see how important the subject has become.

The Role of Joint Flaws in Adhesive Failure

To this point the discussion has focused primarily on so-called "proper" or ideal adhesive joints, assuming intimate contact between components and the absence of flaws and contaminants. In the real world, such conditions are difficult to attain, so that the question of practical joint failure may not be concerned so much with intermolecular forces and entanglement, but with the mechanics of stress propagation in the system. That subject, like so many related to practical applications of surface chemistry, is very extensive and beyond the scope of this book. However, the basic principles involved are such that a few words may serve a useful purpose.

A typical flaw, for purposes of the present discussion, would be an entrapped bubble of air or other contaminant that is itself relatively weak cohesively (Figure 19.6). In the presence of such flaws, little or no energy can be transmitted through the flaw so that the stress becomes concentrated at the junctions of the flaws with the interface or in one of the bulk materials. The "local" stress, therefore is greater than the average value and is more likely to exceed the adhesive or cohesive strength of the system near the flaw. As a result of that situation, the applied stress induces the formation of a crack which continues to propagate along the line of least resistance (with continued application of stress) until joint failure results.

(a)

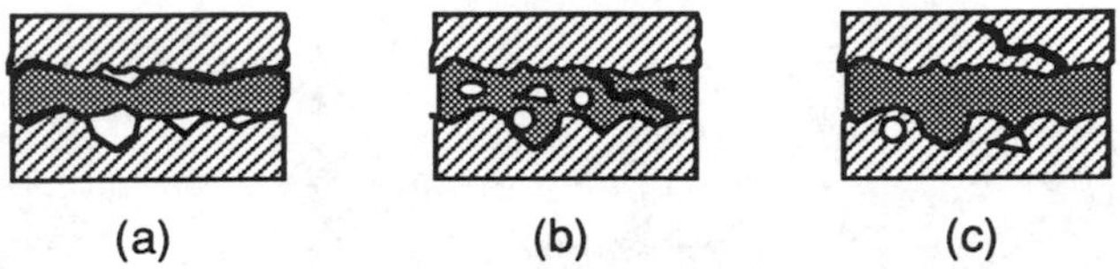

(b)

(c)

Figure 19.6. Schematic illustration of the principle flaws leading to failure in adhesive joints: (a) trapped gas, liquid, or solid contaminants at the interface; (b) cracks, bubbles, or other flaws in the bulk of the adhesive layer; (c) subsurface cracks, bubbles, or other flaws in the adherend near the adhesive interface.

Because there are so many geometries of adhesive joints encountered, and so many types of stress applied (tension, shear, torsion, etc), the analysis of a given system must be tailored to meet the specific application. The processes of experimental design and data analysis, therefore, become quite complicated. It should also be kept in mind that flaws such as those often implicated in adhesive failure can also lead to apparent cohesive failure in the bulk material.

Bibliography

Chapter 1

van Olphen, H.; Mysels, K.J. "Physical Chemistry: Enriching Topics from Colloid and Surface Science." Theorex: La Jolla, Calif., 1975.

Chapter 2

Adamson, A.W. "Physical Chemistry of Surfaces." 3rd ed. Wiley-Interscience: New York, 1984.

Aveyard, R.; Haydon, D.A. "An Introduction to the Principles of Surface Chemistry." Cambridge University Press: Cambridge, 1973.

Jaycock, M.J.; Parfitt, G.D. "Chemistry of Interfaces." Ellis Horwood, Ltd.: Chichester, England, 1981.

Chapter 3

Myers, D.Y. "Surfactant Science and Technology." VCH Publishers, Inc.: New York, 1988.

Rosen, M.J. "Surfactants and Interfacial Phenomena." Wiley-Interscience: New York, 1976.

Shinoda, K.; Nakagawa, T.; Tamamushi, B.; Isemura, T. "Colloidal Surfactants, Some Physico-Chemical Properties." Academic Press: New York, 1963.

Swisher, R.D. "Surfactant Biodegradation." Marcel Dekker, Inc.: New York, 1970.

Chapter 4

Israelachvili, J.N. "Intermolecular and Surface Forces with Applications to Colloidal and Biological Systems." Academic Press: New York, 1985.

Israelachvili, J.N.; Ninham, B.W. in J. Colloid Interface Sci., 1977, 58, 14.

Mahanty, J.; Ninham, B.W. "Dispersion Forces." Academic Press: New York, 1976.

Tanford, C. "The Hydrophobic Effect: Formation of Micelles and Biological Membranes," 2nd edition. Wiley-Interscience: New York, 1980.

Chapter 5

Adamson, A.W. "Physical Chemistry of Surfaces," 3rd edition. Wiley-Interscience: New York, 1984.

Jaycock, M.J.; Parfitt, G.D. "Chemistry of Interfaces." Ellis Horwood, Ltd.: Chichester, England, 1981.

Kruyt, H.R. "Colloid Science." Elsevier, New York, 1965.

Sparnaay, M.J. "The Electrical Double Layer." Pergamon: New York, 1972.

Chapter 6

Aveyard, R.; Haydon, D.A. "An Introduction to the Principles of Surface Chemistry." Cambridge University Press: Cambridge, 1973.

Morrison, S.R.; "The Chemical Physics of Surfaces." Plenum Press: London, 1977.

Chapter 7

Bradley, R.S. "Nucleation in Phase Changes." Q. Rev.(London), 1951, 5, 315.

"Catalysis." Emmett, P.H., Ed.; Reinhold: New York, 1956.

Hayward, D.O.; Trapnell, B.M.W. "Chemisorption." Buttersworth: London, 1964.

"Nucleation." Zettlemoyer, A.C., Ed. Marcel Dekker: New York, 1969.

Somorjai, G.A. "Principles of Surface Chemistry." Prentice-Hall: Englewood Cliffs, New Jersey, 1972.

Chapter 8

Defay, R.; Prigogine, I.; Bellemans, A.; Everett, D.H. "Surface Tension and Adsorption." Longmans, Green and Co.: London, 1966.

Gaines, G.L. "Insoluble Monolayers at Liquid--Gas Interfaces." Interscience: New York, 1966.

"Techniques of Surface and Colloid Chemistry." Good, R.J; Patrick, R.L.; Stromberg, R.R; Eds. Marcel Dekker: 1972.

van Olphen, H.; Mysels, K.J. "Physical Chemistry: Enriching Topics from Colloid and Surface Science." Theorex: La Jolla, Calif., 1975.

Chapter 9

Defay, R.; Prigogine, I.; Bellemans, A.; Everett, D.H. "Surface Tension and Adsorption." Longmans, Green and Co.: London, 1966.

Kipling, J.J. "Adsorption from Solutions of Non-Electrolytes." Academic Press: New York, 1965.

Mittal, K.L. "Adsorption at Interfaces, ACS Symposium Series No. 8." American Chemical Society: Washington, D.C., 1975.

Chapter 10

Adamson, A.W. "Physical Chemistry of Surfaces," 3rd edition. Wiley-Interscience: New York, 1984.

Kruyt, H.R. "Colloid Science." Elsevier, New York, 1965.

Chapter 11

Becher, P. "Emulsions: Theory and Practice," 2nd edition. Reinhold Publishing Corp.: New York, 1965.

Lissant, K.J. "Demulsification, Industrial Applications." Marcel Dekker, Inc.: New York, 1983.

"Emulsions and Emulsion Technology," (in two parts) Lissant, K.J., Ed. Marcel Dekker, Inc.: New York, 1974.

"Encyclopedia of Emulsion Technology," Vol. 1: "Basic Theory," Becher, P., Ed. Marcel Dekker, Inc: New York, 1983.

"Encyclopedia of Emulsion Technology," Vol. 2: "Applications," Becher, P., Ed. Marcel Dekker, Inc: New York, 1985.

"Encyclopedia of Emulsion Technology," Vol. 3: Becher, P., Ed. Marcel Dekker, Inc: New York, 198?.

Chapter 12

Berkman, S.; Egloff, G. "Emulsions and Foams." Reinhold: New York, 1961.

Bikerman, J.J. "Foams." Springer-Verlag: New York, 1973.

Chapter 13

Adam, N.K. "The Physics and Chemistry of Surfaces." 3rd ed. Oxford University Press: London, 1941.

Osipow, L.I. "Surface Chemistry, Theory and Industrial Applications." Krieger: New York, 1977.

Chapter 14

"Surfactants." Tadros, T.F. Ed. Academic Press: London, 1985.

Colloids and Surfaces, 1985, Vol. 13, Goddard, E.D., Ed. a special issue on polymer--surfactant interactions.

Chapter 15

Israelachvili, J.N. "Intermolecular and Surface Forces with Applications to Colloidal and Biological Systems." Academic Press: New York, 1985.

Myers, D.Y. "Surfactant Science and Technology." VCH Publishers, Inc.: New York, 1988.

Rosen, M.J. "Surfactants and Interfacial Phenomena." Wiley-Interscience: New York, 1976.

Shinoda, K.; Nakagawa, T.; Tamamushi, B.; Isemura, T. "Colloidal Surfactants, Some Physico-Chemical Properties." Academic Press: New York, 1963.

Tanford, C. "The Hydrophobic Effect: Formation of Micelles and Biological Membranes," 2nd edition. Wiley-Interscience: New York, 1980.

Chapter 16

Elworthy, P.H.; Florence, A.T.; Macfarlane, C.B. "Solubilization by Surface-Active Agents." Chapman and Hall Ltd.: London, 1968.

Fendler, J.H.; Fendler, E.J. "Catalysis in Micellar and Macromolecular Systems." Academic Press, Inc. New York, 1975.

"Microemulsions Theory and Practice." Prince, L.M., Ed. Academic Press, Inc. New York, 1977.

"Reverse Micelles, Biological and Technological Relevance of Amphiphilic Structures in Apolar Media," Luisi, P.L.; Straub, B.E., Eds. Plenum Press: New York, 1984.

Chapter 17

"Improved Oil Recovery by Surfactant and Polymer Flooding," Shah, D.O.; Schechter, R.S., Eds. Academic Press, Inc.: New York, 1978.

Osipow, L.I. "Surface Chemistry; Theory and Industrial Applications." Krieger: New York, 1977.

Schwartz, A.M. The Physical Chemistry of Detergency, in "Surface abd Colloid Science." Vol. 5, Matijevic, E., Ed. Wiley-Interscience: New York, 1972.

"Wetting, Spreading, and Adhesion." Padday, J.F., Ed. Academic Press Ltd.: London, 1978.

Chapter 18

Bowden, F.P.; Tabor, D. "Friction, An Introduction to Tribology." Anchor Books: New York, 1973.

Moore, D.F. "Principles and Applications of Tribology." Pergamon: New York, 1975.

Chapter 19

Bikerman, J.J. "The Science of Adhesive Joints." Academic Press: New York, 1961.

"Contact Angle, Wettability, and Adhesion. Advances in Chemistry Series No. 43." American Chemical Society: Washington, D.C., 1964.

"Adhesion and Adhesives." Houwink, R.; Salomon, D.; Eds. Elsevier: New York, 1965.

Kaelble, D.H. "Physical Chemistry of Adhesion." Wiley-Interscience: New York, 1971.

REFERENCES

Chapter 1

1. Ostwald, W.; "Der Welt der vernachlassigten Dimensionen." Dresden: Leipzig, 1915.
2. Young, T. "Miscellaneous Works." Vol.1; Peacock, G. Ed. Murry: London, 1855, p.418.
3. Laplace, P.S. de; " Mechanique Celeste." Supplement to Book 10, 1806.
4. Gauss. "Principia Generalia Theorie Figurae Fluidorum." 1830.
5. Poisson. "Nouvelle Theorie de l'action Capillaire." 1831.

Chapter 2

1. Gibbs, J.W. "The Collected Works of J.W. Gibbs." Vol.1. Longmans, Green: New York, 1931.

Chapter 4

1. van der Waals, J.D. Thesis. Leyden, 1873; "Die Kontinuitat des gasformigen und flussigen Zustande - I, II." Leipzig, 1899.
2. London, F. Z. Phys. Chem., B11, 1930, 222.
3. McLachlan, A.D. Mol. Phys., 7, 1963-63, 381.
4. Israelachvili, J.N."Intermolecular and Surface Forces." Academic Press: London, 1985.
5. Hamaker, H.C. Physica, 4, 1937, 1058.

Chapter 5

1. von Helmholtz, H. Weid Ann. Phys. 1879, 7, 337.
2. Gouy, G. J. Phys., 1910, 4, 457.; Chapman, D.L. Phil. Mag., 1913, 25, 475.
3. Verwey, E.J.W.; Overbeek, J. Th. G. "Theory of Stability of Lyophobic Colloids." Elsevier: New York, 1948.
4. Kruyt, H.R. "Colloid Science." Elsevier: New York, 1952.
5. Hiemenz, P.C. "principles of Colloid and Surface Chemistry," 2nd edition; Marcel Dekker: New York, 1986.

Chapter 6

1. Laplace, P.S. de; " Mechanique Celeste." Supplement to Book 10, 1806.

Chapter 8

1. Ramsey, R.; Shields, J. Trans. R. Soc. (London). 1893, A184, 647; J. Chem. Soc., 1893, 1089.

2. Myers, D.Y. "Surfactant Science and Technology." VCH Publishers, Inc.: New York, 1988. (and references cited therein)

Chapter 9

1. Giles, C.H.; MacEwen, T.H.; Nakhwa, S.N.; Smith, D. J. Chem. Soc., 1960, 3973.

Chapter 10

1. Hamaker, H.C. Physica. 4, 1937, 1058.
2. Reerink, H. Thesis. Utrecht, 1951.
3. von Smoluchowski, N. Bull. Int. Acad. Polon. Sci.. Classe Sci. Math. Nat.. 1903, 184.
4. See Langmuir, I. J. Am. Chem. Soc., 1916, 38, 2221.
5. Schultze, H.J. J. prakt. Chem., 1882, 25, 431; 1883, 27, 320.
6. Hardy, W.B. Proc. Roy. Soc. London, 1900, 66, 110; Z. physik. Chem., 1900, 33, 385.

Chapter 11

1. Griffin, W.C. J. Soc. Cosmet. Chem.. 1949, 1, 311.
2. Davies, J.T.; Rideal, E.K. "Interfacial Phenomena," 2nd edition. Academic Press: London, 1963.
3. Little, R.C.; Singleterry, C.R. J. Phys. Chem., 1964, 68, 3453.
4. Israelachvili, J.N.; Mitchell, D.J.; Ninham, B.W. J.Chem. Soc. Faraday Trans. 2. 1976, 72, 1925.
5. Mitchell, D.J.; Ninham, B.W. J.Chem. Soc. Faraday Trans. 2. 1981, 77, 601.
6. Shinoda, K.; Arai, H. J. Phys. Chem.. 1964, 68, 3485.
7. Shinoda, K.; Arai, H.J. Colloid Sci.. 1965, 20, 99.
8. Florence, A.T. Whitehill, D. J. Colloid Interface Sci.. 1981, 79, 243.

Chapter 12

1. Manegold, E. "Schaum, Strassenbau, Chemie und technik." Heidelberg, 1953.
2. Gibbs, J.W. "The Collected Works of J.W. Gibbs." Vol.1. Longmans, Green: New York, 1931, pp. 287,301,307.
3. Myers, D.Y. "Surfactant Science and Technology." VCH Publishers, Inc.: New York, 1988. (and references cited therein)
4. Friberg, S.; Mandell, L.; Larsson, K. J. Colloid Interface Sci.. 1969, 29, 155.

5.   Friberg, S.; Mandell, L. J. Pharm. Sci., 1970, 59, 1001.

Chapter 13

1.   Zettlemoyer, A.C. "Nucleation." Marcel Dekker: New York, 1969.

Chapter 15

1.   Israelachvili, J.N. "Intermolecular and Surface Forces." Academic
      Press: London, 1985.
2.   Fendler, J.H. Acc. Chem. Res., 1980, 13, 7; ibid., 1976, 9, 153.
3.   Tanford, C. "The Hydrophobic Effect: Formation of Micelles and
      Biological Membranes." John Wiley and Sons: New York, 1980.
4.   Myers, D.Y. "Surfactant Science and Technology." VCH Publishers,
      Inc.: New York, 1988. (and references cited therein)
5.   Klevens, H.B. Chem Rev., 1950, 47, 1.

Chapter 17

1.   Wenzel, R.N. Ind. Eng. Chem., 1936, 28, 988; J. Phys. Colloid
      Chem., 1949, 53, 1466.
2.   Baxter, S.; Cassie, A.B.D. J. Text. Inst., 1945, 36, T67; Trans. faraday
      Soc., 1944, 40, 546.
3.   Girifalco, L.A.; Good, R.J. J. Phys. Chem., 1957, 61, 904.
4.   Fowkes, F.M. J. Phys. Chem., 1963, 67, 2538; Adv. Chem. Ser., No.
      43, 1964, 99.

Chapter 18

1.   Moore, D.F. "Principles and Applications of Tribology." Pergamon:
      New York, 1975.

Chapter 19

1.   Good, R.J. in "Treatise on Adhesives and Adhesion," Vol. 1, R.L.
      Patrick, ed. Marcel Dekker: New York, 1967, pp.9-69.)
2.   "Modern Plastics Encyclopedia." McGraw-Hill: New York, 1974.
3.   Sturgeon, D.L.G.; Lacy, R.I. in "Handbook of Fillers and Reinforce-
      ments for Plastics," Katz, H.S.; Milewski, J. V. , Eds., Reinhold: New
      York, 1978, pp. 511-544.
4.   Cottrell, A.H. Proc. Royal Soc. London. A282, 1964, 2.

INDEX